Sampling in Combinatorial and Geometric Set Systems

Mathematical
Surveys
and
Monographs

Volume 265

Sampling in Combinatorial and Geometric Set Systems

Nabil H. Mustafa

AMS AMERICAN MATHEMATICAL SOCIETY
Providence, Rhode Island

2020 *Mathematics Subject Classification.* Primary 68Q87, 52C45, 05D40, 03D32, 11K38.

For additional information and updates on this book, visit
www.ams.org/bookpages/surv-265

Library of Congress Cataloging-in-Publication Data

Names: Mustafa, Nabil H. (Nabil Hassan), 1979– author.
Title: Sampling in combinatorial and geometric set systems / Nabil H. Mustafa.
Description: Providence, Rhode Island: American Mathematical Society, [2022] | Series: Mathematical surveys and monographs, 0076-5376; 265 | Includes bibliographical references and index.
Identifiers: LCCN 2021040893 | ISBN 9781470461560 (paperback) | 9781470468736 (ebook)
Subjects: LCSH: Geometry–Data processing. | AMS: Computer science – Theory of computing – Probability in computer science (algorithm analysis, random structures, phase transitions, etc.). | Convex and discrete geometry – Discrete geometry – Combinatorial complexity of geometric structures. | Combinatorics – Extremal combinatorics – Probabilistic methods in extremal combinatorics, including polynomial methods (combinatorial Nullstellensatz, etc.). | Mathematical logic and foundations – Computability and recursion theory – Algorithmic randomness and dimension. | Number theory – Probabilistic theory: distribution modulo 1; metric theory of algorithms – Irregularities of distribution, discrepancy.
Classification: LCC QA448.D38 M87 2022 | DDC 519.2–dc23
LC record available at https://lccn.loc.gov/2021040893

Visit the AMS home page at `https://www.ams.org/`

10 9 8 7 6 5 4 3 2 1 27 26 25 24 23 22

To my parents: Riffat and Ghulam Mustafa

Contents

Preface ix

Background xiii

Chapter 1. A Probabilistic Averaging Technique 1
 1. Level Sets 4
 2. Concentration Bounds for Sums of Bernoulli Variables 11

Chapter 2. First Constructions of Epsilon-Nets 21
 1. Deterministic 23
 2. Probabilistic 26
 3. Improved Probabilistic Analysis 31

Chapter 3. Refining Random Samples 37
 1. Linear-sized Nets for Disks in $\mathbb{R}^2$ 39
 2. Partitioning Segments in $\mathbb{R}^2$ 44
 3. Application: Forbidden Subgraphs 52

Chapter 4. Complexity of Set Systems 57
 1. VC-dimension 60
 2. Shallow-cell Complexity 67

Chapter 5. Packings of Set Systems 75
 1. Packing Half-Spaces in $\mathbb{R}^d$ 77
 2. A Packing Theorem for Combinatorial Set Systems 80
 3. Applications of the Packing Theorem 85

Chapter 6. Epsilon-Nets: Combinatorial Bounds 89
 1. A First Bound using Ghost Sampling 91
 2. Optimal ϵ-Nets using Packings 96

Chapter 7. Epsilon-Nets: An Algorithm 101
 1. Special Case: Linear-sized Nets 103
 2. General Case 108

Chapter 8. Epsilon-Nets: Weighted Case 113
 1. Special Case: Linear-sized Nets 115
 2. General Case 121
 3. Application: Rounding Linear Programs 128

Chapter 9. Epsilon-Nets: Convex Sets 131
 1. Weak ϵ-nets 132

2. Application: Polytope Approximation 138

Chapter 10. VC-dimension of k-Fold Unions: Basic Case 143
1. Abstract Set Systems 145
2. Axis-Aligned Rectangles in $\mathbb{R}^2$ 150

Chapter 11. VC-dimension of k-Fold Unions: General Case 157
1. Products of Set Systems 158
2. Geometric Set Systems in $\mathbb{R}^d$ 162
3. Lower bounds on Epsilon-Nets 167

Chapter 12. Epsilon-Approximations: First Bounds 171
1. Proof using Induction 174
2. Proof using Discrepancy 178

Chapter 13. Epsilon-Approximations: Improved Bounds 185
1. Improved Analysis via Chaining 186
2. Near-Optimal Bounds via Partitioning 193

Chapter 14. Epsilon-Approximations: Relative Case 199
1. Relative and Sensitive Approximations 202
2. Improved Bounds for Relative Approximations 208

Chapter 15. Epsilon-Approximations: Functional Case 213
1. A Functional View of Approximations 216
2. Application: Sensitivity and Coresets for Clustering 220

Chapter 16. A Summary of Known Bounds 227
1. `LIST`: Complexity of Geometric Set Systems 228
2. `LIST`: Epsilon-Nets 233
3. `LIST`: Epsilon-Approximations 236

Bibliography 241

Index 247

Preface

The use of random sampling in the field of discrete and computational geometry started in the 1980s, motivated by the challenges in designing efficient algorithms for geometric problems. While these earlier uses were tightly coupled with specific geometric scenarios, soon the key problems were formulated abstractly in the framework of combinatorial and geometric set systems. We state one of the principal structures that will be studied in this framework. Let X be a set of n elements and $\mathcal{F}$ a collection of subsets of X; the pair $(X, \mathcal{F})$ forms a set system.

> **Epsilon-nets:** Given a parameter $\epsilon \in (0, 1]$, a set $N \subseteq X$ is an ϵ-net of $(X, \mathcal{F})$ if each $S \in \mathcal{F}$ of size at least $\epsilon |X|$ has non-empty intersection with N.

The goal is to find ϵ-nets of small size; this of course depends on the structure and complexity of $(X, \mathcal{F})$. A classical geometric instance of this question—first studied in 1987 and settled conclusively in 2017—is to determine, given any set P of n points in $\mathbb{R}^d$, the smallest $N \subseteq P$ such that any half-space containing at least ϵn points of P contains at least one point of N.

Selective aspects of ϵ-nets have been presented in earlier texts (*Combinatorial Geometry*, Pach and Agarwal, 1995; *The Discrepancy Method*, Chazelle, 2000; *Lectures on Discrete Geometry*, Matoušek, 2004; *Geometric Approximation Algorithms*, Har-Peled, 2011). However, the last ten years have seen significant progress with many open problems in the area having been resolved during this time. These include optimal lower bounds for ϵ-nets for most geometric set systems, the use of shallow-cell complexity to unify proofs, simpler algorithms to construct ϵ-nets, and the use of ϵ-approximations for construction of coresets via sensitivity analysis, to name a few. This book presents a didactic account of these recent developments. We will revisit classical results, but with new and more elegant proofs which unify earlier work.

> Chapter 1 introduces the two key technical ingredients that lie at the heart of the analysis of random sampling methods in this book: the complexity of certain combinatorial structures arising in geometric configurations and the probability of a random variable deviating far from its expectation. While historically these two have been considered separate statements with entirely different proofs, we present a powerful probabilistic technique from which both of these bounds can be deduced in a uniform way.
>
> Chapters 2 and 3 initiate the study of ϵ-nets for some basic geometric set systems in $\mathbb{R}^2$, delineating the precise geometric properties that are relevant to the construction of ϵ-nets; these are then combined with probabilistic techniques to derive asymptotically optimal bounds on the size of ϵ-nets.

While these will be superseded later by more general and powerful combinatorial machinery, they are important in understanding the intuition, ideas, and analysis at their most elementary level.

The move from geometric to combinatorial set systems requires formulating and proving analogs of geometric properties for combinatorial systems. Chapters 4 and 5 are devoted to building this technical foundation. First, the VC-dimension and the shallow-cell complexity of combinatorial set systems are introduced and studied as measures of complexity of a set system. These are then used to construct combinatorial equivalents of geometric properties relevant to ϵ-net constructions.

Chapters 6 to 8 present the current best bounds for ϵ-nets for combinatorial set systems $(X, \mathcal{F})$, where the bounds depend on the VC-dimension and the shallow-cell complexity of $\mathcal{F}$. Together these chapters contain the insight that one can derive optimal bounds for geometric set systems from combinatorial bounds based on the shallow-cell complexity of the corresponding set system. Chapter 9 studies a geometric case where small ϵ-nets do not exist, set systems induced by convex sets in $\mathbb{R}^d$, and where one has to turn to the notion of a *weak* ϵ-net.

Chapters 10 and 11 are concerned with lower bounds on sizes of ϵ-nets, based on the insight that a lower bound on the size of an ϵ-net for a given set system $(X, \mathcal{F})$ follows from a lower bound on the VC-dimension of a related set system, the k-fold union of $\mathcal{F}$. These lower bounds are then used to show optimality of the ϵ-net bounds presented earlier.

Chapters 12 to 15 study another notion of samples, *ϵ-approximations*, for both geometric and combinatorial set systems. It also includes an application of ϵ-approximations for constructing small coresets for some geometric optimization problems.

Chapter 16 concludes the book with a list of bounds on the VC-dimension and shallow-cell complexity for most commonly studied geometric set systems, as well as on sizes of their ϵ-nets and ϵ-approximations. This will serve as a reference for those looking for the state-of-the-art bounds on these topics.

We now briefly list some topics which are not in this book: algorithms tailored to construct ϵ-nets efficiently for specific geometric set systems, efficient deterministic versions of the probabilistic algorithms, range searching and other classic algorithmic applications, bounds for combinatorial discrepancy of geometric set systems. This choice was guided by two factors. First, the techniques involved in these are rather different, often relying on detailed geometric data-structures; doing justice to this essentially requires another book. Second, parts of it have been covered very nicely in earlier texts (e.g., in *The Discrepancy Method* by Chazelle).

While our key objective is to give a clear account of the ideas (as much as is possible by us), we also hope that reading this book is a pleasant experience (the wonderful texts *Combinatorial Geometry* by Pach and Agarwal and *Lectures in Discrete Geometry* by Matoušek being exemplary in this regard). We have also taken care to make the present text useful for teaching: all calculations are written in sufficient detail; each section begins with an "overview of ideas" which gives intuition into the proof; wherever possible we first present the simplest non-trivial instance of the idea before dealing with the more general case; each chapter can be read mostly independently (though it might use earlier results); additional insights, ideas and calculations that are not crucial to the main text are interspersed throughout in small font size. Each section typically contains one or two results, and is carved up into smaller themes that are delineated by the symbol ☙.

The text should be suitable as a first introduction to sampling aspects for senior undergraduate and graduate students in computer science, mathematics and statistics. While mathematical maturity will certainly help in appreciating the ideas presented here, only a basic familiarity with discrete mathematics, probability and combinatorics is required to understand the material. For background on these topics, the following books are recommended:

H. Tijms. *Understanding Probability: Chance Rules in Everyday Life.* Cambridge University Press, 2007.

J. Matoušek and J. Nešetřil. *Invitation to Discrete Mathematics.* Oxford University Press, 2008.

M. Mitzenmacher and E. Upfal. *Probability and Computing: Randomization and Probabilistic Techniques in Algorithms and Data Analysis.* Cambridge University Press, 2017.

For teaching a course on these topics, we recommend that around 2 hours of class time be devoted to each chapter; this text is suitable for a 30 to 40 hour course on the subject (there is considerable freedom in the choice of topics to cover). We also hope that this book will be useful for researchers in the field as a reference text for looking up specific bounds as well as learning quickly the ideas and techniques behind specific results.

We would be grateful if any errors are reported to `nabilhmustafa@gmail.com`.

Acknowledgments. This text benefited greatly from feedback and discussions with several mentors, colleagues and students. In particular, I am grateful to the following people: Imre Bárány, Victor-Emmanuel Brunel, Jean Cardinal, Timothy Chan, Bernard Chazelle, Mónika Csikós, Kunal Dutta, Fritz Eisenbrand, David Eppstein, Jeff Erickson, Martina Gallato, Arijit Ghosh, Andrey Kupavskii, Jesús De Loera, Frédéric Meunier, Wolfgang Mulzer, Márton Naszódi, János Pach, Dömötör Pálvölgyi, Dominique Perrin, Jeff Phillips, Saurabh Ray, Güntor Rote, Gabor Tardos, Csaba Tóth, Kasturi Varadarajan, and Emo Welzl.

I am also grateful to the Agence Nationale de la Recherche (ANR) for funding my research for the past ten years, and to my colleagues at LIPN, Villetaneuse and LIGM, Marne-la-Vallée.

It was a pleasure to work the people from the AMS Publishing. I would like to especially thank Ina Mette for her great help and patience.

Nabil H. Mustafa
Nogent-sur-Marne, September 2021

Background

Basic notation. The cardinality of a finite set X is denoted by $|X|$. For a real number a, $\lfloor a \rfloor$ denotes the largest integer less than or equal to a; similarly $\lceil a \rceil$ denotes the smallest integer greater than or equal to a. We use the notation $[n]$ for the set $\{1, \ldots, n\}$, where n is a positive integer. We will use the notation $A = B \pm C$, where $A, B \in \mathbb{R}$ and $C \in \mathbb{R}^+$, as a shorthand for $A \in [B - C, B + C]$. We will use $\log n$ for logarithms with base 2, and $\ln n$ for logarithms with base e. The letters $\mathbb{N}$, $\mathbb{R}$, $\mathbb{Z}$ are reserved for the set of all natural numbers, reals and integers, respectively.

Asymptotic notation. For two real valued functions f and g, we say that $f = O(g)$ if there exist large-enough constants $n, C > 0$ such that $f(x) \leq Cg(x)$ for all $x \geq n$. Here the values of n, C might depend on other quantities considered as constants; we will explicitly point out such dependencies when they occur. The notation $f = \Omega(g)$ is equivalent to $g = O(f)$, $f = \Theta(g)$ if and only if $f = O(g)$ and $f = \Omega(g)$, and $f = o(g)$ if $\lim_{x \to \infty} \frac{f(x)}{g(x)} = 0$.

Set systems. Given a finite set X, 2^X will denote the collection of all subsets of X. Similarly, for $0 \leq k|X|$, $\binom{X}{k}$ will denote the collection of all subsets of X of size k, and so $\left|\binom{X}{k}\right| = \binom{|X|}{k}$. A set system is a pair $(X, \mathcal{F})$, where X is a set and $\mathcal{F}$ is a collection of subsets of X. When X is clear from the context, we will simply use $\mathcal{F}$ to denote the set system.

Geometric notions. $\mathbb{R}^d$ will denote the d-dimensional Euclidean space. For a measurable set $X \subseteq \mathbb{R}^d$, $\mathrm{vol}(X)$ denotes the d-dimensional Lebesgue measure of X. The symbol ∂X denotes the boundary of $X \subseteq \mathbb{R}^d$, and $\mathrm{int}(X)$ the interior of X. For $p \in \mathbb{R}^d$ and $r > 0$, $\mathrm{Ball}(p, r)$ denotes the closed ball of radius r centered at p. A set $X \subset \mathbb{R}^d$ is convex if for every $p, q \in X$, the segment pq is contained in X. The set $\mathrm{conv}(X)$ is defined to be the intersection of all convex sets in $\mathbb{R}^d$ containing X. Alternatively, $q \in \mathrm{conv}(X)$ if and only if there exist points $p_1 \in X, \ldots, p_{d+1} \in X$ and nonnegative reals $t_1, \ldots, t_{d+1}$ such that $\sum_i t_i = 1$ and $q = \sum_i t_i p_i$. A finite set $X \subset \mathbb{R}^d$ is said to be 'in convex position' if $p \notin \mathrm{conv}(X \setminus \{p\})$ for all $p \in X$. Radon's theorem states that given any set P of $d+2$ points in $\mathbb{R}^d$, there exists a partition of P into two disjoint sets P_1 and P_2 such that $\mathrm{conv}(P_1) \cap \mathrm{conv}(P_2) \neq \emptyset$.

General position. Throughout the text we will often assume that a configuration of geometric objects is 'in general position'. That is, all properties and corresponding results are invariant to an arbitrarily small perturbation of the configuration. For example, for a set of n points in $\mathbb{R}^d$ in general position, we will assume that no $d+1$ points lie on a common hyperplane, no $d+2$ on a common sphere and so on. The specific properties assumed for a configuration in general position will be explicitly stated where used.

Point-hyperplane duality. The dual of a point $p = (p_1, \ldots, p_d) \in \mathbb{R}^d$ is the hyperplane $p_1x_1 + p_2x_2 + \cdots + p_{d-1}x_{d-1} - x_d = -p_d$. The dual of a 'non-vertical' hyperplane $h\colon a_1x_1 + \cdots + a_{d-1}x_{d-1} + x_d = b$ is the point $(a_1, \ldots, a_{d-1}, b)$. The key property of duality, easy to verify using the above mappings, is that it preserves incidences and sidedness. That is, for any point $p \in \mathbb{R}^d$ and any hyperplane h, p lies above h (with respect to the x_d-coordinate) if and only if the point corresponding to the dual of h lies below hyperplane corresponding to the dual of p.

Graphs. An undirected graph is usually denoted by $G = (V, E)$, where V is the set of its vertices, and $E \subseteq \binom{V}{2}$ is the set of its edges. When the sets V and E are not explicitly defined, they will be denoted by $V(G)$ and $E(G)$. If $\{u, v\} \in E$, we say that u and v are adjacent in G, and that v is a neighbor of u (and vice versa). For any $v \in V$, $N_G(v) \subseteq V$ will be the set of vertices of V which are adjacent to v. The complete graph, where $E = \binom{V}{2}$, on t vertices will be denoted by K_t, and the complete bipartite graph with t_1, t_2 vertices in the two partite sets will be denoted by K_{t_1,t_2}. A subset $V' \subseteq V$ such that there are no edges between any two vertices of V' in G is called an independent set. Any $V' \subseteq V$ such that there is an edge in G between every two vertices of V' is called a clique.

A *drawing* of an undirected graph $G = (V, E)$ in the plane consists of two functions that map V and E to subsets of the plane. The function $\phi_V\colon V \to \mathbb{R}^2$ maps each vertex $v \in V$ to a point $\phi_V(v) \in \mathbb{R}^2$. Then for each edge $e = \{u, v\} \in E$, the continuous function $\phi_e\colon [0,1] \to \mathbb{R}^2$ maps e to a continuous arc in $\mathbb{R}^2$ connecting the images of u and v, i.e., connecting $\phi_e(0) = \phi_V(u)$ to $\phi_e(1) = \phi_V(v)$. We will assume that ϕ_V is injective ($\phi_V(x) = \phi_V(y)$ if and only if $x = y$) and that no arc $\phi_e[0,1]$ passes through the image of any vertex apart from the endpoints of e. A drawing of $G = (V, E)$ is called an *embedding* or a *plane graph* if (the images of) no two edges share an interior point (of course, they may share an endpoint). G is called *planar* if it has an embedding in $\mathbb{R}^2$. A planar graph on n vertices has at most $3n - 6$ edges and at most $2n - 4$ faces in any embedding.

Given a set P of n points in the plane, the Delaunay graph of P has an edge between two points $p, q \in P$ if and only if there is a closed disk containing p and q and no other point of P. The Delaunay graph is planar, and so has at most $3n - 6$ edges.

Probability. $\Pr[A]$ denotes the probability of an event A. The expectation of a random variable X is denoted by $\mathrm{E}[X]$. An indicator random variable is a random variable which can have a value of 0 or 1. For an indicator random variable X, we have $\mathrm{E}[X] = \Pr[X = 1]$. Linearity of expectation states that for two random variables X and Y, we have $\mathrm{E}[X + Y] = \mathrm{E}[X] + \mathrm{E}[Y]$; the usefulness of this statement comes from the fact that it holds regardless of any dependency between X and Y.

Equations and inequalities. Here are some inequalities that will be useful.

Pascal's rule: $$\binom{n}{k} = \binom{n-1}{k} + \binom{n-1}{k-1}, \quad n, k \in \mathbb{Z}^+.$$

Geometric-arithmetic mean inequality: $$\prod_{i=1}^{n} a_i \le \left(\frac{\sum_{i=1}^{n} a_i}{n}\right)^n, \quad a_1, \ldots, a_n \in \mathbb{R}^+.$$

Exponential: $$1 + x \le e^x, \quad \text{for } x \in \mathbb{R}.$$
$$1 - x \ge e^{-2x}, \quad \text{for } x \in [0, 0.79].$$

Binomial theorem: $$(x+y)^n = \sum_{i=0}^{n} \binom{n}{i} x^i \, y^{n-i}, \quad n \in \mathbb{Z}^+, x, y \in \mathbb{R}.$$

CHAPTER 1

A Probabilistic Averaging Technique

This chapter presents a powerful probabilistic proof technique which will be used, in combination with additional ideas, throughout this book. For a simple first application, consider the following problem.

Let $P = \{p_1, \ldots, p_n\}$ be a set of n points in the plane and let $k \geq 0$ be an integer. The k-Delaunay graph of P, denoted by $G_k(P) = (P, E_k)$, is defined as follows:

> $\{p_i, p_j\} \in E_k$ if and only if there exists a closed disk in the plane containing $\{p_i, p_j\}$ and at most k points of $P \setminus \{p_i, p_j\}$.
>
> For each $\{p_i, p_j\} \in E_k$ fix any one such disk and denote it by D_{ij}.

Note that $E_0 \subseteq E_1 \subseteq \cdots \subseteq E_{n-2} = \binom{P}{2}$.

Our goal is to upper bound the number of edges in the k-Delaunay graph of P as a function of n and k. We will prove that $|E_k| = O(n(k+1))$[1].

First, observe that the 0-Delaunay graph of P is simply the Delaunay graph, which is planar and thus E_0 has size at most $3n$. Next, we upper bound $|E_k|$, for any integer $k \geq 1$, by the following argument.

> Let S be a random sample constructed by picking each point of P independently with probability $p = \frac{1}{k+1}$ and let $G_0(S)$ be the 0-Delaunay graph of S. We count the expected number of edges in $G_0(S)$ in two ways.
>
> **Upper bound:** As *any* 0-Delaunay graph on t vertices has at most $3t$ edges, the expected number of edges in $G_0(S)$ is
>
> $$\mathrm{E}\left[3|S|\right] = 3\,\mathrm{E}\left[|S|\right] = 3np = \frac{3n}{k+1}.$$
>
> **Lower bound:** For any $\{p_i, p_j\} \in E_k$, if both p_i and p_j are picked in S and none of the at most k other points of P lying in D_{ij} are picked in S, then $\{p_i, p_j\}$ is an edge in $G_0(S)$. As each point of P was picked independently, the probability that $\{p_i, p_j\}$ is an edge in $G_0(S)$ is at least
>
> $$p^2 \cdot (1-p)^k = \frac{1}{(k+1)^2} \cdot \left(1 - \frac{1}{k+1}\right)^k \geq \frac{1}{(k+1)^2} \cdot \frac{1}{e}, \tag{1.1}$$
>
> where the last step uses the fact that $\left(1 + \frac{1}{k}\right)^k \leq e$, and thus $\frac{1}{e} \leq \left(\frac{k}{k+1}\right)^k = \left(1 - \frac{1}{k+1}\right)^k$.

[1]The '+1' term is there just to take care of the case $k = 0$.

Using linearity of expectation, Equation (1.1) implies that the expected number of edges of E_k that are present in $G_0(S)$ is at least $|E_k| \cdot \frac{1}{(k+1)^2 e}$.

Combining the upper and lower bounds, we get

$$|E_k| \cdot \frac{1}{(k+1)^2 e} \leq \text{expected number of edges in } G_0(S) = \frac{3n}{k+1},$$

implying that $|E_k| = O(n(k+1))$.

Before we move on to other applications of this technique, we make a few remarks.

- The use of random sampling in the above proof is a way to 'implement' a double-counting argument. Essentially we are summing up, over all edges e in $G_k(P)$, the number of subsets $S \subseteq P$ of size $\frac{n}{k+1}$ for which e is an edge in $G_0(S)$[2]. We counted this sum in two ways: iterating over edges of $G_k(P)$ gave a lower bound while iterating over subsets of P gave an upper bound. More precisely, define the set of pairs

$$\mathcal{I} = \{ (e, S) : e \in E_k,\ |S| = \lceil n/(k+1) \rceil,\ e \text{ is in } G_0(S) \}.$$

  Then the above double-counting argument gives

$$|E_k| \cdot \binom{n-2-k}{\lceil n/(k+1) \rceil - 2} \leq |\mathcal{I}| \leq \binom{n}{\lceil n/(k+1) \rceil} \cdot 3\lceil n/(k+1) \rceil.$$

  Solving this for $|E_k|$ gives $|E_k| = O(n(k+1))$, as before.
- The utility of framing the argument probabilistically is that it beautifully captures the intuition behind the key idea: if there are 'too many' edges in $G_k(P)$, then in expectation more than $3|S|$ of these edges will 'filter through' to $G_0(S)$, for a random sample S. This contradicts the fact that for *any* S, $G_0(S)$ has at most $3|S|$ edges.
- The lower bound follows by considering, for each edge $\{p_i, p_j\}$ of $G_k(P)$, a specific event whose occurrence implies that $\{p_i, p_j\}$ appears as an edge in $G_0(S)$. This need not be the *only* event that could cause $\{p_i, p_j\}$ to be an edge in $G_0(S)$—e.g., there could be a disk other than D_{ij} containing p_i and p_j that happens to not contain any other point of S. Thus our lower bound is not necessarily tight.

  In fact, what we actually want to compute is a lower bound on the probability that there exists *some* disk containing $\{p_i, p_j\}$ and no other point of S. However the events for all possible disks containing $\{p_i, p_j\}$ are not independent, which makes computing this probability difficult. Fortunately, we do not lose much by considering *any one* such disk, and in fact the lower bound is optimal up to constant factors for certain point sets. In particular, the bound $|E_k| = O(n(k+1))$ is tight for the instance of n points lying on a line.
- The calculation, when carried out with probability $p \in (0, 1)$ as a parameter, gives $|E_k| = O\left(\frac{n}{p(1-p)^k}\right)$. The value of p is then set to maximize the denominator. Roughly speaking, as the term $(1-p)^k = e^{-\Theta(pk)}$ decreases exponentially with p, it is best to set p so that $e^{-\Theta(pk)}$ is a constant—that is, $p = \Theta\left(\frac{1}{k}\right)$.

[2] A minor technical difference is that in the probabilistic version, the *expected size* is $\frac{n}{k+1}$.

As for the precise value of p that maximizes the denominator, since the derivative of $p\,(1-p)^k$ with respect to p is $(1-kp-p)\,(1-p)^{k-1}$, it can be verified that $p\,(1-p)^k$ is maximized at $p = \frac{1}{k+1}$.

This also makes sense intuitively: for each edge $\{p_i, p_j\} \in E_k$, the disk D_{ij} contains at most k other points of P and so picking each point with probability less than $\frac{1}{k}$ implies that, in expectation, D_{ij} will not contain any of these points.

- Other applications of this technique follow the same 'template'—pick a random sample and calculate the probability of some event due to it in two ways. The main technical work consists in finding good estimates for certain events; this is typically where a variety of other combinatorial and geometric ideas come into play.

1. Level Sets

Counting pairs is the oldest trick in combinatorics ... every time we count pairs, we learn something from it.

Gil Kalai

Our first application is a variant of the k-Delaunay graph problem. Given a finite set P of points in $\mathbb{R}^d$ and an integer $k \geq 1$, the objective is to upper bound the number of subsets of P of size at most k that are 'realizable' by geometric objects in $\mathbb{R}^d$. We first explain the problem for the case of disks in $\mathbb{R}^2$.
For a set P of n points in $\mathbb{R}^2$, define the set system

$$\mathcal{R}(P) = \left\{ D \cap P : \ D \text{ is a disk in } \mathbb{R}^2 \right\}.$$

We call $\mathcal{R}(P)$ the primal set system induced on P by disks. For any integer $k \geq 1$, let $\mathcal{R}_{=k}(P)$ be the sets of $\mathcal{R}(P)$ of size exactly k and let $\mathcal{R}_{\leq k}(P)$ be the sets of $\mathcal{R}(P)$ of size at most k. That is,

$$\mathcal{R}_{=k}(P) = \left\{ R \in \mathcal{R}(P) : \ |R| = k \right\} \quad \text{and} \quad \mathcal{R}_{\leq k}(P) = \left\{ R \in \mathcal{R}(P) : \ |R| \leq k \right\}.$$

The sets of $\mathcal{R}_{\leq k}(P)$ are called the ($\leq k$)-level sets, or simply ($\leq k$)-sets, of $\mathcal{R}(P)$.

> Observe that $\mathcal{R}_{\leq 2}(P)$—the subsets of P of size at most two that are induced by disks—consists of $O(n)$ sets: the sets of size 1 in $\mathcal{R}_{\leq 2}(P)$ are the points of P and the sets of size 2 are precisely the edges of the Delaunay graph of P. At the other end, $\mathcal{R}_{\leq n}(P)$ is just $\mathcal{R}(P)$, with size $O\left(n^3\right)$.

Our first main result of this section implies both of the above two cases.

LEMMA 1.2. *Let P be a set of n points in $\mathbb{R}^2$ and let $\mathcal{R}(P)$ be the primal set system induced on P by disks in the plane. Then for any integer $k \geq 1$,*

$$\left|\mathcal{R}_{\leq k}(P)\right| = O\left(nk^2\right).$$

To simplify the presentation, we will assume that $|P| \geq 3$, and that P is in general position; in particular, no three points lie on a line and no four points lie on a circle.

To prove Lemma 1.2, we will first count a slightly different structure called *canonical disks*, which are disks that are 'fixed' by points of P on their boundary.

DEFINITION 1.3. A canonical disk spanned by $Q \subseteq \mathbb{R}^2$ is a disk whose boundary contains three points of Q.
Furthermore, a canonical disk D spanned by Q is called an empty canonical disk if the interior of D contains no point of Q.

Let $\mathcal{T}(P)$ be the set of all $\binom{n}{3}$ unordered triples of points of P. For a triple $\{p, q, r\} \in \mathcal{T}(P)$, let D_{pqr} be the unique *open* disk whose boundary contains $\{p, q, r\}$; we say that D_{pqr} is spanned by $\{p, q, r\}$. For an integer $k \geq 0$, define the level sets

$$\mathcal{T}_{\leq k}(P) = \left\{ \{p, q, r\} \in \mathcal{T}(P) : \ |D_{pqr} \cap P| \leq k \right\}.$$

We first observe that the size of $\mathcal{R}_{\leq k}(P)$ is bounded, within a constant factor, by that of $\mathcal{T}_{\leq k}(P)$.

CLAIM 1.4. For any integer $k \geq 1$, $|\mathcal{R}_{\leq k}(P)| \leq 8 \cdot |\mathcal{T}_{\leq (k-1)}(P)|$.

PROOF. Take any $R \in \mathcal{R}_{\leq k}(P)$ and let D be a disk realizing R; that is, $R = D \cap P$.
Now D can be scaled and translated—without any point of P 'crossing' the boundary of D—such that it contains three points of P, say $\{p, q, r\}$, on its boundary. Furthermore at least one of p, q or r belongs to R. See figure.
The interior of D_{pqr} contains at most $k-1$ points of P and so $\{p, q, r\} \in \mathcal{T}_{\leq (k-1)}(P)$. By slightly shifting and scaling D_{pqr}, for each of the 8 possible subsets of $\{p, q, r\}$, one can get a disk containing precisely that subset and all the points of P in the interior of D_{pqr}. One of these subsets is R, implying the claim.

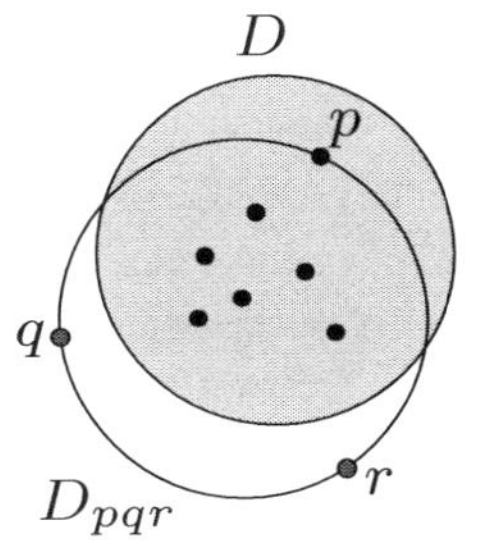

□

We remark here that the constant 8 can be improved with a more careful argument (see discussion).
Now the proof of Lemma 1.2 follows from Claim 1.4 and the following statement.

LEMMA 1.5. *Let P be a set of n points in $\mathbb{R}^2$ and let $k \geq 0$ be an integer. Then*

$$|\mathcal{T}_{\leq k}(P)| = O\left(n(k+1)^2\right).$$

PROOF. First we establish the case $k = 0$.

CLAIM 1.6. For any $S \subseteq P$, $|\mathcal{T}_{\leq 0}(S)| \leq 2|S|$.

PROOF. $\mathcal{T}_{\leq 0}(S)$ consists of unordered triples of S whose corresponding open disks do not contain any point of S in their interior. If the disk D_{pqr}, spanned by $p, q, r \in S$, contains no point of S in its interior, then by slightly shifting D_{pqr}, it follows that each of the three edges $\{p, q\}$, $\{q, r\}$ and $\{p, r\}$ belong to the Delaunay graph of S. In particular, the triangle with vertices $\{p, q, r\}$ is a face of the Delaunay graph of S. Thus $|\mathcal{T}_{\leq 0}(S)|$ is upper bounded by the number of faces in a planar graph on $|S|$ vertices, which is $2|S| - 4$. □

Now consider the case $k \geq 1$. Construct a random sample S by picking each point of P independently with probability $p = \frac{1}{k+1}$.
We count the expected size of $\mathcal{T}_{\leq 0}(S)$ in two ways.

Upper bound: From Claim 1.6,

$$\mathrm{E}\left[\,\left|\mathcal{T}_{\leq 0}(S)\right|\,\right] \leq \mathrm{E}\left[2\left|S\right|\right] = 2\,np.$$

Lower bound: The key is the following observation:

a triple $\{p,q,r\} \in \mathcal{T}(P)$ is present in $\mathcal{T}_{\leq 0}(S)$ *if and only if* $\{p,q,r\} \subseteq S$ and none of the points in $D_{pqr} \cap P$ are picked in S.

As each point of P was picked independently, for any $\{p,q,r\} \in \mathcal{T}(P)$, we have

$$\Pr\left[\,\{p,q,r\} \in \mathcal{T}_{\leq 0}(S)\,\right] = p^3 \cdot (1-p)^{|D_{pqr} \cap P|}.$$

Therefore, by linearity of expectation,

$$\begin{aligned}
\mathrm{E}\left[\,\left|\mathcal{T}_{\leq 0}(S)\right|\,\right] &= \sum_{\{p,q,r\}\in\mathcal{T}(P)} \Pr\left[\,\{p,q,r\} \in \mathcal{T}_{\leq 0}(S)\,\right] \\
&\geq \sum_{\{p,q,r\}\in\mathcal{T}_{\leq k}(P)} \Pr\left[\,\{p,q,r\} \in \mathcal{T}_{\leq 0}(S)\,\right] \\
&= \sum_{\{p,q,r\}\in\mathcal{T}_{\leq k}(P)} p^3 \cdot (1-p)^{|D_{pqr} \cap P|} \\
&\geq \sum_{\{p,q,r\}\in\mathcal{T}_{\leq k}(P)} p^3 \cdot (1-p)^{k} = \left|\mathcal{T}_{\leq k}(P)\right| \cdot p^3 \cdot (1-p)^k.
\end{aligned}$$

Combining the upper and lower bounds,

$$\left|\mathcal{T}_{\leq k}(P)\right| \cdot p^3 \cdot (1-p)^k \;\leq\; \mathrm{E}\left[\,\left|\mathcal{T}_{\leq 0}(S)\right|\,\right] \;\leq\; 2\,np,$$

$$\text{and hence} \quad \left|\mathcal{T}_{\leq k}(P)\right| \leq \frac{2\,n}{p^2 \cdot (1-p)^k} = \frac{2\,n\,(k+1)^2}{\left(1-\frac{1}{k+1}\right)^k} \leq 2e\,n\,(k+1)^2,$$

where the last step follows from the fact that $\left(1-\frac{1}{k+1}\right)^k \geq \frac{1}{e}$. □

We next prove a similar statement for set systems where the elements are geometric objects in $\mathbb{R}^d$ and the sets are induced by points in $\mathbb{R}^d$. We consider the case of disks in the plane.

Given a set $\mathcal{D} = \{D_1, \ldots, D_n\}$ of n distinct closed disks in $\mathbb{R}^2$, define the set system

$$\mathcal{R}(\mathcal{D}) = \left\{\mathcal{D}_p \,:\, p \in \mathbb{R}^2\right\}, \quad \text{where} \quad \mathcal{D}_p = \left\{D \in \mathcal{D} : D \ni p\right\}.$$

We call $\mathcal{R}(\mathcal{D})$ the dual set system induced on $\mathcal{D}$ by $\mathbb{R}^2$. Visually, each cell in the arrangement of $\mathcal{D}$ corresponds to a set in $\mathcal{R}(\mathcal{D})$ (note that different cells may correspond to the same subset).

For simplicity we will assume that $\mathcal{D}$ is in general position—in particular, the intersection of the boundaries of every pair of disks of $\mathcal{D}$ is either empty or consists of two distinct points and the intersection of the boundaries of any three disks of $\mathcal{D}$ is empty.

Our goal is to upper bound, for any integer $k \geq 1$, the size of $\mathcal{R}_{\leq k}(\mathcal{D})$. Our main result is the following.

LEMMA 1.7. *Let $\mathcal{D}$ be a set of n closed disks in $\mathbb{R}^2$ and let $\mathcal{R}(\mathcal{D})$ be the dual set system induced on $\mathcal{D}$. Then for any integer $k \geq 1$,*

$$|\mathcal{R}_{\leq k}(\mathcal{D})| = O(nk).$$

As earlier, it suffices to consider canonical sets, defined as follows. Let $\mathcal{V}(\mathcal{D})$ be the set of at most $2\binom{n}{2}$ points in $\mathbb{R}^2$ that are the intersections of boundaries of the disks of $\mathcal{D}$. For any integer $k \geq 0$, define

$$\mathcal{V}_{\leq k}(\mathcal{D}) = \{v \in \mathcal{V}(\mathcal{D}) : v \text{ is contained in the } \textit{interior} \text{ of at most } k \text{ disks of } \mathcal{D}\}.$$

Similarly one can define $\mathcal{V}_{=k}(\mathcal{D})$. See figure.

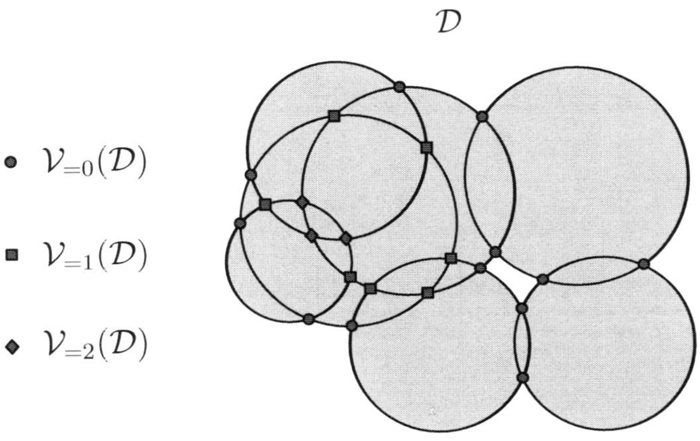

The proof of the following claim is easy and left to the reader.

CLAIM 1.8. For any integer $k \geq 1$, $|\mathcal{R}_{\leq k}(\mathcal{D})| \leq 4 \cdot |\mathcal{V}_{\leq (k-1)}(\mathcal{D})| + |\mathcal{D}|$.

Now the proof of Lemma 1.7 follows from Claim 1.8 and the following statement.

LEMMA 1.9. *For any integer $k \geq 0$, $|\mathcal{V}_{\leq k}(\mathcal{D})| = O(n(k+1))$.*

PROOF. As before, we first upper bound the size of $\mathcal{V}_{\leq 0}(\mathcal{D})$ and then use the averaging technique to upper bound the size of $\mathcal{V}_{\leq k}(\mathcal{D})$ for $k \geq 1$.

CLAIM 1.10. For any $S \subseteq \mathcal{D}$, $|\mathcal{V}_{\leq 0}(S)| \leq 6|S|$.

PROOF. Any $v \in \mathcal{V}_{\leq 0}(S)$ is an intersection point between the boundary of two disks of S and is not contained in the interior of any disk of S. Let $G = (S, E)$ be a graph where there is an edge between two disks of S if and only if a common intersection point of their boundaries belongs to $\mathcal{V}_{\leq 0}(S)$. We now show that G is planar, and so $|\mathcal{V}_{\leq 0}(S)| \leq 2|E| \leq 2(3|S| - 6) \leq 6|S|$.
We claim that the following is a plane drawing of G: draw each edge $\{D_i, D_j\} \in E$ as a line segment between the centers of D_i and D_j. Consider any two edges $\{D_i, D_j\}$, $\{D_k, D_l\}$ and let q_{ij}, q_{kl} be the two corresponding points in $\mathcal{V}_{\leq 0}(S)$. Let l be the bisector of q_{ij} and q_{kl}. As both D_i, D_j contain q_{ij} and do not contain q_{kl}, their centers lie on the side of l containing q_{ij}. Similarly the centers of D_k and D_l lie on the side of l containing q_{kl}. Thus the line segments corresponding to the edges $\{D_i, D_j\}$ and $\{D_k, D_l\}$ cannot intersect. □

Now consider the case $k \geq 1$. Construct a random sample S by picking each disk of $\mathcal{D}$ independently with probability $p = \frac{1}{k+1}$.
We will count the expected size of $\mathcal{V}_{\leq 0}(S)$ in two ways.

Upper bound: From Claim 1.10,

$$\mathrm{E}\left[|\mathcal{V}_{\leq 0}(S)|\right] \leq \mathrm{E}\left[6\,|S|\right] = 6\,\mathrm{E}\left[|S|\right] = 6\,np.$$

Lower bound: This follows by considering the probability of each vertex in $\mathcal{V}_{\leq k}(\mathcal{D})$ ending up as a vertex of $\mathcal{V}_{\leq 0}(S)$. Let $v \in \mathcal{V}_{\leq k}(\mathcal{D})$ and let $D_i, D_j \in \mathcal{D}$ be the two disks such that v is an intersection point of the boundaries of D_i and D_j. Then

$v \in \mathcal{V}_{\leq 0}(S)$ if and only if $\{D_i, D_j\} \subseteq S$ and every disk of $\mathcal{D}$ containing v in its interior is not present in S.

As there are at most k such disks and each disk of $\mathcal{D}$ was picked independently,

$$\Pr\left[v \in \mathcal{V}_{\leq 0}(S)\right] \geq p^2\,(1-p)^k.$$

Therefore,

$$\begin{aligned}\mathrm{E}\left[|\mathcal{V}_{\leq 0}(S)|\right] &= \sum_{v \in \mathcal{V}(\mathcal{D})} \Pr\left[v \in \mathcal{V}_{\leq 0}(S)\right] \\ &\geq \sum_{v \in \mathcal{V}_{\leq k}(\mathcal{D})} \Pr\left[v \in \mathcal{V}_{\leq 0}(S)\right] \geq |\mathcal{V}_{\leq k}(\mathcal{D})| \cdot p^2\,(1-p)^k.\end{aligned}$$

Combining the upper and lower bounds,

$$|\mathcal{V}_{\leq k}(\mathcal{D})| \cdot p^2\,(1-p)^k \leq \mathrm{E}\left[|\mathcal{V}_{\leq 0}(S)|\right] \leq 6\,np,$$

$$\text{and hence}\quad |\mathcal{V}_{\leq k}(\mathcal{D})| \leq \frac{6\,n}{p\,(1-p)^k} = \frac{6\,n\,(k+1)}{\left(1-\frac{1}{k+1}\right)^k} \leq 6e\,n\,(k+1),$$

where the last step used the fact that $\left(1-\frac{1}{k+1}\right)^k \geq \frac{1}{e}$. □

Primal and dual set systems can be defined more generally:

DEFINITION 1.11. Given a set P of points in $\mathbb{R}^d$ and a (possibly infinite) family $\mathcal{R}$ of geometric objects in $\mathbb{R}^d$, the primal set system induced on P by $\mathcal{R}$ is

$$\{O \cap P : O \in \mathcal{R}\}.$$

DEFINITION 1.12. Given a set $\mathcal{R}$ of geometric objects in $\mathbb{R}^d$, the dual set system induced on $\mathcal{R}$ by $\mathbb{R}^d$ is defined as

$$\{\mathcal{R}_p : p \in \mathbb{R}^d\}, \quad \text{where} \quad \mathcal{R}_p = \{R \in \mathcal{R} : R \ni p\}.$$

We now conclude with the case of primal and dual set systems induced by half-spaces in $\mathbb{R}^d$.

Let P be a set of n points in general position in $\mathbb{R}^d$ and $\mathcal{R}(P)$ the primal set system induced on P by downward-facing half-spaces—that is, considering the x_d-axis as vertical, the half-spaces which contain the point that is the 'minus infinity' of the x_d

axis. It can be shown that $|\mathcal{R}(P)| = O(n^d)$. In fact for points in general position there is a precise bound independent of the structure of P (stated without proof):

$$|\mathcal{R}(P)| = \sum_{i=0}^{d} \binom{n}{i}. \tag{1.13}$$

Now let $\mathcal{T}(P)$ be all the $\binom{n}{d}$ subsets of P of size d. For each $e \in \mathcal{T}(P)$, let h_e^+ be the unique downward-facing *open* half-space whose bounding hyperplane contains e. For an integer $k \geq 0$, define the level sets

$$\mathcal{T}_{\leq k}(P) = \left\{e \in \mathcal{T}(P) : \left|h_e^+ \cap P\right| \leq k\right\}.$$

As earlier, the size of $\mathcal{R}_{\leq k}(P)$ can be upper bounded, within a multiplicative factor, by that of $\mathcal{T}_{\leq k}(P)$.
To bound $|\mathcal{T}_{\leq k}(P)$ we again first need a bound on $|\mathcal{T}_{\leq 0}(P)|$. Observe that $|\mathcal{T}_{\leq 0}(P)|$ is simply the number of facets on the lower convex-hull of P. It is well-known, to those who know it, that the Upper Bound Theorem for convex polytopes implies that this is at most $2\sum_{i=0}^{\lfloor \frac{d}{2} \rfloor} \binom{n}{i}$ (see discussion). Now the probabilistic averaging technique of this chapter together with this 0-th level bound implies the following (stated without proof).

THEOREM 1.14. *Given a set P of n points in $\mathbb{R}^d$ and an integer $k \geq 0$,*

$$|\mathcal{T}_{\leq k}(P)| \leq 2\left(\frac{e}{\lceil d/2 \rceil}\right)^{\lceil \frac{d}{2} \rceil} \binom{n}{\lfloor \frac{d}{2} \rfloor} \left(k + \left\lceil \frac{d}{2} \right\rceil\right)^{\lceil \frac{d}{2} \rceil}.$$

The above is $O\left(n^{\lfloor d/2 \rfloor}(k+1)^{\lceil d/2 \rceil}\right)$ when the dimension d is considered a constant.

For $d = 3$, Theorem 1.14 gives a bound of $O(nk^2)$—the same bound, within a multiplicative constant, as the one of Lemma 1.5. This is not a coincidence: there exists a mapping of points in $\mathbb{R}^2$ to $\mathbb{R}^3$, the so-called 'paraboloid lift', with the property that subsets realized by intersection with disks in $\mathbb{R}^2$ can be realized by intersection with half-spaces in $\mathbb{R}^3$. Thus Theorem 1.14 for $d = 3$ implies Lemma 1.5.

For later use, it will be convenient to state Theorem 1.14 in the dual setting.

DEFINITION 1.15. The level of a point $q \in \mathbb{R}^d$ with respect to a set $\mathcal{H}$ of hyperplanes in $\mathbb{R}^d$ is the number of hyperplanes of $\mathcal{H}$ lying strictly below q in the negative x_d direction; that is, the number of hyperplanes intersecting the ray

$$\left\{q + \lambda(0, \ldots, 0, -1) : \lambda > 0\right\}.$$

Given a set $\mathcal{H}$ of hyperplanes in $\mathbb{R}^d$ in general position, *a vertex* in the arrangement of $\mathcal{H}$ is a point lying in the intersection of some d hyperplanes of $\mathcal{H}$. Let $\mathcal{V}_{\leq k}(\mathcal{H})$ be the set of vertices of $\mathcal{H}$ of level at most k. Then by duality, Theorem 1.14 is equivalent to the following statement.

THEOREM 1.16. *Given a set $\mathcal{H}$ of n hyperplanes in $\mathbb{R}^d$ and an integer $k \geq 0$,*

$$|\mathcal{V}_{\leq k}(\mathcal{H})| \leq 2\left(\frac{e}{\lceil d/2 \rceil}\right)^{\lceil \frac{d}{2} \rceil} \binom{n}{\lfloor \frac{d}{2} \rfloor} \left(k + \left\lceil \frac{d}{2} \right\rceil\right)^{\lceil \frac{d}{2} \rceil}.$$

Bibliography and discussion. The influential probabilistic technique in this chapter was used in the seminal paper of Clarkson [**Cla87**], and then Clarkson and Shor [**CS89**], exactly for these problems of bounding the combinatorial complexity of configurations. Indeed, it is sometimes called the 'Clarkson-Shor technique' in the discrete and computational geometry literature. See [**APS08, Wag08**] for surveys on ($\leq k$)-sets for half-spaces and related set systems, where one can find information related to Theorem 1.14. Details on the Upper Bound Theorem for convex polytopes [**McM701a**] and related topics can be found in [**Zie95**]. See [**Mat99**, Section 3.1] for a discussion around Claim 1.4.

[APS08] P. K. Agarwal, J. Pach, and M. Sharir, *State of the union (of geometric objects)*, Surveys on discrete and computational geometry, Contemp. Math., vol. 453, Amer. Math. Soc., Providence, RI, 2008, pp. 9–48, DOI 10.1090/conm/453/08794. MR2405676

[CS89] K. L. Clarkson and P. W. Shor, *Applications of random sampling in computational geometry. II*, Discrete Comput. Geom. **4** (1989), no. 5, 387–421, DOI 10.1007/BF02187740. MR1014736

[Cla87] K. L. Clarkson, *New applications of random sampling in computational geometry*, Discrete Comput. Geom. **2** (1987), no. 2, 195–222, DOI 10.1007/BF02187879. MR884226

[Mat99] J. Matoušek. *Geometric Discrepancy: An Illustrated Guide*. Springer, 1999.

[McM701a] P. McMullen, *The maximum numbers of faces of a convex polytope*, Mathematika **17** (1970), 179–184, DOI 10.1112/S0025579300002850. MR283691

[Wag08] U. Wagner, *k-sets and k-facets*, Surveys on discrete and computational geometry, Contemp. Math., vol. 453, Amer. Math. Soc., Providence, RI, 2008, pp. 443–513, DOI 10.1090/conm/453/08810. MR2405692

[Zie95] G. M. Ziegler, *Lectures on polytopes*, Graduate Texts in Mathematics, vol. 152, Springer-Verlag, New York, 1995, DOI 10.1007/978-1-4613-8431-1. MR1311028

2. Concentration Bounds for Sums of Bernoulli Variables

> *I thought it was a rather trivial lemma, but many things are only trivial once you know them.*
>
> Herman Chernoff

We present an application of the probabilistic technique to computing tail bounds of some common probability distributions. That is, we would like to upper bound the probability that a random variable gets a value far from its expectation. This is a basic technical ingredient in nearly all the constructions and methods that will be seen later.

The setting is the following.

> Let $I = \{1, 2, \dots, n\}$ be a set of n elements from which we will pick a random sample. We aim to pick np elements of I, for a given parameter $p \in [0, 1]$.

The 0-1 valued random variable X_i will be used to indicate whether $i \in I$ is picked in our random sample. Our goal is to estimate the probability that the sum of these n variables, $X = \sum_{i=1}^{n} X_i$, falls far from its expectation $\mathrm{E}[X]$. More precisely, for any $\delta \geq 0$, we are interested in bounding $\Pr\big[X \geq (1+\delta)\,\mathrm{E}[X]\big]$ and $\Pr\big[X \leq (1-\delta)\,\mathrm{E}[X]\big]$.

In fact, we consider the more general case where for a fixed set $J \subseteq I$ with $X_J = \sum_{j \in J} X_j$, we are interested in upper bounds on

$$\Pr\big[X_J \geq (1+\delta) \cdot \mathrm{E}[X_J]\big] \qquad \text{and} \qquad \Pr\big[X_J \leq (1-\delta) \cdot \mathrm{E}[X_J]\big].$$

There are several natural ways to pick a random sample from I. Two basic ones, given a parameter p, are the following.

Binomial distribution: Pick each element of I *independently* with probability p. That is, let $X_1, \dots, X_n$ be n independent 0-1 random variables where

$$X_i = \begin{cases} 1 & \text{with probability } p, \\ 0 & \text{otherwise.} \end{cases}$$

For any $J \subseteq I$, we have

$$\mathrm{E}[X_J] = \mathrm{E}\left[\sum_{j \in J} X_j\right] = \sum_{j \in J} \mathrm{E}[X_j] = \sum_{j \in J} \Pr[X_j = 1] = |J|\, p.$$

One can write the exact equation for the tail bounds using the fact that the value of each X_i was set independently:

$$\Pr\big[X_J \geq (1+\delta) \cdot |J|\, p\big] = \sum_{i=\lceil (1+\delta)\cdot|J|p\rceil}^{|J|} \Pr[X_J = i] \tag{1.17}$$

$$= \sum_{i=\lceil (1+\delta)\cdot|J|p\rceil}^{|J|} \binom{|J|}{i} p^i (1-p)^{|J|-i}.$$

As there is no closed-form formula for this, several methods have been proposed to estimate the right-hand side of the above expression (see discussion).

The fact that $\{X_1, \ldots, X_n\}$ are independent has two advantages: first it makes calculations easier and second, for any $J \subseteq I$ the induced probability distribution on X_J remains the same (that is, each element of J is picked independently with probability p). On the other hand, the number of elements X is not fixed and is a random variable with expectation np.

Sampling without replacement: A second natural way to sample is to choose, out of all $\binom{n}{np}$ np-sized subsets of I, one uniformly at random (assume that np is an integer). This then sets the values of $X_1, \ldots, X_n$, with $\sum_i X_i$ being equal to np. Note that for any $J \subseteq I$, $\mathrm{E}[X_J] = |J|\,p$ since for any i,

$$\Pr[X_i = 1] = \frac{\binom{n-1}{np-1}}{\binom{n}{np}} = \frac{(n-1)!}{(np-1)!(n-np)!} \cdot \frac{(np)!(n-np)!}{n!} = \frac{np}{n} = p.$$

More generally, for any $J \subseteq I$, letting $t = |J|$, the probability that $X_j = 1$ for all $j \in J$, can be upper bounded as

$$\text{(1.18)} \qquad \Pr\left[\left(\prod_{j \in J} X_j\right) = 1\right] = \frac{\binom{n-t}{np-t}}{\binom{n}{np}} = \frac{(n-t)!}{(np-t)!} \cdot \frac{(np)!}{n!}$$
$$= \frac{np}{n}\,\frac{np-1}{n-1} \cdots \frac{np-t+1}{n-t+1} \le p^t,$$

since each term $\frac{np-i}{n-i} \le p$ for $p \le 1$. Similarly, the probability that $X_j = 0$ for all $j \in J$, can be upper bounded as

(1.19)
$$\Pr\left[\left(\prod_{j \in J} (1 - X_j)\right) = 1\right] = \frac{\binom{n-t}{np}}{\binom{n}{np}} = \frac{(n-t)!}{(n-t-np)!} \cdot \frac{(n-np)!}{n!}$$
$$= \frac{n-np}{n}\,\frac{n-np-1}{n-1} \cdots \frac{n-np-t+1}{n-t+1} \le (1-p)^t,$$

since each term $\frac{n-np-i}{n-i} \le (1-p)$ for $i \ge 0$.
We can again write the precise equation for the tail bounds for any $J \subseteq I$:

$$\Pr\left[X_J \ge (1+\delta) \cdot |J|\,p\right] = \sum_{i=\lceil (1+\delta)\cdot|J|p\rceil}^{|J|} \Pr[X_J = i]$$
$$= \sum_{i=\lceil (1+\delta)\cdot|J|p\rceil}^{|J|} \frac{\binom{|J|}{i} \cdot \binom{n-|J|}{np-i}}{\binom{n}{np}}.$$

The advantage of this distribution is that $X = np$ always; however the variables $\{X_1, \ldots, X_n\}$ are no longer independent. Consequently, for a $J \subseteq I$, the induced probability distribution on X_J is *not* the one where a $(|J|p)$-sized subset of J is chosen uniformly at random from the set of all $(|J|p)$-sized subsets of J.

The variables $X_1, \ldots, X_n$ are an example of *negatively associated* random variables. We note that in this case the tail bounds are even better—that is, more sharply concentrated around the expectation—than for binomial distribution. Intuitively, for any $i, j \in I$, the fact that $X_i = 1$ makes it *less* likely that $X_j = 1$ and the fact that $X_i = 0$ makes it *more* likely that $X_j = 1$.

Formally, if $p \in (0,1)$,

$$\Pr\left[X_j = 1 \mid X_i = 1\right] = \frac{\binom{n-2}{np-2}}{\binom{n-1}{np-1}} = \frac{np-1}{n-1} < p, \qquad \text{and}$$

$$\Pr\left[X_j = 1 \mid X_i = 0\right] = \frac{\binom{n-2}{np-1}}{\binom{n-1}{np}} = \frac{np}{n-1} > p.$$

Our main theorem, a multiplicative version of a tail bound for negatively associated random variables, is the following.

THEOREM 1.20. *Let $X_1, \ldots, X_n$ be n indicator random variables and let $\delta > 0$ be a given parameter. Set $X = \sum_{i=1}^{n} X_i$.*

(1) *Let $p_1, \ldots, p_n$ be reals in $[0,1]$, $0 < \sum_{i=1}^{n} p_i < n$, such that*

$$\text{for any } I' \subseteq [n] \qquad \Pr\left[\left(\prod_{i \in I'} X_i\right) = 1\right] \leq \prod_{i \in I'} p_i.$$

Let $\tilde{p} = \frac{\sum_i p_i}{n}$. Then

$$\Pr\left[X \geq (1+\delta)\, n\tilde{p}\right] \leq \left(\frac{\left(1 - \frac{\tilde{p}\delta}{1-\tilde{p}}\right)^{(1+\delta)\tilde{p}-1}}{(1+\delta)^{(1+\delta)\tilde{p}}}\right)^{n}. \tag{1.21}$$

The above expression can be simplified to give

$$\Pr\left[X \geq (1+\delta)\, n\tilde{p}\right] \leq e^{-\frac{\delta^2}{2+\delta} n\tilde{p}}.$$

(2) *Let $r_1, \ldots, r_n$ be reals in $[0,1]$, $0 < \sum_{i=1}^{n} r_i < n$, such that*

$$\text{for any } I' \subseteq [n] \qquad \Pr\left[\left(\prod_{i \in I'} (1 - X_i)\right) = 1\right] \leq \prod_{i \in I'} (1 - r_i).$$

Let $\tilde{r} = \frac{\sum_i r_i}{n}$. Then

$$\Pr\left[X \leq (1-\delta)\, n\tilde{r}\right] \leq \left(\frac{\left(1 + \frac{\delta\tilde{r}}{1-\tilde{r}}\right)^{-1+\tilde{r}-\delta\tilde{r}}}{(1-\delta)^{\tilde{r}(1-\delta)}}\right)^{n}. \tag{1.22}$$

The above expression can be simplified to give

$$\Pr\left[X \leq (1-\delta)\, n\tilde{r}\right] \leq e^{-\frac{\delta^2}{2} n\tilde{r}}.$$

We remark that the two preconditions of the above theorem imply that for any variable X_i, we have

$$\Pr\left[X_i = 1\right] \leq p_i, \quad \text{and}$$
$$\Pr\left[(1 - X_i) = 1\right] \leq 1 - r_i \quad \text{or equivalently,} \qquad \Pr\left[X_i = 1\right] \geq r_i.$$

Thus the variables p_i and r_i are upper and lower bounds on the probability that $X_i = 1$, and therefore $n\tilde{r} \leq \mathrm{E}\left[\sum_i X_i\right] \leq n\tilde{p}$.

The above theorem applies to *both* the two earlier distributions—binomial and sampling without replacement—as it avoids using independence of the X_i's and instead uses an upper bound on $\Pr\left[\left(\prod_{i\in J} X_i\right) = 1\right]$ and $\Pr\left[\left(\prod_{i\in J}(1-X_i)\right) = 1\right]$ for every $J \subseteq I$.

Binomial distribution: Theorem 1.20 and independence implies the following.

COROLLARY 1.23. *Let $I = \{1,\ldots,n\}$ and $p_1,\ldots,p_n \in [0,1]$ be given parameters. Let $R \subseteq I$ be a random sample constructed by picking each $i \in I$ independently with probability p_i. Then for a fixed $J \subseteq I$ and $\delta > 0$,*

$$\begin{aligned}\Pr\Big[\,|J\cap R| \geq (1+\delta)\,\mathrm{E}\left[|J\cap R|\right]\Big] &= \Pr\Big[\,|J\cap R| \geq (1+\delta)\sum_{j\in J} p_j\Big]\\ &\leq e^{-\frac{\delta^2}{2+\delta}\sum_{j\in J} p_j},\\ \Pr\Big[\,|J\cap R| \leq (1-\delta)\,\mathrm{E}\left[|J\cap R|\right]\Big] &= \Pr\Big[\,|J\cap R| \leq (1-\delta)\sum_{j\in J} p_j\Big]\\ &\leq e^{-\frac{\delta^2}{2}\sum_{j\in J} p_j}.\end{aligned}$$

In particular,

$$\Pr\left[\;|J\cap R| \geq (1+\delta)\sum_{j\in J} p_j \quad\bigcup\quad |J\cap R| \leq (1-\delta)\sum_{j\in J} p_j\right] \leq 2\,e^{-\frac{\delta^2}{2+\delta}\sum_{j\in J} p_j}.$$

Sampling without replacement: Theorem 1.20 together with Equations (1.18) and (1.19) implies the following.

COROLLARY 1.24. *Let $I = \{1,\ldots,n\}$ and $t \in [n]$ be a given parameter. Let $R \subseteq I$ be a random sample of size t chosen uniformly from all $\binom{n}{t}$ t-sized subsets of I. Then for any fixed $J \subseteq I$ and $\delta > 0$,*

$$\begin{aligned}\Pr\left[|J\cap R| \geq (1+\delta)\,|J|\frac{t}{n}\right] &\leq e^{-\frac{\delta^2}{2+\delta}|J|\frac{t}{n}},\\ \Pr\left[|J\cap R| \leq (1-\delta)\,|J|\frac{t}{n}\right] &\leq e^{-\frac{\delta^2}{2}|J|\frac{t}{n}}.\end{aligned}$$

In particular,

$$\Pr\left[\;|J\cap R| \geq (1+\delta)\,|J|\frac{t}{n} \quad\bigcup\quad |J\cap R| \leq (1-\delta)\,|J|\frac{t}{n}\right] \leq 2\,e^{-\frac{\delta^2}{2+\delta}|J|\frac{t}{n}}.$$

We remark here that these bounds are tight within constant factors in the exponent for certain ranges of δ. Here is one lower bound (stated without proof; see discussion).

THEOREM 1.25. *Let $I = \{1,\ldots,n\}$ and $p \in \left(0,\frac{1}{2}\right]$. Let $R \subseteq I$ be a random sample constructed by picking each $i \in I$ independently with probability p. Then for $\delta \in \left[\sqrt{\frac{3}{np}},\frac{1}{2}\right]$,*

$$\begin{aligned}\Pr\Big[\,|R| \geq (1+\delta)\,np\Big] &\geq e^{-9\delta^2 np},\\ \Pr\Big[\,|R| \leq (1-\delta)\,np\Big] &\geq e^{-9\delta^2 np}.\end{aligned}$$

Overview of ideas. The proof of Theorem 1.20 will use our probabilistic averaging technique. That is, we will take a random sample of $\{1, 2, \ldots, n\}$ and calculate the probability of a carefully chosen event due to it in two ways.

At first glance, it might seem odd to estimate the probability of a random event—in our case the tail bounds on X—by taking *another* random sample! However it is a mistake to confuse these two separate probability distributions, with very different purposes—one is part of the input problem and the other is part of the averaging proof technique.

Perhaps a more modular way to think about this is to consider the quantity we are bounding—$\Pr[X \geq (1+\delta)\, n\tilde{p}]$—*combinatorially*: the support of the probability distribution consists of 2^n binary strings corresponding to all possible assignments of the 0-1 variables $X_1, \ldots, X_n$. Each string $s \in \{0,1\}^n$ has some probability, say $w(s)$, of being chosen.

> Let $|s|$ denote the number of 1's in s. The precise value of $w(s)$ depends on the probability distribution. For example, when $p_1 = \cdots = p_n = p$ and where np is an integer, $w(s) = p^{|s|}(1-p)^{n-|s|}$ for the binomial distribution. Similarly $w(s) = 1/\binom{n}{np}$ if $|s| = np$, and 0 otherwise, for the sampling without replacement distribution.

Then our goal is to upper bound the combinatorial quantity

$$\sum_{\substack{s \in \{0,1\}^n \\ |s| \geq (1+\delta) n\tilde{p}}} w(s).$$

Seen this way, it is similar to the earlier use of the probabilistic averaging technique to upper bound the sizes of level sets, with one difference being that earlier we were bounding the cardinality instead of a weighted sum.

As a warm-up, we first prove the following weaker bound, called *Markov's inequality*, under the conditions of Theorem 1.20:

$$\Pr\left[X \geq (1+\delta)\, n\tilde{p}\right] \leq \frac{1}{1+\delta}. \tag{1.26}$$

While Markov's inequality has an even simpler direct proof (furthermore, Markov's inequality holds for *any* positive random variable X for which $\mathrm{E}[X]$ exists, with $\mathrm{E}[X]$ replacing $n\tilde{p}$ in the stated bound. That is, X need not be the sum of n indicator variables), the following proof is an easy natural application of the probabilistic averaging technique and gives insight into the proof of Theorem 1.20.

> Let S be a random sample of the index set $I = \{1, 2, \ldots, n\}$ where each index is picked independently with probability q. Note that S is independent of the X_i variables.
>
> We count the following quantity in two ways:
>
> $$\mathrm{E}\left[|S_1|\right], \quad \text{where} \quad S_1 = \{i \in S \colon X_i = 1\}.$$
>
> That is, the expected number of indices $i \in S$ for which $X_i = 1$.

Upper bound: Using linearity of expectation and the fact that we have $\Pr[X_i = 1] \le p_i$,

$$\mathrm{E}[|S_1|] = \sum_{i=1}^{n} \Pr[i \in S \text{ and } X_i = 1] \le \sum_{i=1}^{n} p_i q = n\tilde{p}\, q.$$

The last step used the fact that S and X are independent.

Lower bound: Consider the elements of the event space for the variable $X = X_1 + \cdots + X_n$ for which $X \ge (1+\delta)\, n\tilde{p}$. Note that for each event $\{X_1, \ldots, X_n\}$ with $X = k$, the expected number of indices $i \in S$ with $X_i = 1$ is precisely kq. Thus we have

$$\mathrm{E}\Big[\ |S_1| \ \ \Big|\ \ X \ge (1+\delta)\, n\tilde{p}\ \Big] \ge (1+\delta)\, n\tilde{p}\, q.$$

Summing up over all events,

$$\begin{aligned}
\mathrm{E}[|S_1|] =& \,\mathrm{E}\left[|S_1| \,\middle|\, X \ge (1+\delta)n\tilde{p}\right] \cdot \Pr[X \ge (1+\delta)n\tilde{p}] \ + \\
& \,\mathrm{E}\left[|S_1| \,\middle|\, X < (1+\delta)n\tilde{p}\right] \cdot \Pr[X < (1+\delta)n\tilde{p}] \\
\ge& \,\mathrm{E}\left[|S_1| \,\middle|\, X \ge (1+\delta)n\tilde{p}\right] \cdot \Pr[X \ge (1+\delta)n\tilde{p}] \\
\ge& \,(1+\delta)\, n\tilde{p}\, q \cdot \Pr[X \ge (1+\delta)\, n\tilde{p}]\,.
\end{aligned}$$

Putting the upper and lower bounds together,

$$(1+\delta)\, n\tilde{p}\, q \ \cdot\ \Pr[X \ge (1+\delta)\, n\tilde{p}] \ \le\ \mathrm{E}[|S_1|] \ \le\ n\tilde{p}\, q,$$

$$\text{and hence} \qquad \Pr[X \ge (1+\delta)\, n\tilde{p}] \le \frac{1}{1+\delta}.$$

An astute reader will notice that the proof above is needlessly complicated, as the parameter q does not play any role: the dependence on q is linear in both the upper and lower bounds and thus cancels out. Setting $S = I$ (i.e., $q = 1$) gives the standard proof of Markov's inequality. This will not remain the case for the proof of the main theorem, to which we turn to next.

We now prove our main theorem.

PROOF OF THEOREM 1.20. As before, let S be a random sample where each element in $\{1, 2, \ldots, n\}$ is picked independently with probability q.

We count the following quantity in two ways:

$\Pr\Big[\prod_{i \in S} X_i = 1\Big]$.

That is, the probability that for *each* index $i \in S$, $X_i = 1$.

Note that this probability is over both the choice of S and the choice of X. Furthermore S and X are independent.

Upper bound. It will be instructive to consider it in three, progressively more general, scenarios:

- **each $X_i = 1$ independently with probability p_i:** Then we have

$$\begin{aligned}\Pr\left[\prod_{i\in S} X_i = 1\right] &= \prod_{i=1}^{n}\Big(1 - \Pr\left[i \in S \text{ and } X_i = 0\right]\Big)\\ &= \prod_{i=1}^{n}\left(1 - q\left(1-p_i\right)\right) \le \left(\frac{\sum_{i=1}^{n}\left(1-q\left(1-p_i\right)\right)}{n}\right)^n\\ &= \left(\frac{n - nq + q\sum_{i=1}^{n} p_i}{n}\right)^n = \left(q\tilde{p} + 1 - q\right)^n,\end{aligned}$$

where the third step uses the inequality of arithmetic and geometric means, that $\prod_{i=1}^{n} a_i \le \left(\frac{\sum_{i=1}^{n} a_i}{n}\right)^n$ for any non-negative reals $a_1, \ldots, a_n$.

- **$p_1 = \cdots = p_n = p$:** Then $\tilde{p} = p$ and so

$$\begin{aligned}\Pr\left[\prod_{i\in S} X_i = 1\right] &= \sum_{Q\subseteq[n]} \Pr\Big[S = Q \text{ and } \prod_{i\in Q} X_i = 1\Big]\\ &= \sum_{Q\subseteq[n]} \Pr\left[S = Q\right]\cdot \Pr\left[\prod_{i\in Q} X_i = 1\right] \quad \Big(S,\, X \text{ are independent}\Big)\\ &\le \sum_{i=0}^{n}\binom{n}{i} q^i (1-q)^{n-i}\cdot p^i \quad \Big(\text{by input assumption}\Big)\\ &= (qp + 1 - q)^n \quad \Big(\text{by the binomial theorem}\Big).\end{aligned}$$

- **the general case:**

$$\begin{aligned}\Pr\left[\prod_{i\in S} X_i = 1\right] &= \sum_{Q\subseteq[n]} \Pr\left[S = Q\right]\cdot \Pr\left[\prod_{i\in Q} X_i = 1\right]\\ &\le \sum_{Q\subseteq[n]} q^{|Q|}(1-q)^{n-|Q|}\cdot \prod_{i\in Q} p_i \quad \Big(\text{by input assumption}\Big)\\ &= (1-q)^n \sum_{Q\subseteq[n]}\prod_{i\in Q}\frac{qp_i}{1-q} = (1-q)^n\prod_{i=1}^{n}\left(1 + \frac{qp_i}{1-q}\right),\end{aligned}$$

where the last step uses the fact that $\prod_{i=1}^{n}(1+a_i) = \sum_{Q\subseteq[n]}\prod_{i\in Q} a_i$ (each term in the L.H.S. of this expression, when opened up, corresponds to a choice of either 1 or a from each of the n product terms). Continuing,

$$\begin{aligned}&= (1-q)^n\prod_{i=1}^{n}\left(\frac{qp_i + 1 - q}{1-q}\right) = \prod_{i=1}^{n}\left(qp_i + 1 - q\right)\\ &\le \left(q\tilde{p} + 1 - q\right)^n \quad \Big(\text{as earlier}\Big).\end{aligned}$$

Lower bound. Consider the elements of the event space of $X = X_1 + \cdots + X_n$ for which $X \ge (1+\delta)\, n\tilde{p}$. Note that for each instance of $\{X_1, \ldots, X_n\}$ with $X = k$, the probability that for *each* index $i \in S$ we have $X_i = 1$ is exactly $(1-q)^{n-k}$. In our case $k \geq (1+\delta)\, n\tilde{p}$ and since $(1-q)^{n-k}$ is mononotically increasing with k, we have

$$\Pr\left[\prod_{i \in S} X_i = 1 \mid X \geq (1+\delta)\, n\tilde{p}\right] \geq (1-q)^{n-(1+\delta)n\tilde{p}}.$$

Summing up over all events,

$$\begin{aligned}\Pr\left[\prod_{i \in S} X_i = 1\right] &= \Pr\left[\prod_{i \in S} X_i = 1 \mid X \geq (1+\delta)\, n\tilde{p}\right] \cdot \Pr[X \geq (1+\delta)n\tilde{p}] \; + \\ &\quad \Pr\left[\prod_{i \in S} X_i = 1 \mid X < (1+\delta)\, n\tilde{p}\right] \cdot \Pr[X < (1+\delta)n\tilde{p}] \\ &\geq \Pr\left[\prod_{i \in S} X_i = 1 \mid X \geq (1+\delta)\, n\tilde{p}\right] \cdot \Pr[X \geq (1+\delta)n\tilde{p}] \\ &\geq (1-q)^{n-(1+\delta)n\tilde{p}} \cdot \Pr\left[X \geq (1+\delta)\, n\tilde{p}\right].\end{aligned}$$

Combining the upper and lower bounds,

$$\begin{aligned}(1-q)^{n-(1+\delta)n\tilde{p}} \cdot \Pr\left[X \geq (1+\delta)\, n\tilde{p}\right] \;\leq\; \Pr\left[\prod_{i \in S} X_i = 1\right] \;\leq\; (1-q\,(1-\tilde{p}))^n \\ \Longrightarrow \Pr\left[X \geq (1+\delta)\, n\tilde{p}\right] \leq \left(\frac{1-q\,(1-\tilde{p})}{(1-q)^{1-(1+\delta)\tilde{p}}}\right)^n.\end{aligned}$$

To minimize the R.H.S. of the above expression[3], we set $q = \frac{\delta}{(1-\tilde{p})(1+\delta)}$. Then

$$\begin{aligned}\Pr\left[X \geq (1+\delta)\, n\tilde{p}\right] &\leq \left(\frac{1-\frac{\delta}{(1-\tilde{p})(1+\delta)}(1-\tilde{p})}{\left(1-\frac{\delta}{(1-\tilde{p})(1+\delta)}\right)^{1-(1+\delta)\tilde{p}}}\right)^n \\ &= \left(\frac{\frac{1}{1+\delta}}{\left(\frac{1-\tilde{p}-\tilde{p}\delta}{(1-\tilde{p})(1+\delta)}\right)^{1-(1+\delta)\tilde{p}}}\right)^n = \left(\frac{\left(\frac{1-\tilde{p}-\tilde{p}\delta}{1-\tilde{p}}\right)^{(1+\delta)\tilde{p}-1}}{(1+\delta)^{(1+\delta)\tilde{p}}}\right)^n \\ &= \left(\frac{\left(1-\frac{\tilde{p}\delta}{1-\tilde{p}}\right)^{(1+\delta)\tilde{p}-1}}{(1+\delta)^{(1+\delta)\tilde{p}}}\right)^n,\end{aligned}$$

getting the required bound.

The other direction—an upper bound on the probability that the number of 1's in X is at most $(1-\delta)n\tilde{r}$—is equivalent to upper bounding the probability that the number of 0's in X is at least

$$n - (1-\delta)\, n\tilde{r} = \left(\frac{1-(1-\delta)\,\tilde{r}}{(1-\tilde{r})}\right) n\,(1-\tilde{r}) = \left(1+\frac{\delta\tilde{r}}{1-\tilde{r}}\right) n\,(1-\tilde{r}).$$

---

[3]The partial derivative w.r.t. q is $\frac{(n\tilde{p})\left((1+\delta)(\tilde{p}-1)q+\delta\right)\left(((\tilde{p}-1)q+1)(1-q)^{\delta\tilde{p}+\tilde{p}-1}\right)^n}{(q-1)((\tilde{p}-1)q+1)}$.

Set $Y_i = 1 - X_i$ for $i = 1, \ldots, n$, and let $Y = \sum_i Y_i$. That is, $Y = n - X$ is a random variable denoting the number of 0's in X. Then

$$\Pr\left[X \leq (1-\delta)\, n\tilde{r}\right] = \Pr\left[Y \geq \left(1 + \frac{\delta\tilde{r}}{1-\tilde{r}}\right) n\,(1-\tilde{r})\right].$$

Thus we can apply the previous bound on the variable Y_i's, now with probabilities $(1 - r_i)$ instead of p_i. We have $\frac{\sum_{i=1}^{n}(1-r_i)}{n} = (1-\tilde{r})$ and so from Equation (1.21) with $\delta' = \frac{\delta\tilde{r}}{1-\tilde{r}}$,

$$\begin{aligned}
\Pr\left[Y \geq (1+\delta')\, n\,(1-\tilde{r})\right] &\leq \left(\frac{\left(1 - \frac{(1-\tilde{r})\delta'}{1-(1-\tilde{r})}\right)^{(1+\delta')(1-\tilde{r})-1}}{(1+\delta')^{(1+\delta')(1-\tilde{r})}}\right)^{n} \\
&= \left(\frac{\left(1 - \frac{\delta\tilde{r}}{\tilde{r}}\right)^{(1+\frac{\delta\tilde{r}}{1-\tilde{r}})(1-\tilde{r})-1}}{\left(1 + \frac{\delta\tilde{r}}{1-\tilde{r}}\right)^{(1+\frac{\delta\tilde{r}}{1-\tilde{r}})(1-\tilde{r})}}\right)^{n} \\
&= \left(\frac{(1-\delta)^{-\tilde{r}(1-\delta)}}{\left(1 + \frac{\delta\tilde{r}}{1-\tilde{r}}\right)^{1-\tilde{r}+\delta\tilde{r}}}\right)^{n} = \left(\frac{\left(1 + \frac{\delta\tilde{r}}{1-\tilde{r}}\right)^{-1+\tilde{r}-\delta\tilde{r}}}{(1-\delta)^{\tilde{r}(1-\delta)}}\right)^{n},
\end{aligned}$$

getting the required bound.

The simplifications of these expressions are covered in many places and we refer the reader to existing literature on this (see discussion). □

Bibliography and discussion. The proof given of these tail bounds is from [**IK10**], while Theorem 1.25 is from [**KY15**]. A nice exposition of several proofs of tail bounds similar to the one presented here (Chernoff's bound, Bernstein's inequality, Hoeffding's extension) together with the details of simplification of the expressions in Theorem 1.20 can be found in [**Mul18**] (see also [**DP09**]). A discussion on the differences between sampling with and without replacement can be found in [**FK15**, Section 21.5]. A discussion on the asymmetry between the upper and lower tail bounds in Theorem 1.20 can be found in [**AS16**, Appendix A]. Some approximations for sums of binomial coefficients (such as those of Equation (1.17)) can be found in [**GKP94**, Chapter 5] (see also [**Spi19**]). Many other concentration inequalities can be found in the text [**BLM13**].

[AS16] N. Alon and J. H. Spencer, *The probabilistic method*, 4th ed., Wiley Series in Discrete Mathematics and Optimization, John Wiley & Sons, Inc., Hoboken, NJ, 2016. MR3524748

[BLM13] S. Boucheron, G. Lugosi, and P. Massart, *Concentration inequalities*, Oxford University Press, Oxford, 2013. A nonasymptotic theory of independence; With a foreword by Michel Ledoux, DOI 10.1093/acprof:oso/9780199535255.001.0001. MR3185193

[DP09] D. P. Dubhashi and A. Panconesi, *Concentration of measure for the analysis of randomized algorithms*, Cambridge University Press, Cambridge, 2009, DOI 10.1017/CBO9780511581274. MR2547432

[FK15] A. Frieze and M. Karoński, *Introduction to random graphs*, Cambridge University Press, Cambridge, 2016, DOI 10.1017/CBO9781316339831. MR3675279

[GKP94] R. L. Graham, D. E. Knuth, and O. Patashnik, *Concrete mathematics*, 2nd ed., Addison-Wesley Publishing Company, Reading, MA, 1994. A foundation for computer science. MR1397498

[IK10] R. Impagliazzo and V. Kabanets, *Constructive proofs of concentration bounds*, Approximation, randomization, and combinatorial optimization, Lecture Notes in Comput. Sci., vol. 6302, Springer, Berlin, 2010, pp. 617–631, DOI 10.1007/978-3-642-15369-3_46. MR2755867

[KY15] P. Klein and N. E. Young, *On the number of iterations for Dantzig-Wolfe optimization and packing-covering approximation algorithms*, SIAM J. Comput. **44** (2015), no. 4, 1154–1172, DOI 10.1137/12087222X. MR3390154

[Mul18] W. Mulzer, *Five proofs of Chernoff's bound with applications*, Bull. Eur. Assoc. Theor. Comput. Sci. EATCS **124** (2018), 59–76. MR3793013

[Spi19] M. Z. Spivey, *The art of proving binomial identities*, Discrete Mathematics and its Applications (Boca Raton), CRC Press, Boca Raton, FL, 2019, DOI 10.1201/9781351215824. MR3931743

CHAPTER 2

First Constructions of Epsilon-Nets

Consider the *minimum hitting set* problem for disks:

> Let P be a set of n points in $\mathbb{R}^2$ and $\mathcal{R}$ a collection of m subsets of P induced by disks in the plane. Call a set $Q \subseteq P$ a hitting set of $\mathcal{R}$ if Q contains at least one point from each set of $\mathcal{R}$. Then our goal is to compute a hitting set of $\mathcal{R}$ of minimum cardinality.

Let OPT be the size of a minimum hitting set of $\mathcal{R}$. As computing such a set is NP-hard, we present a polynomial-time algorithm that computes a hitting set of size at most $c \cdot \mathrm{OPT}$, where $c > 1$ is a constant independent of n. We will do so by choosing a random sample from P. However a set $R \in \mathcal{R}$ of small cardinality is unlikely to be hit by a *uniform* random sample—unless one samples many points of P, possibly much larger than OPT. To circumvent this problem, our plan will be to sample *non-uniformly*, so that points in small sets have a higher probability of being picked. This will be done by assigning weights to points of P using a linear program, as follows.

We formulate the minimum hitting set problem as an integer program, with a boolean variable for each point of P and one constraint for each set of $\mathcal{R}$. Relaxing this integer program gives a linear program (LP) on n variables, with a variable $x_p \in [0, 1]$ for each $p \in P$. This LP, shown below, can be solved in polynomial time.

$$\text{Minimize} \quad \sum_{p \in P} x_p$$

subject to

$$\text{(C1)} \quad \text{for each } R \in \mathcal{R}\colon \sum_{p \in R} x_p \geq 1,$$

$$\text{(C2)} \quad \text{for each } p \in P\colon 0 \leq x_p \leq 1.$$

Fix values of $\{x_p : p \in P\}$ that realize the optimal solution, and set

$$W^* = \sum_{p \in P} x_p.$$

Note that $W^* \leq \mathrm{OPT}$, since the more constrained integer program—when each variable x_p is either 0 or 1—has value OPT. Call x_p the weight of the point p, and define the weight of a set to be the sum of the weights of its elements.

The LP constraint (C1) forces the sum of the variables in each set $R \in \mathcal{R}$ to be least 1. In other words,

> the weights $\{x_p : p \in P\}$ are set such that each set of $\mathcal{R}$ has weight at least $\frac{1}{W^*}$-th of the total weight.

The LP-assigned x_p can be viewed as quantifying the 'importance' of p—points lying in sets of small cardinality will generally get assigned higher values. Here the role of the LP ends and we will now use the weights x_p to guide us in constructing our hitting set.

The key is that our earlier problem—that sets of $\mathcal{R}$ with small cardinality were unlikely to be hit by a uniform random sample—is now resolved if we sample according to weights, as each set of $\mathcal{R}$ is now guaranteed to have weight at least $\frac{1}{W^*}$-th of the total weight. In particular, say we construct a random sample N by picking each $p \in P$ independently with probability x_p. Then we have

$$\mathrm{E}\left[|N|\right] = \sum_{p \in P} x_p = W^*,$$

and further, for each $R \in \mathcal{R}$, LP constraint (C1) implies that

$$\mathrm{E}\left[|N \cap R|\right] = \sum_{p \in R} x_p \geq 1.$$

So far we have not used the fact that $\mathcal{R}$ is induced by disks in the plane. That now comes into play due to the following theorem (proof presented later in this chapter) that "implements" our probabilistic intuition.

THEOREM. *There exists an absolute constant $c \geq 1$ such that the following is true. Let P be a set of n points in $\mathbb{R}^2$ and let $w\colon P \to [0,1]$ be a weight function on P. Then for any $\epsilon > 0$, there exists a set $N \subseteq P$ of cardinality at most $\frac{c}{\epsilon}$ such that N hits any disk containing points of weight at least an ϵ-th fraction of the total weight.*

We apply the above theorem to P, with $w(p) = x_p$ and $\epsilon = \frac{1}{W^*}$, to get a set N with

$$|N| \leq \frac{c}{\epsilon} = c \cdot W^* \leq c \cdot \mathrm{OPT}.$$

Finally, N is a hitting set of $\mathcal{R}$ since each set of $\mathcal{R}$ has weight $\frac{1}{W^*}$-th of the total weight. This concludes the proof.

The general task of converting the solution of a linear program into an integer solution is called *rounding* in optimisation and algorithms.

We now turn to the main object of study—the set N of the above theorem. This notion is defined for general set systems as follows.

DEFINITION 2.1. Given a set system $(X, \mathcal{F})$ and a parameter $\epsilon > 0$, a set $N \subseteq X$ is an ϵ-net of $(X, \mathcal{F})$ if for each $F \in \mathcal{F}$ with $|F| \geq \epsilon \cdot |X|$, we have $N \cap F \neq \emptyset$.

Note that in the definition above, one seeks to hit sets whose *cardinality* is at least an ϵ-th fraction of the total *number* of points, while for the hitting set problem we wanted to hit all sets with *weight* an ϵ-th fraction of the total weight. But they are essentially equivalent: due to the fact that the coefficients of our LP are all rational numbers, each weight x_p is a rational number and so by scaling up the weights to be integers and replacing each $p \in P$ with correspondingly many distinct new elements, the weighted statement can be reduced to the cardinality one (this will be formally presented later, in Theorem 8.20).

Our goal in this chapter is to show the existence of ϵ-nets of small size when the set system is derived from configurations of geometric objects.

1. Deterministic

Many people have an impression that mathematics is an austere and formal subject concerned with complicated and ultimately confusing rules for the manipulation of numbers, symbols, and equations, rather like the preparation of a complicated income tax return. Good mathematics is quite opposite to this. Mathematics is an art of human understanding.

William Thurston

Given a set P of n points in the plane, let $\mathcal{R}$ be the primal set system induced on P by disks. That is,

$$\mathcal{R} = \left\{ D \cap P \colon D \text{ is a disk in } \mathbb{R}^2 \right\}.$$

What is the size of the smallest ϵ-net N of $\mathcal{R}$?

See the figure (a) for an example with $n = 16$ and a $\frac{1}{4}$-net N consisting of 6 points (shaded). In this case, any disk containing at least $\frac{1}{4} \cdot 16 = 4$ points must contain at least one of the six points of N.

It is easy to see that there are point sets where every ϵ-net must have size at least $\lfloor \frac{1}{\epsilon} \rfloor$. For example, arrange n points in groups of size ϵn, place the points in each group inside a small disk and place these disks disjoint from each other. See figure (b). Clearly, N must contain at least one point from each such disk and there are $\lfloor \frac{1}{\epsilon} \rfloor$ of them.

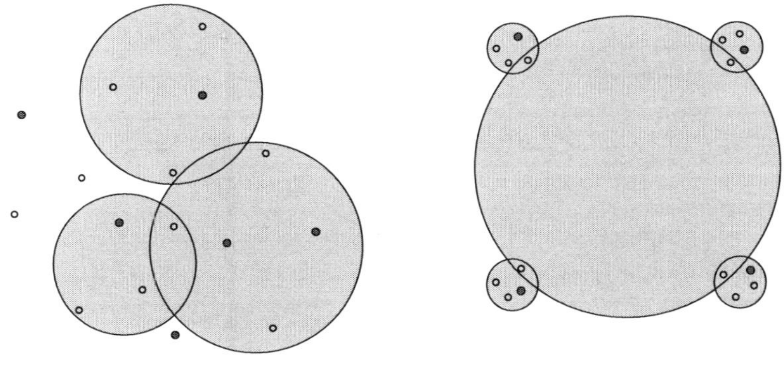

(a) (b)

On the other hand, in the above example, constructing N by arbitrarily picking one point from each disk is not sufficient, as there could exist a disk containing ϵn points of P but not containing any point of N.

Surprisingly, as we will see in Chapter 3, for any point set P in $\mathbb{R}^2$ and $\epsilon > 0$, there *does* exist an ϵ-net of size $O\left(\frac{1}{\epsilon}\right)$ of the primal set system induced on P by disks. In this section we deal with some simpler set systems, showing $O\left(\frac{1}{\epsilon}\right)$-sized ϵ-nets via ad-hoc geometric constructions.

Intervals in $\mathbb{R}$. Given a set P of n points in $\mathbb{R}$, our goal is to pick a set $N \subseteq P$ such that any interval that contains at least ϵn points of P contains some point of N. This is easy: sort the points of P in increasing order and simply pick every

$\lceil \epsilon n \rceil$-th point in this order. As any interval contains a contiguous subset of P with respect to this ordering, every interval containing at least ϵn points of P will be hit by N. The size of N is $\frac{n}{\lceil \epsilon n \rceil} \leq \frac{n}{\epsilon n} \leq \frac{1}{\epsilon}$.

Anchored rectangles in $\mathbb{R}^2$. Let P be a set of $n \geq 3$ points in $\mathbb{R}^2$, each with positive y-coordinate. We seek an ϵ-net N of the primal set system induced on P by axis-aligned rectangles 'anchored' at the x-axis—that is, rectangles with one horizontal edge lying on the x-axis. To construct N, assume the points of $P = \{p_1, \ldots, p_n\}$ are sorted by increasing x-coordinates. Partition P into sets $P_1, \ldots, P_t$ of contiguous points, with each P_i containing $\lceil \frac{\epsilon n}{2} \rceil$ points of P (except the last set P_t which could contain fewer points). For each $i = 1, \ldots, t$ add a point with the lowest y-coordinate in P_i—denote this point by q_i—to N. Note that $|N| = t \leq \frac{2}{\epsilon} + 1$. See the figure.

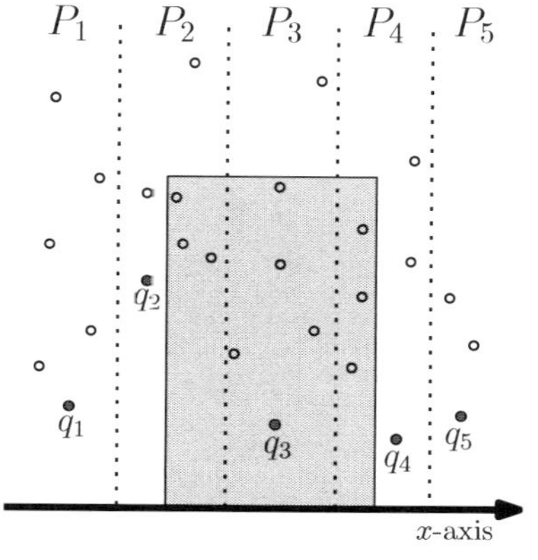

To see why N is an ϵ-net, consider any anchored rectangle R containing at least ϵn points of P. If R contains points from two sets of our partition, then it must contain one of those sets completely[1] and will be hit by N. Otherwise R must contain points from at least 3 sets in our partition, say the sets P_i, P_j and P_k where $i < j < k$. Then R must contain the point $q_j \in N$ with the lowest y-coordinate in P_j.

Half-spaces in $\mathbb{R}^2$. Our final result of this section is the following.

THEOREM 2.2. *Given a set P of n points in the plane and a parameter $\epsilon > 0$, there exists an ϵ-net of size at most $2\lceil \frac{1}{\epsilon} \rceil + 1$ of the primal set system induced on P by half-spaces in $\mathbb{R}^2$.*

PROOF. We assume that P is in general position: no three points of P lie on a common line.

First consider the easier case where P is in convex position. Then any half-space must contain a contiguous subset, with respect to the cyclic order in which the points appear on the convex-hull, of P. Thus picking every $\lceil \epsilon n \rceil$-th point along the convex-hull gives an ϵ-net of size $\lceil \frac{1}{\epsilon} \rceil$ (this is essentially the interval case).

[1] This was the reason for setting the cardinalities of the P_i's as we did.

Otherwise, if P is not in convex position, we will first map the points in P to a set P' which *is* in convex position, construct an ϵ-net N' of P', and from N' reconstruct an ϵ-net N of P.

Pick any point lying in the interior of the convex-hull conv (P) of P, say $o \in P$. For each $p_i \in P \setminus \{o\}$, trace a ray from o through p_i till this ray intersects the boundary of conv(P) at some point, say denoted by p_i'. Map p_i to p_i' (note that if p_i was on conv (P), then $p_i' = p_i$).

Let P' be the resulting set of $n-1$ points in convex position. Pick an ϵ-net N' of P' of size at most $\lceil \frac{1}{\epsilon} \rceil$.

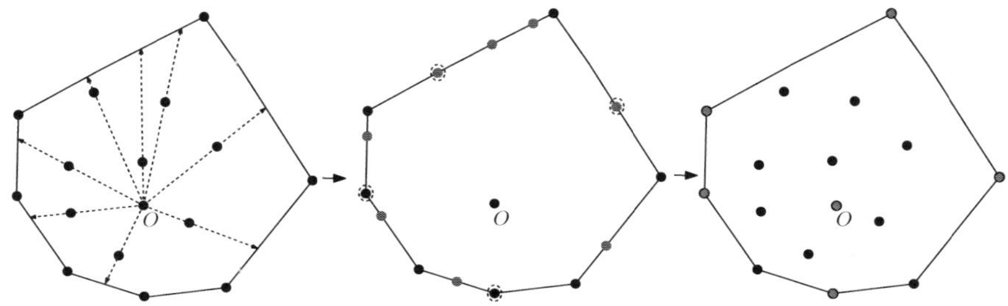

Now we show how to construct an ϵ-net N from N'. For each $p_i' \in N'$ there are two possibilities:

$p_i' = p_i$**:** That is, p_i was already on conv (P). Add p_i to N.

$p_i' \neq p_i$**:** Then p_i' lies on some edge of conv(P) that is spanned by some two points of P. Add both these points to N.

See the figure. Finally, add the point o to N. Clearly,

$$|N| \leq 2 \cdot |N'| + 1 = 2\left\lceil \frac{1}{\epsilon} \right\rceil + 1.$$

We claim that N is an ϵ-net. Consider any half-space h containing at least ϵn points of P. If h contains o, we are done as $o \in N$. Otherwise, we have the property that if a point p_i lies in h then the point p_i' also lies in h. Thus h contains at least ϵn points of P', and so it must contain a point $p' \in N'$. This implies that h must also contain at least one of the endpoints of the edge of conv(P) on which p' lies, both of which were added to N. □

Bibliography and discussion. The proof for half-spaces that we presented was invented here for didactic purposes; the original proof in [**KPW922a**] is by an extremal configuration argument.

[KPW922a] J. Komlós, J. Pach, and G. Woeginger, *Almost tight bounds for ϵ-nets*, Discrete Comput. Geom. **7** (1992), no. 2, 163–173, DOI 10.1007/BF02187833. MR1139078

2. Probabilistic

A modern mathematical proof is not very different from a modern machine—the simple fundamental principles are hidden and almost invisible under a mass of technical details.

Hermann Weyl

Our first theorem gives an upper bound on the size of ϵ-nets for general set systems.

THEOREM 2.3. *Let $(X, \mathcal{F})$ be a set system with $|X| = n$ and $|\mathcal{F}| = m$. Then given a parameter $\epsilon > 0$, there exists an ϵ-net N of $\mathcal{F}$ of size at most $\frac{\ln(m+1)}{\epsilon}$. Such a set can be computed by a greedy deterministic algorithm in time $O(nm)$.*
Furthermore, given a parameter $\gamma \in \left(0, \frac{1}{2}\right]$, a uniform random sample $N \subseteq X$ of size $\frac{1}{\epsilon} \ln \frac{m}{\gamma}$ is an ϵ-net of $\mathcal{F}$ with probability at least $1-\gamma$. The same holds if N is constructed by picking each point of X independently with probability $\frac{1}{\epsilon n} \ln \frac{m}{\gamma}$.

If the set system $\mathcal{F}$ is arbitrary, then the above bound is tight within constant factors.
Theorem 2.3 does not assume anything about the structure of $\mathcal{F}$. Obviously set systems derived from geometry have additional properties and hence one can expect to get better bounds in many cases. Consider next the following case of a basic geometric set system where this bound can be improved.
Let P be a set of n points in $\mathbb{R}^d$ and $\mathcal{R}$ the primal set system induced by axis-aligned boxes. That is,

$$\mathcal{R} = \left\{ B \cap P \colon B \text{ is an axis-aligned box in } \mathbb{R}^d \right\}.$$

First observe that $|\mathcal{R}| = O\left(n^{2d}\right)$:

> take any set $R \in \mathcal{R}$, and let B be an axis-aligned box such that $R = B \cap P$. Now let B' be another axis-aligned box constructed by translating each of the $2d$ axis-parallel bounding hyperplanes of B 'inwards' till they each contain a point of P. Then clearly B' contains precisely the points of R, and further, each of its $2d$ axis-parallel bounding hyperplanes contains a point of P. As there are n choices for each such hyperplane, there are at most n^{2d} possible B''s, implying the same bound for the size of $\mathcal{R}$.

Theorem 2.3 now implies the existence of an ϵ-net of size $\frac{1}{\epsilon} \log\left(|\mathcal{R}|+1\right) = O\left(\frac{d}{\epsilon} \log n\right)$ of $\mathcal{R}$. Our second main result of this section shows that this bound can improved by a simple geometric observation.

THEOREM 2.4. *Let P be a set of n points in $\mathbb{R}^d$ and let $\epsilon \in \left(0, \frac{1}{2}\right]$ be a given parameter. Then, with probability at least $\frac{1}{2}$, a uniform random sample of X of size $\Theta\left(\frac{d}{\epsilon} \log \frac{d}{\epsilon}\right)$ is an ϵ-net of the primal set system induced by axis-aligned boxes in $\mathbb{R}^d$.*

The fact that the above bound of $\Theta\left(\frac{d}{\epsilon} \log \frac{d}{\epsilon}\right)$ is *independent* of the size of X and the size of $\mathcal{R}$ is the key in many applications of ϵ-nets.

PROOF OF THEOREM 2.3. Note that for our purposes we can assume that each set in $\mathcal{F}$ has size at least ϵn. Let $\mathcal{F} = \{F_1, \ldots, F_m\}$.
There are several, essentially equivalent, ways of viewing the proof.

Iterative view: We construct N by a greedy algorithm. Set $\mathcal{F}_1 = \mathcal{F}$. Pick an element of X that hits the maximum number of sets of $\mathcal{F}_1$, say $p_1 \in X$. Add p_1 to N, remove the sets hit by p_1 from $\mathcal{F}_1$ and denote this smaller collection by $\mathcal{F}_2$. Now re-iterate this procedure on $\mathcal{F}_2$; we continue these iterations until there are no remaining unhit sets.
Let $p_i \in X$ be the point added to N in the i-th iteration, and let $\mathcal{F}_{i+1}$ be the unhit sets after i iterations. By the pigeonhole principle, at the i-th iteration there exists an element of X hitting at least

$$\frac{\sum_{F \in \mathcal{F}_i} |F|}{n} \geq \frac{\sum_{F \in \mathcal{F}_i} \epsilon n}{n} = \epsilon \cdot |\mathcal{F}_i|$$

sets of $\mathcal{F}_i$. Thus for any $i \geq 1$, we have

$$|\mathcal{F}_{i+1}| \leq (1-\epsilon)\,|\mathcal{F}_i| \leq (1-\epsilon)^2\,|\mathcal{F}_{i-1}| \leq \cdots \leq (1-\epsilon)^i \cdot |\mathcal{F}|.$$

After the t-th iteration, we have in total added t elements to N and the number of unhit sets remaining is at most $(1-\epsilon)^t\, m$. Setting $t = \frac{\ln(m+1)}{\epsilon}$, we get

$$|\mathcal{F}_{t+1}| \leq (1-\epsilon)^t \cdot m \leq m\, e^{-\epsilon\, t} \leq m\, e^{-\ln(m+1)} = \frac{m}{m+1} < 1.$$

In other words, $\mathcal{F}_{t+1}$ is empty and so N is a hitting set of $\mathcal{F}$, of size at most $\frac{\ln(m+1)}{\epsilon}$.
This algorithm can be implemented in deterministic $O(nm)$ time:

> Let A be an array of size n, such that $A[j]$ is the number of sets of $\mathcal{F}$ containing the j-th element of X; A can be constructed in $O(nm)$ time simply by iterating over all sets of $\mathcal{F}$.
> In each iteration $i = 1, \ldots,$ we can compute p_i in $O(n)$ time by scanning A to find a maximum value. Next we iterate over all sets of $\mathcal{F}$, and for each $F \in \mathcal{F}$ that is hit by p_i, we
> a) remove F from $\mathcal{F}$, and
> b) update A by decrementing each $A[j]$ if F contains the j-th element.
>
> If t_i sets are hit by p_i, then the total time taken is $O(m + nt_i)$.
> As the total number of iterations is at most $\min\{n, m\}$, the total time is
>
> $$O(mn) + \sum_{i=1}^{\min\{n,m\}} O(n + m + nt_i) = O(nm) + O(n) \sum_i t_i = O(nm).$$

Combinatorial view: Note that the pigeonholing in the above proof essentially states that, on average, each element of X hits ϵm sets of $\mathcal{F}$. Of course, this average goes down with the number of iterations, which is what results in the multiplicative logarithmic term. This argument can be viewed succinctly combinatorially, as follows. Let t be a positive integer whose value will be fixed later. Then, we

> count the number of subsets of X of size t that are *not* ϵ-nets of $\mathcal{F}$.

For each $F \in \mathcal{F}$, there are $\binom{n-|F|}{t} \leq \binom{n-\epsilon n}{t}$ subsets of X of size t that do not hit F. Hence the number of subsets of X that do not hit at least one set of $\mathcal{F}$ can be upper bounded as

$$\sum_{F \in \mathcal{F}} \Big(\text{\# of } t\text{-sized subsets of } X \text{ that do not hit } F\Big) \leq m \cdot \binom{n-\epsilon n}{t}.$$

If this is less than the total number of subsets of size t, then clearly there exists a set of size t that hits all sets of $\mathcal{F}$. A calculation now shows that for $t = \frac{\ln(m+1)}{\epsilon}$, this is indeed the case:

$$\begin{aligned}
\frac{m \cdot \binom{n-\epsilon n}{t}}{\binom{n}{t}} &= m \cdot \frac{(n-t)!}{(n-\epsilon n-t)!} \cdot \frac{(n-\epsilon n)!}{n!} \\
&= m \cdot \frac{(n-t)}{n} \frac{(n-t-1)}{(n-1)} \cdots \frac{(n-\epsilon n-t+1)}{(n-\epsilon n+1)} \\
&= m \cdot \left(1 - \frac{t}{n}\right)\left(1 - \frac{t}{n-1}\right) \cdots \left(1 - \frac{t}{n-\epsilon n+1}\right) \\
&\leq m \cdot \left(1 - \frac{t}{n}\right)^{\epsilon n} \leq m \cdot e^{-\epsilon t} = m \cdot e^{-\ln(m+1)} < 1.
\end{aligned}$$

Probabilistic view: Perhaps the simplest view is the probabilistic one. Take a random sample N in the following way.

N: choose an element uniformly at random from X t times (with replacement).

Then

$$\Pr\big[\text{a fixed set } F \in \mathcal{F} \text{ is not hit by } N\big] = \left(1 - \frac{|F|}{n}\right)^t \leq (1-\epsilon)^t \leq e^{-\epsilon t}. \tag{2.5}$$

By the union bound over all sets in $\mathcal{F}$,

$$\Pr\big[N \text{ fails to be an } \epsilon\text{-net of } \mathcal{F}\big] \leq m e^{-\epsilon t}. \tag{2.6}$$

For $t = \frac{\ln(m+1)}{\epsilon}$, this is less than 1. In particular there is a non-zero probability that N will hit all sets and so there has to exist at least one such set. Furthermore, setting $t = \frac{1}{\epsilon} \ln \frac{m}{\gamma}$ implies that N is an ϵ-net of $\mathcal{F}$ with probability at least $1 - \gamma$. If N is constructed by picking each point of X independently with probability $\frac{1}{\epsilon n} \ln \frac{m}{\gamma}$, the probability that N fails to be an ϵ-net of $\mathcal{F}$ can be upper bounded as

$$\sum_{F \in \mathcal{F}} \left(1 - \frac{1}{\epsilon n} \ln \frac{m}{\gamma}\right)^{|F|} \leq m \left(1 - \frac{1}{\epsilon n} \ln \frac{m}{\gamma}\right)^{\epsilon n} \leq m \exp\left(-\ln \frac{m}{\gamma}\right) = \gamma,$$

as desired.

□

Observe that in the probabilistic view of the proof of Theorem 2.3, to get a constant probability of hitting any *fixed* set of size at least ϵn, we only need to take a random sample of size $\Theta\left(\frac{1}{\epsilon}\right)$ (Equation (2.5)). The additional multiplicative $\ln m$ factor arose due to the requirement of hitting *all* m sets of $\mathcal{F}$ simultaneously, and for which we used the union bound to upper bound the failure probability (Equation (2.6)).

Our second main result improves this $\ln m$ factor when the set system is induced by axis-aligned boxes in $\mathbb{R}^d$, by showing that the number of sets over which to apply the union bound can be made much smaller.

THEOREM 2.4. *Let P be a set of n points in $\mathbb{R}^d$ and let $\epsilon \in \left(0, \frac{1}{2}\right]$ be a given parameter. Then, with probability at least $\frac{1}{2}$, a uniform random sample of X of size $\Theta\left(\frac{d}{\epsilon}\log\frac{d}{\epsilon}\right)$ is an ϵ-net of the primal set system induced by axis-aligned boxes in $\mathbb{R}^d$.*

PROOF. Given P and $\epsilon > 0$, we will show the existence of a small set of boxes—in fact $O\left(\left(\frac{4d}{\epsilon}\right)^{2d}\right)$, independent of n—such that it suffices to construct an $\frac{\epsilon}{2}$-net for these boxes. The next lemma proves their existence.

LEMMA 2.7. *Given a set P of n points in $\mathbb{R}^d$ and a parameter $\epsilon > 0$, there exists a set $\mathcal{B} = \{B_1, \ldots, B_t\}$ of $t = O\left(\left(\frac{4d}{\epsilon}\right)^{2d}\right)$ axis-aligned boxes such that*

- *each box of $\mathcal{B}$ contains at least $\frac{\epsilon n}{2}$ points of P, and*
- *for any axis-aligned box B containing at least ϵn points of P, there exists an index $j \in \{1, \ldots, t\}$ such that $B_j \subseteq B$.*

PROOF. For each $i \in [d]$, let $\mathcal{H}_i$ be a set of hyperplanes orthogonal to the i-th axis such that there are at most $\frac{\epsilon n}{4d}$ points of P between two consecutive, as ordered along the i-th axis, hyperplanes of $\mathcal{H}_i$. The same also holds for the two half-infinite slabs bounded by the first and last hyperplane of $\mathcal{H}_i$. Note that

$$|\mathcal{H}_i| \leq \left\lceil \frac{n}{(\epsilon n/4d)} \right\rceil = \left\lceil \frac{4d}{\epsilon} \right\rceil.$$

Set $\mathcal{H} = \mathcal{H}_1 \cup \cdots \cup \mathcal{H}_d$. Note that the hyperplanes of $\mathcal{H}$ induce a non-uniform grid in $\mathbb{R}^d$. See the figure for an illustration in $\mathbb{R}^2$.

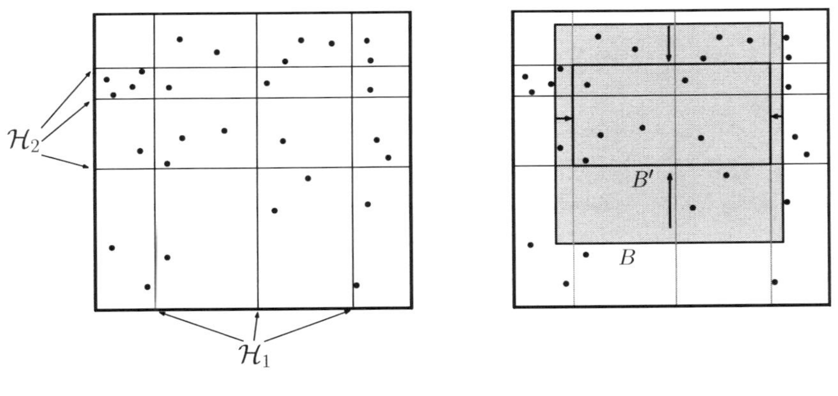

Now for any box B containing ϵn points of P, one can move each of B's $2d$ bounding hyperplanes 'inwards' till each coincides with a hyperplane of $\mathcal{H}$; let B' be the resulting smaller box. As there are at most $\frac{\epsilon n}{4d}$ points of P between any two consecutive hyperplanes of $\mathcal{H}_i$, we have

$$|B' \cap P| \geq |B \cap P| - 2d \cdot \frac{\epsilon n}{4d} \geq \epsilon n - \frac{\epsilon n}{2} = \frac{\epsilon n}{2}.$$

Thus our required set $\mathcal{B}$ is simply the set of all possible boxes formed by picking $2d$ hyperplanes from $\mathcal{H}$, where exactly two hyperplanes are picked from each $\mathcal{H}_i$,

$i = 1, \ldots, d$. We only keep those boxes which contain at least $\frac{\epsilon n}{2}$ points of P. We have

$$|\mathcal{B}| \leq \binom{|\mathcal{H}_1|}{2} \cdot \binom{|\mathcal{H}_2|}{2} \cdots \binom{|\mathcal{H}_{d-1}|}{2} \cdot \binom{|\mathcal{H}_d|}{2} = O\left(\left(\frac{4d}{\epsilon}\right)^{2d}\right).$$

□

To compute an ϵ-net of the set system induced on P by axis-aligned boxes in $\mathbb{R}^d$, it suffices to compute an $\frac{\epsilon}{2}$-net N of the primal set system induced on P by the boxes $\mathcal{B} = \{B_1, \ldots, B_t\}$ given by Lemma 2.7. Now the bound follows by applying Theorem 2.3:

$$|N| \leq \frac{\ln(|\mathcal{B}| + 1)}{\epsilon/2} = \frac{\ln\left(O\left(\left(\frac{4d}{\epsilon}\right)^{2d}\right)\right)}{\epsilon/2} = O\left(\frac{d}{\epsilon} \log \frac{d}{\epsilon}\right).$$

□

Bibliography and discussion. The general set system proof is folklore, though typically only the probabilistic or greedy view is presented; see [**Cha00**] for the idea in a more general weighted setting, as well as for deterministic algorithms to compute such nets. The proof for axis-aligned rectangles given here was invented for didactic purposes, mainly to illustrate the surprising fact that the size of an ϵ-net need not depend on the number of elements or sets of a given set system.

[Cha00] B. Chazelle, *The discrepancy method*, Cambridge University Press, Cambridge, 2000. Randomness and complexity, DOI 10.1017/CBO9780511626371. MR1779341

3. Improved Probabilistic Analysis

He [Lefschetz] liked to repeat, as an example of mathematical pedantry, the story of one of E. H. Moore's visits to Princeton, when Moore started a lecture by saying, 'Let a be a point and let b be a point.' 'But why don't you just say, Let a and b be points!' asked Lefschetz. 'Because a may equal b,' answered Moore.

Lefschetz got up and left the lecture room.

Gian-Carlo Rota

Theorem 2.4 stated the existence of ϵ-nets of size $O\left(\frac{d}{\epsilon}\log\frac{d}{\epsilon}\right)$ of the primal set system induced by axis-aligned boxes in $\mathbb{R}^d$. We now incorporate the key idea present in that proof into a general technique for proving ϵ-net bounds for a broader category of geometric set systems. As a warm-up illustration of the technique, we first re-prove the bound for axis-aligned rectangles in $\mathbb{R}^2$.

THEOREM 2.8. *Let P be a set of n points in $\mathbb{R}^2$ and $\epsilon \in \left(0, \frac{1}{2}\right]$ a given parameter. Then there exists an ϵ-net of size $O\left(\frac{1}{\epsilon}\log\frac{1}{\epsilon}\right)$ of the primal set system induced on P by axis-aligned rectangles in $\mathbb{R}^2$.*

Then we present another application of this technique.

THEOREM 2.9. *Let P be a set of n points in $\mathbb{R}^2$ and $\epsilon \in \left(0, \frac{1}{2}\right]$ a given parameter. Then there exists an ϵ-net of size $O\left(\frac{1}{\epsilon}\log\frac{1}{\epsilon}\right)$ of the primal set system induced on P by disks in $\mathbb{R}^2$.*

The construction in Theorems 2.8 and 2.9 remains the same as that in Theorem 2.4: given a set P of n points in $\mathbb{R}^2$, we construct a random sample S by choosing each point of P independently with probability p. We show that for $p = \Theta\left(\frac{1}{\epsilon n}\log\frac{1}{\epsilon}\right)$, there is a constant probability that S is an ϵ-net.

PROOF OF THEOREM 2.8. We will assume that P is in general position. In particular, that no two points of P have the same x or y coordinate.
The key notion is that of *canonical rectangles*, similar to Definition 1.3. Let $\mathcal{U}$ be an axis-aligned square containing P in its interior.

DEFINITION 2.10. An axis-aligned rectangle R is called a canonical rectangle spanned by $Q \subset \mathcal{U}$ if each bounding edge of R contains a point of Q or lies on $\partial\mathcal{U}$.
Furthermore, a canonical rectangle R spanned by Q is called an empty canonical rectangle if the interior of R contains no point of Q.

Note that there are $O\left(n^4\right)$ canonical rectangles spanned by any set of n points in the plane. Furthermore, there are four types of canonical rectangles, 'fixed' by 4, 3, 2 or 1 points of Q on the boundary. See the figure for some examples of empty canonical rectangles.

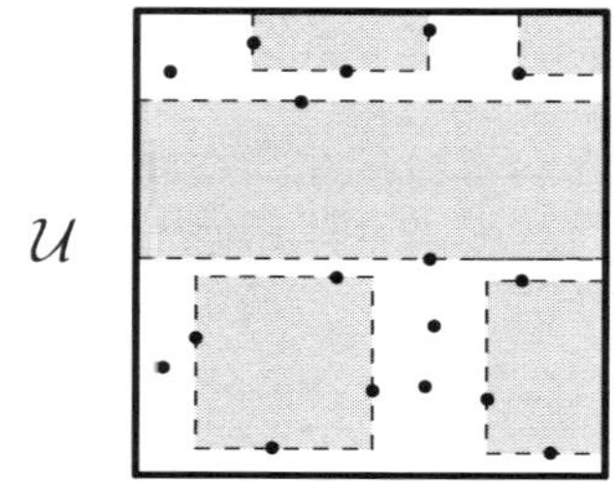

For the rest of the proof, we only consider canonical rectangles fixed by 4 points; the other cases are very similar and treated *mutatis mutandis.*

Let S be a random sample constructed by picking each point of P independently with probability p, for a parameter p to be fixed later.

The following key claim shows the relevance of canonical rectangles to ϵ-nets.

CLAIM 2.11. If the interior of each empty canonical rectangle spanned by S has less than ϵn points of P, then S is an ϵ-net.

PROOF. We prove the contrapositive. Assume that S is not an ϵ-net and let R be an axis-aligned rectangle containing at least ϵn points of P and no point of S. Then expand R by translating its left, right, top and bottom edges 'outwards' to transform it to an empty canonical rectangle R' spanned by S. As $R \subseteq \operatorname{int}(R')$, the interior of R' contains at least ϵn points of P. $\square$

Thus it suffices to show that for an appropriate value of p, with non-zero probability, the interior of each empty canonical rectangle spanned by the random sample S contains less than ϵn points of P.

We first sketch a rough calculation that, while technically incorrect, conveys the right idea.

> There are $O\left(\mathrm{E}\left[|S|\right]^4\right)$ possible canonical rectangles spanned by the points of S. Let R be an *empty* canonical rectangle spanned by S. If R contains at least ϵn points of P in its interior, then the probability that R contains no point of S is at most $(1-p)^{\epsilon n}$ (circular reasoning, as we already assumed that R was empty!). Thus by the union bound, the probability that there exists an empty canonical rectangle spanned by S and containing at least ϵn points of P is at most
>
> $$O\left(\mathrm{E}\left[|S|\right]^4\right) \cdot (1-p)^{\epsilon n} = O\left((np)^4 \cdot e^{-p\epsilon n}\right) < \frac{1}{2},$$
>
> by setting $p = \frac{c}{\epsilon n} \ln \frac{1}{\epsilon}$ for a sufficiently large constant c.

The correct way of doing the probability calculations will use the following observation, stated without proof.

CLAIM 2.12. A canonical rectangle R spanned by four points of P ends up as an empty canonical rectangle in S if and only if

(1) the four points spanning R are picked into S, and
(2) none of the points contained in the interior of R are picked into S.

Let $\mathcal{R}_{can}$ be the set of *all* canonical rectangles spanned by 4 points of P. We have $|\mathcal{R}_{can}| = O\left(n^4\right)$.

For each $R \in \mathcal{R}_{can}$ let $\mathcal{E}_R$ be the event that R ends up as an empty canonical rectangle spanned by S. As each point of P was picked independently into S with probability p, Claim 2.12 implies that

$$\Pr\left[\mathcal{E}_R\right] = p^4 \cdot (1-p)^{|\mathrm{int}(R)\cap P|}.$$

Setting $p = \frac{c}{\epsilon n}\ln\frac{1}{\epsilon}$ for a constant c, and using the union bound over all rectangles in $\mathcal{R}_{can}$, the probability that a rectangle of $\mathcal{R}_{can}$ containing at least ϵn points of P in its interior ends up as an empty canonical rectangle spanned by S, is

$$\begin{aligned}\Pr\left[\bigcup_{\substack{R\in\mathcal{R}_{can}\\ |\mathrm{int}(R)\cap P|\geq\epsilon n}}\mathcal{E}_R\right] &\leq \sum_{\substack{R\in\mathcal{R}_{can}\\ |\mathrm{int}(R)\cap P|\geq\epsilon n}}\Pr\left[\mathcal{E}_R\right] = \sum_{\substack{R\in\mathcal{R}_{can}\\ |\mathrm{int}(R)\cap P|\geq\epsilon n}} p^4\cdot(1-p)^{|\mathrm{int}(R)\cap P|}\\ &= O\left(n^4\right)\cdot O\left(p^4\cdot e^{-p\,\epsilon n}\right) = O\left(n^4\cdot\left(\frac{c}{\epsilon n}\ln\frac{1}{\epsilon}\right)^4\cdot e^{-c\ln\frac{1}{\epsilon}}\right)\\ &= O\left(\left(c\ln\frac{1}{\epsilon}\right)^4\cdot\epsilon^{c-4}\right)\leq\frac{1}{16},\end{aligned}$$

for a sufficiently large constant c (e.g., as $\epsilon \leq 1/2$, taking $c = 25$ gives $(25\ln 2)^4 \cdot \frac{1}{2^{21}} \leq \frac{1}{2^2}$).

Similarly, consider the set of all canonical rectangles containing at least ϵn points of P and fixed by 3 points of P. The probability that one of them ends up as an empty canonical rectangle spanned by S is at most

$$O\left(n^3\right)\cdot O\left(p^3 e^{-p\epsilon n}\right) = O\left(\left(c\ln\frac{1}{\epsilon}\right)^3\cdot\epsilon^{c-3}\right)\leq\frac{1}{16}.$$

Likewise for the canonical rectangles fixed by 2 and 1 points of P. Therefore, with probability at least $1-\frac{4}{16}=\frac{3}{4}$, the interior of each empty canonical rectangle spanned by S has less than ϵn points of P and so by Claim 2.11, S is an ϵ-net.

Finally we have

$$\mathrm{E}\left[|S|\right] = np = \frac{c}{\epsilon}\ln\frac{1}{\epsilon}.$$

By Markov's inequality (Equation (1.26)), the probability that $|S| \geq 4\,\mathrm{E}\left[|S|\right]$ is at most $\frac{1}{4}$. Thus by the union bound,

$$\begin{aligned}\Pr\left[S \text{ is not an } \epsilon\text{-net } \textbf{or } |S|\geq 4np\right] &\leq \Pr\left[S \text{ is not an } \epsilon\text{-net }\right] + \Pr\left[|S|\geq 4np\right]\\ &\leq \frac{1}{4}+\frac{1}{4}=\frac{1}{2}.\end{aligned}$$

and so with probability at least $\frac{1}{2}$, S is an ϵ-net and has size at most $4np = O\left(\frac{1}{\epsilon}\log\frac{1}{\epsilon}\right)$, completing the proof. □

To avoid having to use Markov's inequality, sometimes the sampling is done with a different distribution, either by letting S be a uniform random subset of size $t = \Theta\left(\frac{1}{\epsilon}\log\frac{1}{\epsilon}\right)$ or by picking t elements from P with replacement. Then the size of S is fixed, though the probability calculations become slightly more involved. The final result is the same within constant factors.

The reader will notice that the only property of axis-aligned rectangles used in the proof of Theorem 2.8 is that a canonical rectangle R is 'fixed' by a constant number of points of P and that R is an empty canonical rectangle spanned by S if and only if these points are picked into S and none of the points in the interior of R are picked into S. This is a very general idea and we now outline how a similar method of analysis shows the existence of an ϵ-net of size $O\left(\frac{1}{\epsilon}\log\frac{1}{\epsilon}\right)$ of the primal set system induced by disks.

PROOF OF THEOREM 2.9. We will assume that P is in general position. In particular, that no three points of P lie on a common line and no four points of P lie on a common circle.
As before, let S be a random sample constructed by picking each point of P independently with probability p, with the parameter p to be fixed later.
Consider the notion of canonical halfplanes and canonical disks.

DEFINITION 2.13. A canonical halfplane spanned by $Q \subseteq \mathbb{R}^2$ is a halfplane whose boundary line contains two points of Q.
Furthermore, a canonical halfplane H spanned by Q is called an empty canonical halfplane if the interior of H contains no point of Q.

DEFINITION 1.3. A canonical disk spanned by $Q \subseteq \mathbb{R}^2$ is a disk whose boundary contains three points of Q.
Furthermore, a canonical disk D spanned by Q is called an empty canonical disk if the interior of D contains no point of Q.

We will need the following independent geometric fact, an analog of Claim 2.11.

CLAIM 2.14. Let $Q \subset \mathbb{R}^2$, $|Q| \geq 3$, be a finite set of points and let D be a closed disk not containing any point of Q. Then D lies in the interior of the union of at most two empty canonical objects—disks/halfplanes—spanned by Q.

PROOF. Let c be the center of D. Keeping c fixed, increase the radius of D until it becomes tangent to some point of Q, say q_1. Denote this new disk by D_1; clearly $D \subset D_1$. Next, move the center c of D_1 along the direction $\vec{q_1 c}$, while keeping D_1 tangent to q_1, until it becomes tangent to another point of Q[2], say q_2. Denote this disk by D_2 and its center by c_2. Again we have $D_1 \subset D_2$. See the figure.
Note that c_2 lies on the perpendicular bisector of the segment q_1q_2. Transform D_2, while keeping it tangent to both q_1 and q_2, by moving its center c_2 along this bisector in each of the two directions till it becomes tangent to a third point of Q. This gives two empty canonical disks or halfplanes whose union contains D_2 in its interior, completing the proof. □

[2]It could happen that D_1 becomes a halfplane, in which case we can rotate it clockwise and anti-clockwise around q_1 to get two empty canonical halfplanes spanned by Q such that the interior of their union contains D_1.

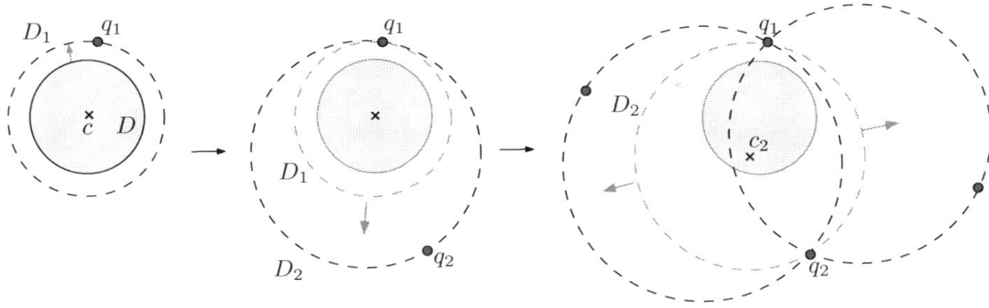

Claim 2.14 implies that

> if the interior of each empty canonical disk and each empty canonical halfplane spanned by S contains less than $\frac{\epsilon n}{2}$ points of P, then any closed disk containing no point of S contains less than ϵn points of P, and thus S is an ϵ-net.

A canonical disk D spanned by three points of P is an empty canonical disk spanned by S if and only if the three points spanning D are present in S and no point of P lying in the interior of D is picked into S. The probability that this happens for a fixed canonical disk D is precisely

$$p^3 \cdot (1-p)^{|\mathrm{int}(D) \cap P|}.$$

There are $O(n^3)$ canonical disks spanned by P. Thus by the union bound, the probability that some canonical disk spanned by P and containing at least $\frac{\epsilon n}{2}$ points of P in its interior ends up as an empty canonical disk spanned by S can be upper bounded by

$$O(n^3) \cdot p^3 \cdot (1-p)^{\frac{\epsilon n}{2}} = O\left(n^3 \cdot p^3 \cdot e^{-\frac{p\epsilon n}{2}}\right) = O\left(n^3 \cdot \left(\frac{2c}{\epsilon n} \ln \frac{1}{\epsilon}\right)^3 \cdot e^{-c \ln \frac{1}{\epsilon}}\right)$$
$$= O\left(\left(c \ln \frac{1}{\epsilon}\right)^3 \cdot \epsilon^{c-3}\right) \leq \frac{1}{8},$$

by setting $p = \frac{2c}{\epsilon n} \ln \frac{1}{\epsilon}$ for a sufficiently large constant c.

A similar analysis shows that the probability that there exists an empty canonical halfplane spanned by Q and containing at least $\frac{\epsilon n}{2}$ points of P is at most $\frac{1}{8}$. Markov's inequality implies that with probability at least $\frac{3}{4}$, we have $|S| \leq \frac{8c}{\epsilon} \ln \frac{1}{\epsilon}$. Thus one can conclude that with probability at least $\frac{1}{2}$, S is an ϵ-net of size $O\left(\frac{1}{\epsilon} \log \frac{1}{\epsilon}\right)$. □

Bibliography and discussion. The construction and analysis of ϵ-nets using the framework of canonical structures is implicit in [**Cla88**]. The ideas presented in this section apply more generally in the framework of so-called LP-type problems [**SW92**] and their generalization, violator spaces [**Gär+08**].

[Cla88] K. L. Clarkson, *A randomized algorithm for closest-point queries*, SIAM J. Comput. **17** (1988), no. 4, 830–847, DOI 10.1137/0217052. MR953296

[Gär+08] B. Gärtner, J. Matoušek, L. Rüst, and P. Škovroň, *Violator spaces: structure and algorithms*, Discrete Appl. Math. **156** (2008), no. 11, 2124–2141, DOI 10.1016/j.dam.2007.08.048. MR2437006

[SW92] M. Sharir and E. Welzl, *A combinatorial bound for linear programming and related problems*, STACS 92 (Cachan, 1992), Lecture Notes in Comput. Sci., vol. 577, Springer, Berlin, 1992, pp. 569–579. MR1255620

CHAPTER 3

Refining Random Samples

Let $\mathcal{R}$ be the primal set system induced by disks on a set P of n points in the plane. An earlier probabilistic construction—Theorem 2.3—of an ϵ-net N of $\mathcal{R}$ consisted of taking a uniform random sample $N \subseteq P$. For N to be an ϵ-net with non-zero probability, calculations using the union bound required that

$$|N| = \Omega\left(\frac{1}{\epsilon}\log|\mathcal{R}|\right) = \Omega\left(\frac{1}{\epsilon}\log n\right).$$

We were able to improve this bound by showing, using geometric properties of disks, that the number of 'bad' events that cause a $N \subseteq P$ to *not* be an ϵ-net can be upper bounded by a polynomial function of $\frac{1}{\epsilon}$. This implied that a uniform random sample N of size $\Theta\left(\frac{1}{\epsilon}\log\frac{1}{\epsilon}\right)$ is an ϵ-net of $\mathcal{R}$ with probability at least $\frac{1}{2}$. If one is limited to taking a *single uniform* random sample, trivial considerations show that this cannot be improved; that is, for a uniform random sample to be an ϵ-net with probability at least $\frac{1}{2}$ (or any absolute constant), we must sample $\Omega\left(\frac{1}{\epsilon}\ln\frac{1}{\epsilon}\right)$ points.

> Consider the case when $\mathcal{R}$ consists of $\frac{1}{\epsilon}$ disjoint sets of ϵn points each. Let N be a uniform random sample of P of size $\frac{1}{2\epsilon}\ln\frac{1}{\epsilon}$, constructed by repeatedly choosing a point of P uniformly at random (with replacement). Then for each $S \in \mathcal{R}$,
> $$\Pr[S \text{ is not hit by } N] = (1-\epsilon)^{|N|} \geq e^{-2\,\epsilon|N|} \geq \epsilon,$$
> where the second step uses the fact that $1 - x \geq e^{-2x}$ for $x \in [0, 0.79]$. As the sets of $\mathcal{R}$ are disjoint,
> $$\Pr[N \text{ is an } \epsilon\text{-net of } \mathcal{R}] = \prod_{S\in\mathcal{R}}\left(1 - \Pr[S \text{ is not hit by } N]\right) \leq (1-\epsilon)^{\frac{1}{\epsilon}} \leq \frac{1}{e}.$$

If we are interested in the *existence* of small ϵ-nets, we need not restrict ourselves to only taking random samples. The goal of this chapter is to improve the bounds on ϵ-nets to $o\left(\frac{1}{\epsilon}\log\frac{1}{\epsilon}\right)$ by the following general idea.

> To improve upon the $\log t$ factor that is needed to take care of t bad events using the union bound, we allow *some* bad events to happen and then 'fix' these bad events with additional work. This involves a trade-off: we pick a smaller initial random sample, but pay an additional price to fix the problems caused by the failed events.

This idea is captured, in the parlance of our times, by the quote 'the biggest risk is to not take any risk'. It is also called the *alterations technique* in the study of probabilistic methods. We illustrate the idea in our context by slightly improving the upper bound of Theorem 2.3, which we first recall.

THEOREM 2.3. *Let* $(X, \mathcal{F})$ *be a set system with* $|X| = n$ *and* $|\mathcal{F}| = m$. *Then given a parameter* $\epsilon > 0$, *there exists an* ϵ*-net* N *of* $\mathcal{F}$ *of size at most* $\frac{\ln(m+1)}{\epsilon}$. *Such a set can be computed by a greedy deterministic algorithm in time* $O(nm)$. *Furthermore, given a parameter* $\gamma \in \left(0, \frac{1}{2}\right]$, *a uniform random sample* $N \subseteq X$ *of size* $\frac{1}{\epsilon} \ln \frac{m}{\gamma}$ *is an* ϵ*-net of* $\mathcal{F}$ *with probability at least* $1 - \gamma$. *The same holds if* N *is constructed by picking each point of* X *independently with probability* $\frac{1}{\epsilon n} \ln \frac{m}{\gamma}$.

Here is the improved statement.

THEOREM 3.1. *Let* $(X, \mathcal{F})$ *be a set system with* $|X| = n$ *and* $|\mathcal{F}| = m$. *Then given a parameter* $\epsilon > 0$, *there exists an* ϵ*-net* N *of* $\mathcal{F}$ *of size at most* $\frac{\ln(\epsilon m)}{\epsilon} + \frac{1}{\epsilon}$.

PROOF. Let N_0 be constructed by choosing an element, uniformly at random from X, t times (with replacement). Let $\mathcal{F}^!$ be the sets of $\mathcal{F}$ *not* hit by N_0. Add an arbitrary point from each set of $\mathcal{F}^!$ to N_0 to get an ϵ-net N. Then

$$\begin{aligned} \mathrm{E}\left[|N|\right] &\leq |N_0| + \mathrm{E}\left[|\mathcal{F}^!|\right] \\ &\leq t + \sum_{\substack{S \in \mathcal{F}, \\ |S| \geq \epsilon n}} \Pr\left[S \cap N_0 = \emptyset\right] \\ &= t - \sum_{\substack{S \in \mathcal{F}, \\ |S| \geq \epsilon n}} \left(1 - \frac{|S|}{n}\right)^t \\ &\leq t + m\left(1 - \epsilon\right)^t \leq t + me^{-\epsilon t}. \end{aligned}$$

Setting $t = \frac{\ln(\epsilon m)}{\epsilon}$ gives the required bound. □

While this improvement is modest in general, it is useful in cases where $|\mathcal{F}|$ can be related to $\frac{1}{\epsilon}$. Here is an example.

LEMMA 3.2. *Let* $(X, \mathcal{F})$ *be a set system where each set in* $\mathcal{F}$ *has size at least* $\epsilon|X|$, *and let* $D \geq 2$ *be an integer such that each element of* X *belongs to at most* D *sets of* $\mathcal{F}$. *That is,* $\deg_{\mathcal{F}}(p) \leq D$ *for all* $p \in X$. *Then there exists an* ϵ*-net of* $\mathcal{F}$ *of size* $O\left(\frac{1}{\epsilon} \ln D\right)$.

PROOF. We first upper bound $|\mathcal{F}|$ by a double-counting argument:

$$|\mathcal{F}| \cdot \epsilon|X| \;\leq\; \sum_{F \in \mathcal{F}} |F| \;=\; \sum_{p \in X} \deg_{\mathcal{F}}(p) \;\leq\; |X| \cdot D,$$

$$\implies |\mathcal{F}| \leq \frac{D}{\epsilon}.$$

The proof now follows by applying Theorem 3.1. □

The rest of this chapter is essentially variations on this theme. When combined with properties of geometric set systems, this is sufficient to deduce optimal bounds for ϵ-nets of several geometric set systems in $\mathbb{R}^2$.

1. Linear-sized Nets for Disks in $\mathbb{R}^2$

If you understand something, you understand that it is obvious.

Israel Gelfand

The main theorem of this chapter is the following.

THEOREM 3.3. *Let P be a set of n points in $\mathbb{R}^2$ and $\epsilon > 0$ a given parameter. Then there exists an ϵ-net N of size $O\left(\frac{1}{\epsilon}\right)$ for the primal set system induced on P by disks in $\mathbb{R}^2$.*

Overview of ideas. We first recall the probabilistic argument of Theorem 2.9 for the existence of an ϵ-net of size $O\left(\frac{1}{\epsilon}\log\frac{1}{\epsilon}\right)$ of the primal set system induced by disks.

Let $S \subseteq P$ be a uniform random sample constructed by choosing each point of P independently with probability $p = \frac{2c}{\epsilon n}\ln\frac{1}{\epsilon}$, for a constant $c \geq 1$. The analysis used the following geometric fact:

if the interior of every empty canonical disk and empty canonical halfplane spanned by S contains fewer than $\frac{\epsilon n}{2}$ points of P, then S is an ϵ-net of the primal set system induced on P by disks.

From now onwards we only consider the case of empty canonical disks; the analysis for empty canonical halfplanes is similar.

The total number of canonical disks spanned by P is $O\left(n^3\right)$ and the probability that a canonical disk with at least $\frac{\epsilon n}{2}$ points of P in its interior ends up as an empty canonical disk spanned by S is at most $p^3 \cdot (1-p)^{\frac{\epsilon n}{2}}$. Thus the expected number of empty canonical disks spanned by S and whose interior contains at least $\frac{\epsilon n}{2}$ points of P can be upper bounded by

$$O\left(n^3\right)\cdot p^3\cdot(1-p)^{\frac{\epsilon n}{2}} = O\left(n^3\cdot\left(\frac{2c}{\epsilon n}\ln\frac{1}{\epsilon}\right)^3\cdot e^{-c\ln\frac{1}{\epsilon}}\right) = O\left(\left(\frac{c}{\epsilon}\ln\frac{1}{\epsilon}\right)^3\cdot\epsilon^c\right).$$

The above is less than 1 for c sufficiently large, implying that S is an ϵ-net with constant probability, with $\mathrm{E}\left[|S|\right] = np = \Theta\left(\frac{1}{\epsilon}\log\frac{1}{\epsilon}\right)$.

Now consider a uniform random sample S constructed by picking each point of P independently with the smaller probability $p = \frac{1}{\epsilon n}$.

At first glance, the above argument prohibits setting $p = \frac{c}{\epsilon n}$ for any constant c, as then we only get an upper bound of

$$O\left(n^3\cdot\left(\frac{c}{\epsilon n}\right)^3\cdot\left(1-\left(\frac{c}{\epsilon n}\right)\right)^{\frac{\epsilon n}{2}}\right) = O\left(\frac{c^3}{\epsilon^3}e^{-\frac{c}{2}}\right) = O\left(\frac{1}{\epsilon^3}\right) \tag{3.4}$$

for the expected number of empty canonical disks spanned by S and containing at least $\frac{\epsilon n}{2}$ points of P. This is not less than 1 for small-enough ϵ, and so S need not be an ϵ-net.

To further improve this analysis will require the following two new ideas.

First idea: The probability of a canonical disk D spanned by P ending up as an empty canonical disk spanned by S is exactly

$$p^3\left(1-p\right)^{|\mathrm{int}(D)\cap P|}.$$

Earlier we upper bounded this by $p^3\left(1-p\right)^{\frac{\epsilon n}{2}}$, as we are only interested in canonical disks containing at least $\frac{\epsilon n}{2}$ points of P. However if $|\mathrm{int}\,(D)\cap P|$ is much larger than $\frac{\epsilon n}{2}$, this probability becomes considerably smaller, decreasing exponentially with $|\text{int}(D) \cap P|$. Thus the 'hard case' is when $|\text{int}(D) \cap P|$ is around $\frac{\epsilon n}{2}$. However, the total number of such canonical disks—say those containing at least $\frac{\epsilon n}{2}$ and less than ϵn points of P—is much smaller than the naive bound of $O(n^3)$ we used earlier:

> Lemma 1.5 states that the number of canonical disks spanned by P and containing at most k points of P in their interior is $O(nk^2)$.

Thus for any $t \geq 1$, we have

$$\begin{aligned} \mathrm{E}\left[\begin{smallmatrix}\text{\# of empty canonical disks } D \text{ spanned} \\ \text{by } S, \text{ with } |\text{int}(D)\cap P| \in \left[t\frac{\epsilon n}{2}, t\epsilon n\right)\end{smallmatrix}\right] &\leq O\left(n\,(t\epsilon n)^2\right) \cdot p^3 \cdot (1-p)^{\frac{t\epsilon n}{2}} \\ &= O\left(n\left(t^2\epsilon^2 n^2\right) \cdot \frac{1}{\epsilon^3 n^3} \cdot e^{-\frac{t}{2}}\right) \\ &= O\left(\frac{1}{\epsilon} \cdot t^2 e^{-\frac{t}{2}}\right). \end{aligned} \tag{3.5}$$

For $t = 1$, we get $O\left(\frac{1}{\epsilon}\right)$, improving the earlier estimate of $O\left(\frac{1}{\epsilon^3}\right)$ of Equation (3.4). In particular, the key thing to note here is that the expected number of empty canonical disks spanned by S, and containing between $t\frac{\epsilon n}{2}$ and $t\epsilon n$ points of P, decreases exponentially with t.

Second idea: If S fails to be an ϵ-net, there exists a disk D' that contains at least ϵn points of P and is not hit by S. By Claim 2.14, D' must contain at least $\frac{\epsilon n}{2}$ points from the interior of some empty canonical disk D spanned by S. See the figure, where the points of the random sample S are drawn in solid.

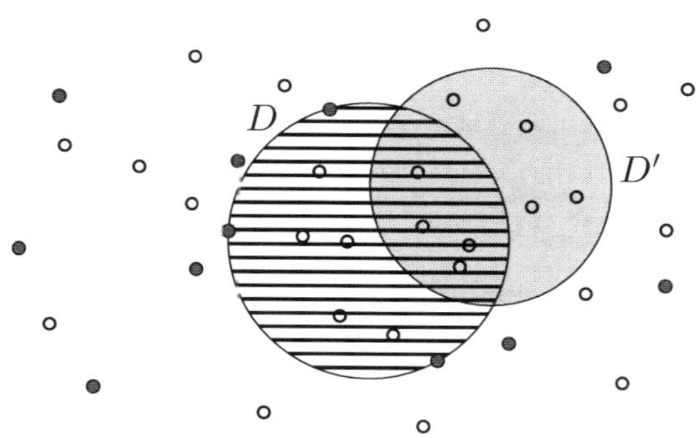

For simplicity, assume that all empty canonical disks spanned by S contain between $\frac{\epsilon n}{2}$ and ϵn points of P. Then D' contains at least $\frac{\epsilon n}{2}$ points out of the (at most) ϵn points of P lying in the interior of D—that is, D' contains at least half of the points of $\text{int}(D) \cap P$. Thus

> *a $\frac{1}{2}$-net S_D of the primal set system induced by disks on* $\text{int}(D) \cap P$ *hits D'.*

Using the sub-optimal $O(\frac{1}{\epsilon}\log\frac{1}{\epsilon})$ bound of Theorem 2.9, $|S_D| = O(2\log 2)$. Therefore, the set S together with the additional sets S_D constructed for all such empty canonical disks D spanned by S, forms an ϵ-net. Using Equation (3.5), its expected size can be upper bounded by

$$\mathrm{E}\left[|S|\right] + \mathrm{E}\left[\substack{\#\text{ of empty canonical disks } D \text{ spanned by } S,\\ \text{with } |\mathrm{int}(D)\cap P| \in \left[\frac{\epsilon n}{2}, \epsilon n\right)}\right] \cdot O(2\log 2) = O\left(\frac{1}{\epsilon}\right),$$

as desired.

We're almost done—it just remains to complete this analysis for all empty canonical disks spanned by S whose interior contains at least ϵn points of P. As the expected number of empty canonical disks spanned by S, and that contain between $t\cdot\frac{\epsilon n}{2}$ and $t\cdot\epsilon n$ points of P decreases exponentially with t, the worst case is the one shown above.

PROOF OF THEOREM 3.3. Let S be a random sample constructed by picking each point of P independently with probability $p = \frac{1}{\epsilon n}$.
Let $\mathcal{D}_{can}$ be the set of *all* canonical disks spanned by the points of P. For each $i = 0, \ldots, \lceil \log \frac{1}{\epsilon} \rceil$, let

$$\mathcal{D}^i_{can} = \left\{ D \in \mathcal{D}_{can} \ : \ |\mathrm{int}(D) \cap P| \in \left[2^{i-1}\cdot \epsilon n, 2^i \cdot \epsilon n\right) \right\}.$$

Lemma 1.5 implies that $|\mathcal{D}^i_{can}| = O\left(n\left(2^i \epsilon n\right)^2\right)$.
Set $\epsilon_i = \frac{1}{2^{i+1}}$. For each $i \geq 0$ and each $D \in \mathcal{D}^i_{can}$,
let S_D be an ϵ_i-net, given by Theorem 2.9, of the primal set system induced on $\mathrm{int}(D) \cap P$ by disks.

Note that

$$|S_D| = O\left(\frac{1}{\epsilon_i}\log\frac{1}{\epsilon_i}\right) = O\left(2^{i+1}\log 2^{i+1}\right). \tag{3.6}$$

We return $N = S \cup M$ as our ϵ-net, where

$$M = \bigcup_{i=0}^{\lceil \log \frac{1}{\epsilon}\rceil} \ \bigcup_{\substack{D\in\mathcal{D}^i_{can}\,:\, D \text{ is an empty}\\ \text{canonical disk spanned by } S}} S_D.$$

CLAIM 3.7. *N is an ϵ-net of the primal set system induced on P by disks.*

PROOF. Let D' be any disk in the plane containing at least ϵn points of P. Then either D' contains a point of S or, by Claim 2.14, there a $D \in \mathcal{D}^i_{can}$ for some i such that D' contains at least $\frac{\epsilon n}{2}$ points from $\mathrm{int}(D) \cap P$, and D is an empty canonical disk spanned by S. Thus we have

$$|D' \cap (\mathrm{int}(D) \cap P)| \geq \frac{\epsilon n}{2} > \frac{|\mathrm{int}(D) \cap P|/2^i}{2} = \frac{|\mathrm{int}(D) \cap P|}{2^{i+1}} = \epsilon_i \cdot |\mathrm{int}(D) \cap P|,$$

and so the ϵ_i-net S_D of the set system induced by disks on $\mathrm{int}(D) \cap P$ must hit D'. □

CLAIM 3.8.

$$\mathrm{E}\left[|M|\right] = O\left(\frac{1}{\epsilon}\right).$$

PROOF. For each $D \in \mathcal{D}_{can}$, let I_D be the indicator random variable which is 1 if D ends up as an empty canonical disk spanned by S and 0 otherwise. Then

$$\begin{aligned}
\mathrm{E}\left[|M|\right] &= \mathrm{E}\left[\sum_{i=0}^{\lceil \log \frac{1}{\epsilon} \rceil} \sum_{D \in \mathcal{D}_{can}^i} I_D \cdot |S_D|\right] = \sum_{i=0}^{\lceil \log \frac{1}{\epsilon} \rceil} \sum_{D \in \mathcal{D}_{can}^i} \mathrm{E}\left[I_D\right] \cdot |S_D| \\
&= \sum_{i=0}^{\lceil \log \frac{1}{\epsilon} \rceil} \sum_{D \in \mathcal{D}_{can}^i} \Pr\left[\begin{smallmatrix} D \text{ is an empty canonical} \\ \text{disk spanned by } S \end{smallmatrix}\right] \cdot |S_D| \\
&\leq \sum_{i=0}^{\lceil \log \frac{1}{\epsilon} \rceil} |\mathcal{D}_{can}^i| \cdot p^3 \cdot (1-p)^{2^{i-1}\epsilon n} \cdot O\left(2^{i+1} \log 2^{i+1}\right) \\
&= \sum_{i=0}^{\lceil \log \frac{1}{\epsilon} \rceil} O\left(n\left(2^{2i}\epsilon^2 n^2\right) \cdot \frac{1}{\epsilon^3 n^3} \cdot e^{-2^{i-1}}\right) \cdot O\left(2^{i+1}\,(i+1)\right) \\
&= \frac{1}{\epsilon} \sum_{i=0}^{\lceil \log \frac{1}{\epsilon} \rceil} O\left(\frac{(i+1)\,2^{3i}}{e^{2^{i-1}}}\right) = O\left(\frac{1}{\epsilon}\right),
\end{aligned}$$

since the last summation can be upper bounded by a decreasing geometric series. That is, there exists a sufficiently large constant c such that for all $i \geq 0$ we have $\frac{(i+1)\,2^{3i}}{e^{2^{i-1}}} \leq \frac{c}{2^i}$. □

Finally we have $\mathrm{E}\left[|N|\right] = \mathrm{E}\left[|S|\right] + \mathrm{E}\left[|M|\right] = O\left(\frac{1}{\epsilon}\right)$.
A similar analysis shows that the expected number of points added due to empty canonical halfplanes is $O\left(\frac{1}{\epsilon}\right)$, concluding the proof. □

Bibliography and discussion. The idea of sampling and refinement for ϵ-nets was used by Chazelle and Friedman [**CF90**], though in a more abstract setting. The proof of optimal ϵ-nets of disks given above was constructed for didactic purposes; a more detailed construction and analysis gives the current-best bound of $\frac{13.4}{\epsilon}$ [**Bus+16**]. This sampling refinement idea can also be done for dual set systems [**CV07, AES10**], though later results (see Chapter 6) will prove such statements in greater generality. For other applications and uses of the alterations technique, we refer the reader to [**AS16**, Chapter 3].

[AES10] B. Aronov, E. Ezra, and M. Sharir, *Small-size ϵ-nets for axis-parallel rectangles and boxes*, SIAM J. Comput. **39** (2010), no. 7, 3248–3282, DOI 10.1137/090762968. MR2678074

[AS16] N. Alon and J. H. Spencer, *The probabilistic method*, 4th ed., Wiley Series in Discrete Mathematics and Optimization, John Wiley & Sons, Inc., Hoboken, NJ, 2016. MR3524748

[Bus+16] N. Bus, S. Garg, N. H. Mustafa, and S. Ray, *Tighter estimates for ϵ-nets for disks*, Comput. Geom. **53** (2016), 27–35, DOI 10.1016/j.comgeo.2015.12.002. MR3454542

[CF90] B. Chazelle and J. Friedman, *A deterministic view of random sampling and its use in geometry*, Combinatorica **10** (1990), no. 3, 229–249, DOI 10.1007/BF02122778. MR1092541

[CV07] K. L. Clarkson and K. Varadarajan, *Improved approximation algorithms for geometric set cover*, Discrete Comput. Geom. **37** (2007), no. 1, 43–58, DOI 10.1007/s00454-006-1273-8. MR2279863

2. Partitioning Segments in $\mathbb{R}^2$

The price of metaphor is eternal vigilance.

Norbert Wiener

The following main theorem of this section illustrates an important spatial aspect of ϵ-nets.

THEOREM 3.9. *Let S be a set of n line segments in general position in $\mathbb{R}^2$ and let m be the number of pairs of intersecting segments of S. Let $1 \leq r \leq n$ be a given parameter. Then there exists a decomposition of $\mathbb{R}^2$ into $O\left(r + \frac{mr^2}{n^2}\right)$ interior-disjoint triangles (some of which are unbounded) such that the interior of each triangle in this decomposition intersects at most $\frac{n}{r}$ segments of S.*
Such a decomposition is called a $\frac{1}{r}$-cutting of S.

Before we present the proof of Theorem 3.9, we make two remarks.

Optimality: The bound of Theorem 3.9 is optimal up to constant factors. Given S, assume that there exists a decomposition Ξ of $\mathbb{R}^2$ with t triangles such that the interior of each triangle intersects at most $\frac{n}{r}$ segments of S. First, the boundary of each triangle of Ξ can contain at most 3 segments of S by the general position assumption on S. The remaining $n - 3t$ segments of S must intersect the interior of at least one triangle of Ξ and thus we have

$$t \geq \frac{n-3t}{(n/r)} \quad \Longrightarrow \quad \frac{tn}{r} \geq n - 3t$$
$$\Longrightarrow \quad t \geq \frac{n}{(n/r)+3} = \frac{n}{n+3r} \cdot r = \Omega(r).$$

Next, the interior of each triangle of Ξ can contain at most $\binom{n/r}{2} = O\left(\frac{n^2}{r^2}\right)$ intersection points between segments of S. As S is in general position, there can be at most $6t$ intersection points lying on the boundary (edges and vertices) of the triangles of Ξ and so

$$m = O\left(t \cdot \frac{n^2}{r^2} + 6t\right) \qquad \Longrightarrow \qquad t = \Omega\left(\frac{mr^2}{n^2}\right).$$

General position: For simplicity of exposition we have assumed that S is in general position, though Theorem 3.9 is true even if that is not the case. The fact that S need not be in general position is the reason for requiring that the *interior* of each triangle of the decomposition intersect at most $\frac{n}{r}$ segments of S. The statement is not true if we also require the boundary of each triangle to intersect at most $\frac{n}{r}$ segments of S—this happens, e.g., when all the segments of S have a non-empty intersection.

For the rest of this section, S will denote a set of n line segments in the plane in general position. Let $\mathcal{U}$ be a large-enough rectangle containing all the segments of S in its interior. It will suffice to construct a $\frac{1}{r}$-cutting of S inside $\mathcal{U}$.
For any set R of segments in the plane in general position, let $\mathrm{I}(R)$ denote the set of intersection points between segments of R and set $m_R = |\mathrm{I}(R)|$.

Overview of ideas. Given S, let $(S, \mathcal{F})$ be the set system induced on S by intersection with open triangles in the plane. That is,

$$\mathcal{F} = \Big\{ \{ s \in S \,:\, s \cap \operatorname{int}(\Delta) \neq \emptyset \} \,:\, \Delta \text{ is a triangle in } \mathbb{R}^2 \Big\}.$$

The key observation is that a $\frac{1}{r}$-net R of $\mathcal{F}$ can be used to construct a $\frac{1}{r}$-cutting of size $O(|R| + m_R)$, as follows.

Given R, decompose $\mathcal{U}$ into a set $\mathcal{T}$ of $O(|R| + m_R)$ interior-disjoint triangles such that no triangle $\Delta \in \mathcal{T}$ intersects any segment of R in its interior. This can be done by 'cutting up' the segments of R into $O(|R| + m_R)$ interior-disjoint segments which then serve as edges of the triangles in $\mathcal{T}$. See the figure.

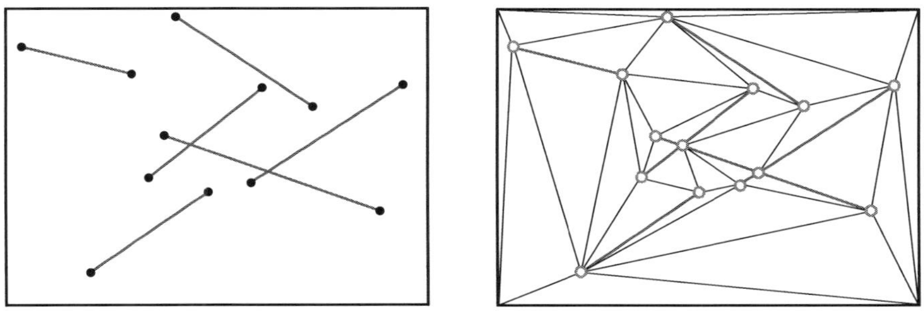

Now for any $\Delta \in \mathcal{T}$, $\mathrm{int}(\Delta)$ must intersect less than $\frac{n}{r}$ segments of S—otherwise the segments of S intersecting $\mathrm{int}(\Delta)$ form a set in $\mathcal{F}$ of size at least $\frac{n}{r}$ that is not hit by the $\frac{1}{r}$-net R, a contradiction.

Note that we view ϵ-nets contrapositively here: if R is an ϵ-net of $\mathcal{F}$, then the interior of any triangle Δ in the plane that does *not* intersect any segment of R must intersect less than ϵn segments of S.

Construct a random sample R by picking each segment of S independently with probability $p = c \cdot \frac{r}{n} \log r$, for a sufficiently large constant c. Let $\mathcal{T}$ be the triangulation of R as above.

It can be shown, with a proof similar to that of Theorem 2.8, that with constant probability R is a $\frac{1}{r}$-net of $\mathcal{F}$. It then follows from the above discussion that $\mathcal{T}$ is a $\frac{1}{r}$-cutting for S. We have $|\mathcal{T}| = O(|R| + m_R)$, where

$$\text{(3.10)} \qquad \mathrm{E}\left[|R|\right] = np \; = \; cr\log r \quad \text{and} \quad \mathrm{E}\left[m_R\right] = mp^2 \; = \; m\frac{c^2r^2\log^2 r}{n^2},$$

where the bound on $\mathrm{E}[m_R]$ uses the fact that each segment of S was picked independently. By applying Markov's inequality to the random variables $|R|$ and m_R, we have $\Pr\left[|R| \geq 10\,np\right] \leq \frac{1}{10}$ and $\Pr\left[m_R \geq 10\,mp^2\right] \leq \frac{1}{10}$. Thus with probability at least $\frac{8}{10}$,

$$\text{(3.11)} \qquad |\mathcal{T}| \leq 10\left(np + mp^2\right) = O\left(r\log r + \frac{mr^2\log^2 r}{n^2}\right).$$

This already gets us to the upper bound of Theorem 3.9 within logarithmic factors. As in the previous section, the optimal bound will follow by taking a smaller initial sample R and 'fixing' any triangles whose interior intersect more than $\frac{n}{r}$ segments of S. The proof is quite similar to the one seen for disks and we encourage the reader to keep this correspondence in mind when reading the proof.

There is one subtlety here—unlike the case of disks, improving upon the size of a $\frac{1}{r}$-net $R \subseteq S$ of $\mathcal{F}$ from $O(r\log r)$ to $O(r)$ is not possible, as there is a lower bound of $\Omega\left(r\left(\frac{\log r}{\log\log r}\right)^{\frac{2}{2}}\right)$ (see Chapter 16)! However, the refinement ideas that we saw for the case of disks *do* go through, as the goal is to compute a $\frac{1}{r}$-cutting of S and not a $\frac{1}{r}$-net of $\mathcal{F}$. In particular, we do not need a $R \subseteq S$ hitting *every* open triangle that intersects at least $\frac{n}{r}$ segments of S—only that this property holds for all open triangles in the specific triangulation $\mathcal{T}$ that we will construct using R. This will lead to a proof similar to the one seen for disks.

Trapezoidal decompositions. We first briefly review a spatial decomposition technique called *trapezoidal decompositions.* Given any set $R \subseteq S$, decompose $\mathcal{U}$ with respect to R as follows.

From each of the $2|R|$ endpoints of segments in R and each of the m_R intersection points between two segments of R, trace a vertical ray upwards and downwards until it hits another segment of R or the bounding rectangle $\mathcal{U}$. The union of all these vertical segments, together with R, decomposes $\mathcal{U}$ into a set of regions. Each such region is called a *trapezoidal region* (or a trapezoid) and the decomposition is called a *trapezoidal decomposition.* See the figure.

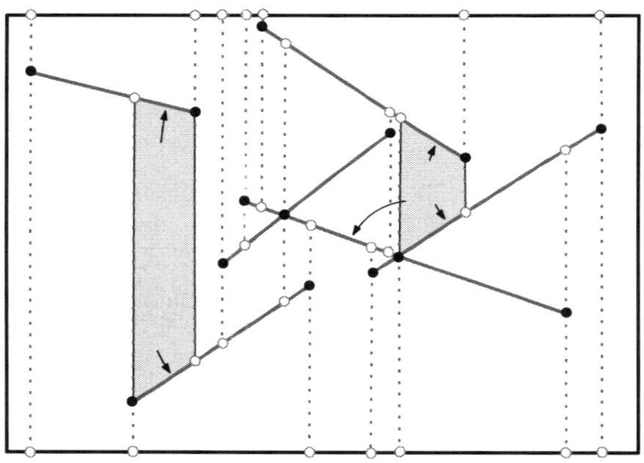

Denote by $\mathcal{T}(R)$ this set of trapezoids for R and let $|\mathcal{T}(R)|$ denote its size. As each trapezoid can be decomposed into two triangles, it suffices to show the existence of a $\frac{1}{r}$-cutting of the desired size consisting of triangles and trapezoids.

Each of the $2|R| + m_R$ points, consisting of endpoints and intersections, produce two additional points from the vertical and horizontal rays traced from them. The trapezoidal decomposition can be seen as a planar graph on these $3(2|R| + m_R)$ points. Then $|\mathcal{T}(R)|$ is simply the number of faces of this graph and so

$$|\mathcal{T}(R)| \leq 2 \cdot 3(2|R| + m_R) = O(|R| + m_R). \tag{3.12}$$

The trapezoids in $\bigcup_{R\subseteq S}\mathcal{T}(R)$ form the set of *canonical trapezoids.*

DEFINITION 3.13. Given S, the set of trapezoids present in the trapezoidal decompositions of *all possible* $R \subseteq S$ are called the canonical trapezoids of S and denoted by $\Xi(S)$.

Furthermore, let $\Xi_{\leq k}(S) \subseteq \Xi(S)$ be the set of canonical trapezoids of S whose interior intersects at most k segments of S.

For a canonical trapezoid Δ, let S_Δ denote the set of segments of S intersecting the interior of Δ, and let $n_\Delta = |S_\Delta|$. The crucial fact that will be needed later is that each trapezoid $\Delta \in \Xi(S)$ is 'determined' by a constant—1, 2, 3 or 4—number of segments of S (recall that we assume S to be in general position). These are called the *determining segments* of Δ and have the following property (stated without proof).

FACT 3.14. A canonical trapezoid $\Delta \in \Xi(S)$ belongs to $\mathcal{T}(R)$ *if and only if* its determining segments are present in R and R does not contain any of the segments of S_Δ.

See the figure where the determining segments of two canonical trapezoids are indicated.

We will need the following level-set bound, whose proof is a routine application of the technique of Chapter 1 together with Fact 3.14 (we defer its proof to the end).

LEMMA 3.15. *For any integer $k \geq 1$, we have $|\Xi_{\leq k}(S)| = O(nk^3 + mk^2)$.*

For the rest of the proof, we only work with canonical trapezoids determined by 4 segments. The case for canonical trapezoids determined by $3, 2$ or 1 segments is similar.

The reader will notice that all of this pretty much mirrors the notion of canonical disks; indeed we are following, in a different setting, the ideas in the proof for the case of disks.

We now restate and prove our main theorem.

THEOREM 3.9. *Let S be a set of n line segments in general position in $\mathbb{R}^2$ and let m be the number of pairs of intersecting segments of S. Let $1 \leq r \leq n$ be a given parameter. Then there exists a decomposition of $\mathbb{R}^2$ into $O\left(r + \frac{mr^2}{n^2}\right)$ interior-disjoint triangles (some of which are unbounded) such that the interior of each triangle in this decomposition intersects at most $\frac{n}{r}$ segments of S.*
Such a decomposition is called a $\frac{1}{r}$-cutting of S.

PROOF. The set of triangles in the required $\frac{1}{r}$-cutting will be denoted by $\mathcal{T}$.

Set $p = \frac{cr}{n}$ for a sufficiently large constant c to be fixed later and construct R by picking each segment of S independently with probability p.

Construct the trapezoidal decomposition $\mathcal{T}(R)$ and for each $\Delta \in \mathcal{T}(R)$, define

$$S'_\Delta = \left\{ s \cap \text{int}(\Delta) \ : \ s \in S_\Delta \right\}.$$

Now we iterate over each trapezoid in $\mathcal{T}(R)$ and construct our final cutting $\mathcal{T}$. Set $\mathcal{T} = \emptyset$ and for each $\Delta \in \mathcal{T}(R)$, do the following:

Case $n_\Delta \leq \frac{n}{r}$: Add Δ to $\mathcal{T}$.

Case $n_\Delta > \frac{n}{r}$: Let $t_\Delta > 1$ be such that $n_\Delta = t_\Delta \cdot \frac{n}{r}$. Applying Equation (3.11) to the set system induced on S'_Δ by intersection with open triangles, there exists a

$\frac{1}{t_\Delta}$-cutting $\mathcal{T}_\Delta$ inside Δ of size

$$O\left(t_\Delta \log t_\Delta + \frac{m'_\Delta \cdot t_\Delta^2 \log^2 t_\Delta}{n_\Delta^2}\right) = O\left(t_\Delta^2 \log^2 t_\Delta\right), \tag{3.16}$$

where $m'_\Delta \leq \binom{n_\Delta}{2}$ is the number of intersections between segments of S'_Δ. By construction, the interior of each triangle in $\mathcal{T}_\Delta$ intersects at most $\frac{n_\Delta}{t_\Delta} = \frac{n}{r}$ segments of S'_Δ and hence of S.
Add all the triangles in $\mathcal{T}_\Delta$ to $\mathcal{T}$.

CLAIM 3.17.

$$\mathrm{E}\left[|\mathcal{T}|\right] = O\left(r + \frac{mr^2}{n^2}\right).$$

PROOF. We have

$$\mathrm{E}\left[|\mathcal{T}|\right] \leq \mathrm{E}\left[\left|\mathcal{T}(R)\right|\right] + \sum_{\substack{\Delta\in\Xi(S)\\ n_\Delta > \frac{n}{r}}} \Pr\left[\Delta \in \mathcal{T}(R)\right] \cdot |\mathcal{T}_\Delta|.$$

The first term can be upper bounded using Equation (3.12) by

$$\mathrm{E}\left[\left|\mathcal{T}(R)\right|\right] = O\Big(\mathrm{E}\left[|R|\right] + \mathrm{E}\left[m_R\right]\Big) = O\Big(np + mp^2\Big) = O\left(r + \frac{mr^2}{n^2}\right).$$

The second term can be upper bounded using Fact 3.14 by

$$\sum_{\substack{\Delta\in\Xi(S)\\ n_\Delta > \frac{n}{r}}} \Pr\left[\Delta \in \mathcal{T}(R)\right] \cdot |\mathcal{T}_\Delta| = \sum_{\substack{\Delta\in\Xi(S)\\ n_\Delta > \frac{n}{r}}} p^4 (1-p)^{n_\Delta} \cdot |\mathcal{T}_\Delta|$$

Using the fact that $n_\Delta = t_\Delta \frac{n}{r}$ and using Equation (3.16),

$$\leq \sum_{i=0}^{\lceil \log r \rceil} \sum_{\substack{\Delta\in\Xi(S)\\ t_\Delta\in(2^i, 2^{i+1}]}} p^4 (1-p)^{\frac{2^i n}{r}} \cdot O\left(t_\Delta^2 \log^2 t_\Delta\right)$$

$$= \sum_{i=0}^{\lceil \log r \rceil} \left|\Xi_{\leq \frac{2^{i+1}n}{r}}(S)\right| \cdot \left(\frac{cr}{n}\right)^4 e^{-c\,2^i} \cdot O\left(2^{2(i+1)} \log^2 2^{i+1}\right)$$

Using Lemma 3.15,

$$= \sum_{i=0}^{\lceil \log r \rceil} O\left(n\left(\frac{2^{i+1}n}{r}\right)^3 + m\left(\frac{2^{i+1}n}{r}\right)^2\right)\left(\frac{cr}{n}\right)^4 e^{-c\,2^i} O\left(2^{2i}(i+1)^2\right)$$

$$= r\sum_{i=0}^{\lceil \log r \rceil} O\left(2^{5i} c^4 e^{-c\,2^i} (i+1)^2\right) + \frac{mr^2}{n^2}\sum_{i=0}^{\lceil \log r \rceil} O\left(2^{4i} c^4 e^{-c\,2^i} (i+1)^2\right)$$

$$= O\left(r + \frac{mr^2}{n^2}\right),$$

where the last step follows from the fact that both the summations can be upper bounded by a geometric series. In particular, for c sufficiently large, we have $\frac{2^{5i} c^4 (i+1)^2}{e^{c 2^i}} \leq \frac{1}{2^i}$ for all $i \geq 0$. □

$\mathcal{T}$ is a $\frac{1}{r}$-cutting of S by construction, and so the above claim completes the proof of Theorem 3.9. □

The last remaining step is the proof of the level-set bound, which is a routine application of the technique of Chapter 1.

LEMMA 3.15. *For any integer $k \geq 1$, we have* $|\Xi_{\leq k}(S)| = O\left(nk^3 + mk^2\right)$.

PROOF. We give the proof for the case of trapezoids in $\Xi_{\leq k}(S)$ determined by 4 segments of S; the other cases are similar.

Construct a random sample T by picking each segment of S independently with probability p_0, where p_0 will be fixed later.

We will count the expected size of $\mathcal{T}(T)$ in two ways.

Upper bound: Using Equation (3.12),

$$\mathrm{E}\left[|\mathcal{T}(T)|\right] = \mathrm{E}\left[O\left(|T| + m_T\right)\right] = O\left(np_0 + mp_0^2\right).$$

Lower bound: By Fact 3.14, any fixed canonical trapezoid $\Delta \in \Xi(S)$ is present in $\mathcal{T}(T)$ with probability $p_0^4 \cdot (1-p_0)^{n_\Delta}$. Therefore

$$\begin{aligned} \mathrm{E}\left[|\mathcal{T}(T)|\right] &= \sum_{\Delta \in \Xi(S)} p_0^4 \cdot (1-p_0)^{n_\Delta} \geq \sum_{\Delta \in \Xi_{\leq k}(S)} p_0^4 \cdot (1-p_0)^{n_\Delta} \\ &\geq \sum_{\Delta \in \Xi_{\leq k}(S)} p_0^4 \cdot (1-p_0)^k = |\Xi_{\leq k}(S)| \cdot p_0^4 \cdot (1-p_0)^k. \end{aligned}$$

Combining the upper and lower bounds,

$$|\Xi_{\leq k}(S)| \cdot p_0^4 \cdot (1-p_0)^k \leq \mathrm{E}[|\mathcal{T}(T)|] = O\left(np_0 + mp_0^2\right),$$

and hence
$$|\Xi_{\leq k}(S)| = O\left(\frac{np_0 + mp_0^2}{p_0^4\,(1-p_0)^k}\right) = O\left(nk^3 + mk^2\right),$$

by setting $p_0 = \frac{1}{k+1}$. □

We had to refine our original random sample $R \subseteq S$ to guarantee that the interior of each triangle in our decomposition intersected at most $\frac{n}{r}$ segments of S. In particular,

(1) To ensure that the interior of each trapezoid in the trapezoidal decomposition intersected, in expectation, at most $\frac{n}{r}$ segments of S, we pick each segment into R with probability $\Omega\left(\frac{r}{n}\right)$.

(2) The resulting size of the spatial decomposition was a function of the complexity of arrangement induced by the random sample. It was linear in both $|R|$ and the number of intersections between segments of R, with expected size $O\left(n\left(\frac{r}{n}\right) + m\left(\frac{r^2}{n^2}\right)\right) = O\left(r + \frac{mr^2}{n^2}\right)$.

These ideas work in a number of related settings, some of which we list below; we omit the proofs which are similar to that of Theorem 3.9.

Higher dimensions: Given a set $\mathcal{S}$ of $(d-1)$-dimensional simplices in $\mathbb{R}^d$ and a parameter r, the goal is to decompose $\mathbb{R}^d$ into regions of constant complexity such that the interior of each region intersects at most $\frac{n}{r}$ elements of $\mathcal{S}$. Here the 'complexity' of $\mathcal{S}$ is captured by the number of d-tuples of $\mathcal{S}$ which have a non-empty intersection; assume there are m such d-tuples. Then the expected complexity of the spatial decomposition induced by a random sample $R \subseteq \mathcal{S}$, where each element is picked with probability $\frac{r}{n}$, is $O\left(r^{d-1+\delta} + m \cdot \left(\frac{r}{n}\right)^d\right)$, where $\delta > 0$ is any small constant and the constant in the asymptotic notation depends on δ. This spatial decomposition method is a generalization of trapezoidal decompositions to $\mathbb{R}^d$. However unlike the two-dimensional case, precise exponents are not known and hence the additional δ in the exponent. The second term is the expected number of vertices in the arrangement induced by R, while the first term bounds the number of all higher-dimensional faces of the arrangement induced by R. Thus we get the following theorem [**Pel97**] (stated without proof).

THEOREM 3.18. *Let $\mathcal{S}$ be an n-element set of $(d-1)$-dimensional simplices in $\mathbb{R}^d$, $d \geq 3$, and let m denote the number of d-tuples of simplices in $\mathcal{S}$ having a point in common. Then for any $\delta > 0$ and any $r \leq n$, there is a $\frac{1}{r}$-cutting with respect to $\mathcal{S}$ of size at most*

$$C_{d,\delta} \cdot \left(r^{d-1+\delta} + m\frac{r^d}{n^d}\right).$$

Here the constant $C_{d,\delta}$ depends only on d and δ.

Cuttings inside a simplex: Given a set $\mathcal{H}$ of n hyperplanes in $\mathbb{R}^d$ and a simplex Δ, we have the following bound for a $\frac{1}{r}$-cutting *inside* Δ [**Cha12**] (stated without proof).

THEOREM 3.19. *Let $\mathcal{H}$ be a set of n hyperplanes in $\mathbb{R}^d$ and let Δ be a simplex containing X vertices of the arrangement induced by $\mathcal{H}$. Then there exists a $\frac{1}{r}$-cutting inside Δ of size $O\left(r^{d-1} + X\frac{r^d}{n^d}\right)$.*

When Δ is large-enough to contain all vertices of the arrangement of $\mathcal{H}$, we get the bound of $O\left(r^d\right)$ for the size of $\frac{1}{r}$-cuttings for a set of n hyperplanes in $\mathbb{R}^d$.

Shallow cuttings: The notion of the level of a point with respect to a set of hyperplanes (recall Definition 1.15) leads naturally to the following definition.

DEFINITION 3.20. Given a set $\mathcal{H}$ of n hyperplanes in $\mathbb{R}^d$, a real parameter $r > 0$ and an integer $k \geq 0$, a $\left(\frac{1}{r}, k\right)$-shallow cutting is a set Ξ of interior-disjoint simplices (some of which are unbounded) such that

- the union of the simplices in Ξ covers all points in $\mathbb{R}^d$ of level at most k with respect to $\mathcal{H}$, and
- the interior of each simplex in Ξ intersects at most $\frac{n}{r}$ hyperplanes of $\mathcal{H}$.

The number of vertices of level at most k in the arrangement induced by $\mathcal{H}$ is $O\left(n^{\lfloor \frac{d}{2} \rfloor} k^{\lceil \frac{d}{2} \rceil}\right)$ (see Theorem 1.16) where the constant in the asymptotic notation depends on d. Thus the arrangement of a random sample of $\mathcal{H}$ of size r has an expected $O\left(n^{\lfloor \frac{d}{2} \rfloor} k^{\lceil \frac{d}{2} \rceil} \cdot \frac{r^d}{n^d}\right)$ vertices, which leads to the following theorem [**Mat92**] (stated without proof).

THEOREM 3.21. *Let $\mathcal{H}$ be a set of n hyperplanes in $\mathbb{R}^d$, $r > 0$ a real and $k \geq 0$ an integer. Then there exists a $\left(\frac{1}{r}, k\right)$-shallow cutting of $\mathcal{H}$ of size*

$$O\left(r^{\lfloor \frac{d}{2} \rfloor}\left(\frac{kr}{n}+1\right)^{\lceil \frac{d}{2} \rceil}\right).$$

Given a set $\mathcal{H}$ of n hyperplanes in $\mathbb{R}^3$ and $\epsilon > 0$, apply Theorem 3.21 with $k = 2\epsilon n$ and $r = \frac{2}{\epsilon}$, to get a set $\mathcal{T}$ of $O\left(\frac{1}{\epsilon}\right)$ simplices whose union covers all points in $\mathbb{R}^3$ of level at most $2\epsilon n$ with respect to $\mathcal{H}$. For each $\Delta \in \mathcal{T}$, choose a hyperplane of $\mathcal{H}$ that lies below Δ (in the x_d direction), if such a hyperplane exists. Let N be the set of these $O\left(\frac{1}{\epsilon}\right)$ hyperplanes.

Consider any point $p \in \mathbb{R}^3$ with level ϵn, and let $\Delta \in \mathcal{T}$ be the simplex containing p. As the interior of Δ intersects at most $\frac{n}{r} = \frac{\epsilon n}{2}$ hyperplanes of $\mathcal{H}$, there are at least $\frac{\epsilon n}{2}$ hyperplanes of $\mathcal{H}$ lying below Δ, at least one of which was picked in N. Thus for every $p \in \mathbb{R}^3$ with level ϵn, there exists a hyperplane in N lying below p. In the dual, this leads to the existence of $O\left(\frac{1}{\epsilon}\right)$ sized ϵ-net of the primal set system induced by half-spaces in $\mathbb{R}^3$.

Bibliography and discussion. The key idea of refining an initial random sample is from [**CF90**], and was used to give an optimal bound on the size of cuttings for hyperplanes in $\mathbb{R}^d$. The proof of the main theorem is from [**BS95**]. While we only considered the existence of cuttings, there has been extensive work on algorithms (e.g., [**Cha93, Mat91**]) which can be read from [**Cha00**]. The use of shallow cuttings to construct an ϵ-net of the primal set system induced by half-spaces in $\mathbb{R}^3$ was shown in [**Mat92**]. For details on partitioning $\mathbb{R}^d$ with respect to geometric objects, we refer the reader to the nice survey of Agarwal and Sharir [**AS00**].

[AS00] P. K. Agarwal and M. Sharir, *Arrangements and their applications*, Handbook of computational geometry, North-Holland, Amsterdam, 2000, pp. 49–119, DOI 10.1016/B978-044482537-7/50003-6. MR1746675

[BS95] M. de Berg and O. Schwarzkopf, *Cuttings and applications*, Internat. J. Comput. Geom. Appl. **5** (1995), no. 4, 343–355, DOI 10.1142/S0218195995000210. MR1359425

[CF90] B. Chazelle and J. Friedman, *A deterministic view of random sampling and its use in geometry*, Combinatorica **10** (1990), no. 3, 229–249, DOI 10.1007/BF02122778. MR1092541

[Cha00] B. Chazelle, *The discrepancy method*, Cambridge University Press, Cambridge, 2000. Randomness and complexity, DOI 10.1017/CBO9780511626371. MR1779341

[Cha12] T. M. Chan, *Optimal partition trees*, Discrete Comput. Geom. **47** (2012), no. 4, 661–690, DOI 10.1007/s00454-012-9410-z. MR2901245

[Cha93] B. Chazelle, *Cutting hyperplanes for divide-and-conquer*, Discrete Comput. Geom. **9** (1993), no. 2, 145–158, DOI 10.1007/BF02189314. MR1194032

[Mat91] J. Matoušek, *Cutting hyperplane arrangements*, Discrete Comput. Geom. **6** (1991), no. 5, 385–406, DOI 10.1007/BF02574697. MR1115098

[Mat92] J. Matoušek, *Reporting points in halfspaces*, Comput. Geom. **2** (1992), no. 3, 169–186, DOI 10.1016/0925-7721(92)90006-E. MR1190599

[Pel97] M. Pellegrini, *On counting pairs of intersecting segments and off-line triangle range searching*, Algorithmica **17** (1997), no. 4, 380–398, DOI 10.1007/BF02523679. MR1429210

3. Application: Forbidden Subgraphs

The real satisfaction from mathematics is in learning from others and sharing with others.

William Thurston

A classic problem in extremal graph theory is bounding the number of edges in graphs and hypergraphs not containing certain forbidden subgraphs. The following is a special case of the *Zarankiewicz problem.*

> Let $G = (V, E)$ be a graph on n vertices, and $t \geq 1$ be a given integer. What is the maximum size of E if G does not contain $K_{t,t}$ as a subgraph?

Note that this is different from the forbidden *induced* subgraph problem—the Zarankiewicz problem is more restrictive as it forbids the presence of $K_{t,t}$ as a subgraph of G, and not just as an induced subgraph.

In this section we consider a geometric instance of this problem, when G is the intersection graph of a set S of n line segments in the plane in general position. Denote by $G_I = (S, E_I)$ the intersection graph of S. That is,

$$E_I = \Big\{ \{s, s'\} : s, s' \in S \text{ and } s \cap s' \neq \emptyset \Big\}.$$

Note that $|E_I|$ is simply the number of intersections between the segments of S.

The main result of this section, an application of $\frac{1}{r}$-cuttings, is the following.

THEOREM 3.22. *Let S be a set of n line segments in general position in the plane and let $t \geq 2$ be an integer. If the intersection graph $G_I = (S, E_I)$ of S does not contain $K_{t,t}$ as a subgraph, then $|E_I| = O(n)$, where the constant in the asymptotic notation depends exponentially on t.*

An early bound—and still the best known for general graphs—is given by the Kövári-Sós-Turán theorem, which implies the following.

THEOREM 3.23. *Let $G = (V, E)$ be a simple graph on n vertices and let $t \geq 2$ be an integer such that G does not contain $K_{t,t}$ as a subgraph. Then $|E| \leq n^{2-\frac{1}{t}}$.*

PROOF. For any $v \in V$, let $N_G(v)$ denote the set of neighbors of v in G. We will count, in two ways, the size of the following set:

$$\mathcal{I} = \Big\{ (v, U) : v \in V, \ \ |U| = t, \text{ and } \ U \subseteq N_G(v) \Big\}.$$

Upper bound: Each fixed subset $U \subseteq V$ with $|U| = t$ can be present in at most $(t-1)$ tuples of $\mathcal{I}$, as otherwise a $K_{t,t}$ would exist in G. So we have

$$|\mathcal{I}| \leq \binom{n}{t} \cdot (t-1).$$

Lower bound: Counting $|\mathcal{I}|$ vertex by vertex,

$$|\mathcal{I}| = \sum_{v \in V} \binom{|N_G(v)|}{t} \geq n \binom{\left\lfloor \frac{2|E|}{n} \right\rfloor}{t},$$

where the last step follows from the fact that, as $\sum_v |N_G(v)| = 2|E|$, the above summation is minimized when the terms $|N_G(v)|$ are as equal as possible.

This fact is true more generally for convex functions and is called Jensen's inequality. In our case it follows immediately from the fact that for any two integers a_1 and a_2 with $a_1 \geq a_2 + 1$,

$$\binom{a_1}{t} + \binom{a_2}{t} \geq \binom{a_1 - 1}{t} + \binom{a_2 + 1}{t}$$

$$\binom{a_1}{t} - \binom{a_1 - 1}{t} \geq \binom{a_2 + 1}{t} - \binom{a_2}{t}$$

$$\binom{a_1 - 1}{t - 1} \geq \binom{a_2}{t - 1}, \quad \text{which holds if } a_1 \geq a_2 + 1.$$

The last step used Pascal's identity, that $\binom{n}{k} = \binom{n-1}{k} + \binom{n-1}{k-1}$.

Combining the upper and lower bounds,

$$n \binom{\left\lfloor \frac{2|E|}{n} \right\rfloor}{t} \leq \binom{n}{t} \cdot (t - 1).$$

Re-arranging the terms and simplifying,

$$\frac{\left\lfloor \frac{2|E|}{n} \right\rfloor}{n} \cdot \frac{\left\lfloor \frac{2|E|}{n} \right\rfloor - 1}{n - 1} \cdots \frac{\left\lfloor \frac{2|E|}{n} \right\rfloor - t + 1}{n - t + 1} \leq \frac{t - 1}{n}$$

$$\Longrightarrow \left(\frac{\frac{2|E|}{n} - t}{n - t + 1} \right)^t \leq \frac{t - 1}{n},$$

since $\frac{2|E|}{n} \leq n$. This gives the required upper bound:

$$|E| \leq \frac{n}{2} \left(\frac{(t - 1)^{\frac{1}{t}} (n - t + 1)}{n^{\frac{1}{t}}} + t \right) \leq n^{2 - \frac{1}{t}}.$$

□

The above proof is perhaps more natural when seen inductively—we give a sketch for the case $t = 2$. Let $T(n)$ denote the maximum number of edges in any n-vertex graph with no $K_{2,2}$ subgraph. If every vertex of V has degree less than $\sqrt{n}$, then already $|E| \leq n^{\frac{3}{2}}$. Otherwise let v be a vertex with $|N_G(v)| \geq \sqrt{n}$. Now any vertex $w \in V \setminus \{v\}$ can have at most one neighbor in $N_G(v)$, as otherwise $\{v, w\}$ together with any two of their common neighbors in $N_G(v)$ contains a $K_{2,2}$. Therefore there are at most $|N_G(v)| + (n-1) \leq 2n$ edges of E that are incident to the vertices of $N_G(v)$. Removing the vertices $\{v\} \cup N_G(v)$ from G and recursively upper bounding the edges in the remaining graph, we get $T(n) \leq 2n + T(n - \sqrt{n})$, which solves to $T(n) = O\left(n^{3/2}\right)$.

Theorem 3.22 shows that the above bound can be significantly improved if G is the intersection graph of segments in the plane.

PROOF OF THEOREM 3.22. We will need the following consequences of S being in general position.

- the intersection of every two segments of S is either empty or is a point lying in the interior of both segments, and

- the three supporting lines of any three segments of S do not have a non-empty intersection.

Let $m = |E_I|$ be the number of intersections between the segments of S. Apply Theorem 3.9 to S, with parameter r to be fixed later, to get a decomposition of the plane into $s \leq c \cdot \left(r + \frac{mr^2}{n^2}\right)$ triangles, where c is an absolute constant. Denote the triangles in this decomposition by

$$\mathcal{T}(S) = \{\Delta_1, \ldots, \Delta_s\}.$$

By increasing it if necessary, we can assume that $c \geq 2$.

For each $i = 1, \ldots, s$, let $S_i \subseteq S$ be the set of segments that either intersect the interior of Δ_i, or which contain a vertex of Δ_i, or which lie on an edge of Δ_i. Note that for each i,

- at most $\frac{n}{r}$ segments intersect the interior of Δ_i by Theorem 3.9, and
- there are at most 6 segments passing through the vertices of Δ_i by general position assumption, and
- at most one segment can coincide with any edge of Δ_i by general position assumption, and thus there are at most 3 segments containing an edge of Δ_i.

Thus for each $i = 1, \ldots, s$,

$$|S_i| \leq \frac{n}{r} + 9 \leq \frac{2n}{r},$$

assuming that $9 \leq \frac{n}{r}$ (our value of r will satisfy this).

Now consider any intersection point q between two segments of S. Then there must exist an index $i \in \{1, \ldots, s\}$ such that q lies in the interior or on the boundary of Δ_i. As q lies in the interior of both segments defining it, both of these segments must be present in S_i. Upper bounding the number of intersections within each triangle of $\mathcal{T}(S)$ by Theorem 3.23, we get

$$\begin{aligned} m &\leq \sum_{i=1}^{s} \Big(\#\text{ of intersections of } S \text{ lying in the interior or boundary of } \Delta_i\Big) \\ &\leq \sum_{i=1}^{s} |S_i|^{2-\frac{1}{t}} \leq \sum_{i=1}^{s} \left(\frac{2n}{r}\right)^{2-\frac{1}{t}} \leq 4\,c\left(r + \frac{mr^2}{n^2}\right) \cdot \left(\frac{n}{r}\right)^{2-\frac{1}{t}}. \end{aligned}$$

Note that if $\frac{n}{(8c)^t} \leq 2$, then trivially $m \leq \frac{n^2}{2} \leq (8c)^t \cdot n$. Otherwise setting $r = \frac{n}{(8c)^t}$,

$$\begin{aligned} m &\leq 4c\left(\frac{n}{(8c)^t} + \frac{m\left(\frac{n}{(8c)^t}\right)^2}{n^2}\right) \cdot \left((8c)^t\right)^{2-\frac{1}{t}} = 4c\left(\frac{n}{(8c)^t} + \frac{m}{(8c)^{2t}}\right) \cdot \left((8c)^t\right)^{2-\frac{1}{t}} \\ &= 4c\left(\frac{n}{(8c)^t} \cdot \left((8c)^t\right)^{2-\frac{1}{t}} + \frac{m}{8c}\right) \leq 4 \cdot 8^t c^{t+1} n + \frac{m}{2}. \end{aligned}$$

Thus $m \leq 4 \cdot 8^t c^{t+1} n + \frac{m}{2}$, implying that $m \leq (8c)^{t+1} n$. This completes the proof. □

We remark here that the proof we presented gives an upper bound with an exponential dependence on t, but with the advantage that it generalizes immediately to higher dimensions using Theorem 3.18.

Bibliography and discussion. A different proof using planar separators gives a better bound of $O(nt \log t)$ but works only in $\mathbb{R}^2$ [**FP08**]. The proof we presented is from [**MP16**] and follows the influential technique of using cuttings for divide-and-conquer, the seminal paper being [**Cla+90**]. While we used this technique to prove a combinatorial bound, it has had many applications in the design of algorithms for range searching, point location and other problems in computational geometry (see [**Cha00**] for some of these applications). We refer the reader to [**FS13**] for a history of the Zarankiewicz problem.

[Cha00] B. Chazelle, *The discrepancy method*, Cambridge University Press, Cambridge, 2000. Randomness and complexity, DOI 10.1017/CBO9780511626371. MR1779341

[Cla+90] K. L. Clarkson, H. Edelsbrunner, L. J. Guibas, M. Sharir, and E. Welzl, *Combinatorial complexity bounds for arrangements of curves and spheres*, Discrete Comput. Geom. **5** (1990), no. 2, 99–160, DOI 10.1007/BF02187783. MR1032370

[FP08] J. Fox and J. Pach, *Separator theorems and Turán-type results for planar intersection graphs*, Adv. Math. **219** (2008), no. 3, 1070–1080, DOI 10.1016/j.aim.2008.06.002. MR2442062

[FS13] Z. Füredi and M. Simonovits, *The history of degenerate (bipartite) extremal graph problems*, Erdös centennial, Bolyai Soc. Math. Stud., vol. 25, János Bolyai Math. Soc., Budapest, 2013, pp. 169–264, DOI 10.1007/978-3-642-39286-3_7. MR3203598

[MP16] N. H. Mustafa and J. Pach, *On the Zarankiewicz problem for intersection hypergraphs*, J. Combin. Theory Ser. A **141** (2016), 1–7, DOI 10.1016/j.jcta.2016.02.001. MR3479234

CHAPTER 4

Complexity of Set Systems

The goal of this chapter is to build the technical foundations that will enable us to generalize the constructions of Chapters 2 and 3 to combinatorial set systems. We now briefly revisit the key ideas behind two earlier constructions—that of Theorem 2.9 and Theorem 3.3—to see the precise properties that are needed and their possible generalizations to combinatorial set systems.

Let P be a set of n points in $\mathbb{R}^2$ and let $\mathcal{R}$ be the primal set system induced on P by disks.

The first key idea forms the basis of the following theorem.

Theorem 2.9: *A uniform random sample of P of size $\Omega\left(\frac{1}{\epsilon}\log\frac{1}{\epsilon}\right)$ is an ϵ-net of $\mathcal{R}$ with constant probability.*

An important fact used in this proof is that $|\mathcal{R}| = O\left(\binom{n}{3}\right) = O\left(n^3\right)$, and that this is true in an 'hereditary' or 'local' manner:

(4.1) for any set $P' \subseteq P$,

$$\text{the number of distinct subsets of } P' \text{ induced by disks is } O\left(|P'|^3\right).$$

It is important to note here that this bound depends *polynomially* and *only* on $|P'|$. The proof of this follows from the fact that disks are 'fixed' by three points (see Claim 1.4). The same idea shows that there are $O\left(n^4\right)$ sets induced on P by axis-aligned rectangles in $\mathbb{R}^2$, $O\left(n^d\right)$ sets induced on n points in $\mathbb{R}^d$ by half-spaces and so on.

The number of sets induced on any subset of $P' \subseteq P$ by intersection with disks corresponds, when dealing with an abstract set system $(X, \mathcal{F})$, to the number of subsets of $Y \subseteq X$ that can be obtained by intersection with sets of $\mathcal{F}$. Formally, define the *projection* of $\mathcal{F}$ onto any $Y \subseteq X$ as the set system

$$\mathcal{F}|_Y = \{S \cap Y \,:\, S \in \mathcal{F}\}.$$

Note that the set system induced on $P' \subseteq P$ by disks is precisely $\mathcal{R}|_{P'}$. Furthermore, just as the bound in Equation (4.1) of $O\left(|P'|^3\right)$ on the number of subsets of $P' \subseteq P$ induced by disks follows from the fact that a disk is 'fixed' by three points, one can derive a bound on the size of the projection $\mathcal{F}|_Y$ for any $Y \subseteq X$ by assuming a similar property for general set systems. It is not immediately clear how to generalize the property of a disk being fixed by three points to combinatorial set systems (see the remark after the proof of Lemma 4.3). The key, however, is to note that

> if a disk D is fixed by the set $Y = \{p, q, r\}$ of three points on its boundary, then by slightly shifting and scaling D, one can obtain all subsets of Y—$\emptyset$, $\{p\}$, $\{q\}$, $\{r\}$, $\{p, q\}$, $\{p, r\}$, $\{q, r\}$, $\{p, q, r\}$—by intersection with disks.

This property, stated abstractly, is the notion of the Vapnik-Chervonenkis dimension (*VC-dimension* for short) of a set system[1].

DEFINITION 4.2. The *VC-dimension* of a set system $(X, \mathcal{F})$, denoted by VC-dim$(\mathcal{F})$, is the size of the largest $Y \subseteq X$ for which $|\mathcal{F}|_Y| = 2^{|Y|}$. We say that such a Y is *shattered* by $\mathcal{F}$.
For a family $\mathcal{R}$ of geometric objects in $\mathbb{R}^d$—e.g., the family of all half-spaces in $\mathbb{R}^d$—the VC-dimension of $\mathcal{R}$ is defined to be the VC-dimension of the primal set system $(\mathbb{R}^d, \mathcal{R})$.

The following key lemma confirms that a statement analogous to Equation (4.1) holds for any set system $(X, \mathcal{F})$ of VC-dimension d.

LEMMA 4.3 (Primal shatter lemma). *Given a set system* $(X, \mathcal{F})$ *with* VC-dim$(\mathcal{F}) \leq d$ *and any* $Y \subseteq X$, *we have*

$$|\mathcal{F}|_Y| \leq \sum_{i=0}^{d} \binom{|Y|}{i}.$$

When $|Y| \geq d$, *the above summation can be upper bounded by* $\left(\frac{e|Y|}{d}\right)^d = O\left(|Y|^d\right)$.

> It is not hard to see that the set system $\mathcal{R}$ induced by disks has VC-dimension three; in other words, it is not possible to find four points in the plane such that one can get all subsets of these four points by intersection with disks. This together with Lemma 4.3 allows us to recover Equation (4.1) for disks, but in an abstract way!

The first goal of this chapter is to study the VC-dimension of set systems.
We now turn to the second key idea which was already used in the following theorem.

Theorem 3.3: *There exists an* ϵ*-net of size* $O\left(\frac{1}{\epsilon}\right)$ *of* $\mathcal{R}$.
The improvement of Theorem 3.3 is due to an additional property of $\mathcal{R}$: the number of sets of size at most k induced by disks is bounded by $O\left(nk^2\right)$ (Lemma 1.2); for small values of k this is much smaller than the total number $O\left(n^3\right)$ of sets induced by disks. This difference turns out to be important for sampling purposes.

Unfortunately VC-dimension alone gives us no information about this *growth* in the number of sets as a function of the set size k. To this end, we will use a finer complexity measure of a set system, called the *shallow-cell complexity*.

DEFINITION 4.4. A set system $(X, \mathcal{F})$ has shallow-cell complexity $\varphi_{\mathcal{F}}(\cdot, \cdot)$ if for any positive integer k and any finite $Y \subseteq X$, the number of sets in $\mathcal{F}|_Y$ of size at most k is upper bounded by $|Y| \cdot \varphi_{\mathcal{F}}(|Y|, k)$.

[1]There are several other related notions of complexity, e.g., compression complexity, Littlestone dimension, teaching dimension.

For a family $\mathcal{R}$ of geometric objects in $\mathbb{R}^d$—e.g., the family of all half-spaces—the shallow-cell complexity of $\mathcal{R}$ is defined to be the shallow-cell complexity of the primal set system $(\mathbb{R}^d, \mathcal{R})$.

Throughout the rest of the text we will assume that the function bounding the shallow-cell complexity of a set system is *well-behaved*. Roughly, we want $\varphi_{\mathcal{F}}(m,k)$ to not increase exponentially with m and k. There are many ways this can be specified; for our purposes the following will suffice.

DEFINITION 4.5. A function $\varphi(\cdot,\cdot)$ is (a,b)-well-behaved, for $a \in (1,2)$ and $b \in \mathbb{R}^+$, if it is non-decreasing in both arguments and for all positive integers $m \geq k \geq b \geq 2$,

$$\varphi(m,k) \leq \left(\varphi\left(\frac{m}{2}, \frac{k}{2}\right)\right)^a.$$

> The constant b in Definition 4.5 can be seen as analogous to the VC-dimension, in the sense that $m \cdot \varphi(m,k)$ can be $\binom{m}{k}$ for $k < b$.

The second goal is to study the shallow-cell complexity of set systems. The first part of this chapter studies the VC-dimension of geometric set systems, and the second part is devoted to the study of shallow-cell complexity.

1. VC-dimension

Very little [of mathematics] is easily accessible. But I think a lot more of it can be explained so that a lot more people understand it ... I like to try to make mathematics easy, not to make it hard. I think there is a tendency among mathematicians to try to make it hard. I try to combat that when I see people wrap up their mathematics in formal fancy theories that make it less accessible.

William Thurston

A set system $(X, \mathcal{F})$ with VC-dimension d behaves in many respects like a set system induced by hyperplanes or half-spaces in $\mathbb{R}^d$. For example, consider the fact that the number of subsets induced by hyperplanes on a set $P \subset \mathbb{R}^d$ of n points in general position is $\sum_{i=0}^{d} \binom{n}{i}$.[2] The first of the two main theorems of this section shows that the same is true for set systems with VC-dimension d, and illustrates why VC-dimension is such a nice way to abstractly model a geometric phenomenon.

LEMMA 4.3 (Primal shatter lemma). *Given a set system $(X, \mathcal{F})$ with* VC-dim$(\mathcal{F}) \leq d$ *and any $Y \subseteq X$, we have*

$$\left|\mathcal{F}|_Y\right| \leq \sum_{i=0}^{d} \binom{|Y|}{i}.$$

When $|Y| \geq d$, the above summation can be upper bounded by $\left(\frac{e|Y|}{d}\right)^d = O\left(|Y|^d\right).$

Lemma 4.3 is what makes VC-dimension a useful parameter of set system complexity and justifies the analogy of such set systems with those induced by geometric objects. We make three remarks.

Optimality: The above bound is optimal, in the sense that for every positive integers n and d, there exists a set system $(X, \mathcal{F})$ with $|X| = n$, VC-dim$(\mathcal{F}) = d$ and for which any $Y \subseteq X$ has $\left|\mathcal{F}|_Y\right| = \sum_{i=0}^{d} \binom{|Y|}{i}$. This is the case, for example, when $\mathcal{F}$ is the primal set system induced by hyperplanes on a set X of n points in general position in $\mathbb{R}^d$.

The precise bound in Lemma 4.3: It might strike the reader a little odd that we state the upper bound first as $\left(\frac{e|Y|}{d}\right)^d$ instead of just $O\left(|Y|^d\right)$. As we will see, this precision allows us to derive optimal bounds, as a function of d, on sizes of ϵ-nets and ϵ-approximations. Just using $O\left(|Y|^d\right)$ as an upper bound on $\left|\mathcal{F}|_Y\right|$ would give slightly worse bounds, with an additional factor of d in the logarithmic terms.

From $\left|\mathcal{F}|_Y\right|$ to VC-dimension: The other direction is true as well—if for each $Y \subseteq X$ we have $\left|\mathcal{F}|_Y\right| = O\left(|Y|^d\right)$, then the VC-dimension of $\mathcal{F}$ cannot be too large.

[2] A similar bound holds for the primal set system induced by half-spaces; see Equation (1.13).

LEMMA 4.6. *Given a set system $(X, \mathcal{F})$, if there exist constants c, d with $c \leq \frac{d}{2}$, such that*

$$\text{for any } Y \subseteq X \text{ with } |Y| \geq d, \text{ we have } \quad |\mathcal{F}|_Y| \leq \left(\frac{|Y|}{c}\right)^d,$$

then $\text{VC-dim}(\mathcal{F}) \leq (1 + o(1))\, d \log \frac{d}{c}$.

PROOF. Let $t = \text{VC-dim}(\mathcal{F})$ and $Y \subseteq X$ be any set realizing the VC-dimension of $\mathcal{F}$. If $t \leq d$, there is nothing to prove. So assume that $t > d$, and $|\mathcal{F}|_Y| = 2^{|Y|}$. Then we have

$$2^t = |\mathcal{F}|_Y| \leq \left(\frac{t}{c}\right)^d \quad \Longrightarrow \quad t \leq d \log \frac{t}{c}. \tag{4.7}$$

There does not exist a closed-form solution for t in the above expression. However, one can simply verify that any value of t larger than the stated one contradicts Equation (4.7). Alternatively, one can use a common trick, that of applying Equation (4.7) repeatedly, to derive an approximate upper bound:

$$\begin{aligned} t &\leq d \cdot \log \frac{t}{c} \leq d \cdot \log \left(\frac{d \log \frac{t}{c}}{c}\right) = d \log \frac{d}{c} + d \log \log \frac{t}{c} \\ &\leq \cdots \leq d \log \frac{d}{c} + d \left(\log \log \frac{d}{c} + \log \log \log \frac{d}{c} + \cdots\right). \end{aligned}$$

Upper bounding the second term by a geometric series gives the desired bound. □

To state our second main result we need the notion of the *unit distance graph* of a set system.

Define the symmetric difference of two sets S, S' to be

$$\Delta(S, S') = (S \setminus S') \cup (S' \setminus S).$$

DEFINITION 4.8. Given a set system $\mathcal{F} = \{S_1, \ldots, S_m\}$ on X, the unit distance graph $G_U(\mathcal{F}) = (\mathcal{F}, E_{\mathcal{F}})$ is a graph on the vertex set $\mathcal{F}$ such that $\{S_i, S_j\} \in E_{\mathcal{F}}$ if and only if $|\Delta(S_i, S_j)| = 1$.

Our second main result of this section is the following.

LEMMA 4.9. *Given a set system $\mathcal{F} = \{S_1, \ldots, S_m\}$ on X, let $G_U(\mathcal{F}) = (\mathcal{F}, E_{\mathcal{F}})$ be its unit distance graph. If* $\text{VC-dim}(\mathcal{F}) \leq d$, *then* $|E_{\mathcal{F}}| \leq d \cdot |\mathcal{F}|$.

We now present the proof of Lemma 4.3. It uses an operation called *shifting*, whose application to a set system $(X, \mathcal{F})$ consists of applying it to each set of $\mathcal{F}$. The goal is to get a 'simpler' set system without changing the number of sets or the VC-dimension.

Given a set system $\mathcal{F} = \{S_1, \ldots, S_m\}$ on the set X and an element $a \in X$, let $\mathcal{F}_a$ be the set system formed by removing the element a from each set of $\mathcal{F}$ *as long as it does not create duplicate sets.*

Formally, for each $S_i \in \mathcal{F}$ define the set S_i' as

$$S_i' = \begin{cases} S_i & \text{if } S_i \setminus \{a\} \in \mathcal{F}, \\ S_i \setminus \{a\} & \text{otherwise.} \end{cases}$$

We say that S_i' is derived by shifting S_i with a (and S_i is shifted with a) and let $\mathcal{F}_a = \{S_1', \ldots, S_m'\}$ be the set system derived by shifting each set of $\mathcal{F}$ with a. Note that the order in which we shift the sets of $\mathcal{F}$ does not matter—the resulting set system is the same.

Let $\mathcal{F} = \{S_1, \ldots, S_m\}$ be the original set system and $\mathcal{F}_a = \{S_1', \ldots, S_m'\}$ the set system shifted with $a \in X$. There are two key features of this operation. First, that $|\mathcal{F}| = |\mathcal{F}_a|$; this follows directly from the definition of shifting. Second, that shifting does not increase the VC-dimension of the set system, as we now prove.

LEMMA 4.10. *For any $a \in X$, we have* VC-dim$(\mathcal{F}_a) \leq$ VC-dim$(\mathcal{F})$.

PROOF. Fix a set $Y \subseteq X$ that is shattered by $\mathcal{F}_a$. We show that it is also shattered by $\mathcal{F}$. If $a \notin Y$, then the intersections $S_i \cap Y$ and $S_i' \cap Y$ are identical for all $i \in \{1, \ldots, m\}$ and the statement follows. Thus assume that $a \in Y$.
Fix any set $B \subseteq Y$ and let $S_i' \in \mathcal{F}_a$ be such that $B = Y \cap S_i'$. We now exhibit a set $S_k \in \mathcal{F}$ such that $B = Y \cap S_k$. We distinguish two cases:

$a \in B$: As $B = Y \cap S_i'$, $a \in S_i'$ and so the set S_i must not have been shifted. Then $S_i = S_i'$ and $B = Y \cap S_i$.

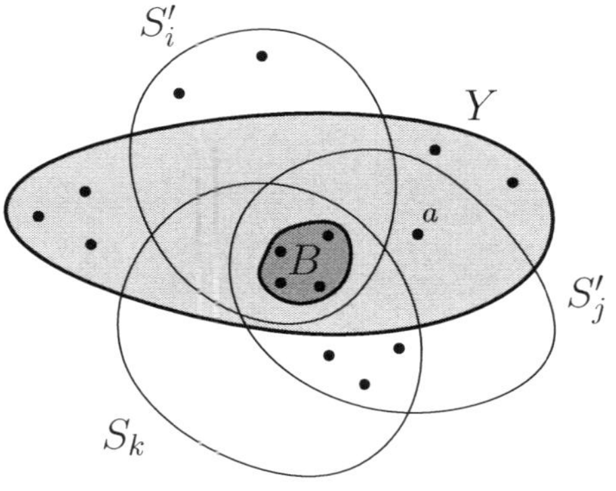

$a \notin B$: Then $a \notin S_i'$. See the figure. Now S_i need not be equal to S_i', as a could have been in S_i, in which case $Y \cap S_i$ would contain the additional element a. However, as Y is shattered by $\mathcal{F}_a$, there exists some other set $S_j' \in \mathcal{F}_a$ such that $B \cup \{a\} = Y \cap S_j'$. Furthermore, the set $S_k = S_j' \setminus \{a\}$ *must* be in $\mathcal{F}$—otherwise S_j would have been shifted to remove the element a and then S_j' would not contain a. And so $B = Y \cap S_k$.

□

We remark here that to show that any fixed $B \subseteq Y$ that can be realized using a set of $\mathcal{F}_a$ can also be realized using a set of $\mathcal{F}$, we needed to use the fact that Y was shattered by $\mathcal{F}_a$. Simply the fact that an individual B can be realized by a set of $\mathcal{F}_a$ is not sufficient.

Proof of Lemma 4.3. Repeatedly apply shifting on $\mathcal{F}|_Y$ with each element of Y. Let $\mathcal{F}'|_Y$ be the resulting set system where shifting with *any* element of Y does not change any set of $\mathcal{F}'|_Y$. This process must terminate as each shifting operation that changes the set system decreases the sum of cardinalities of the sets. First observe that $\mathcal{F}'|_Y$ is downward closed—that is, if $A \in \mathcal{F}'|_Y$, then all subsets of A must be present in $\mathcal{F}'|_Y$. Next, by Lemma 4.10, we have that VC-dim $(\mathcal{F}'|_Y) \le$ VC-dim $(\mathcal{F}|_Y) \le d$. Together they imply that the size of the largest set in $\mathcal{F}'|_Y$ is at most d. Now the proof follows by summing up the number of sets in $\mathcal{F}'|_Y$ by their sizes:

$$|\mathcal{F}|_Y| = |\mathcal{F}'|_Y| \le \sum_{i=0}^{d} \binom{|Y|}{i}.$$

There is no closed-form expression for the above sum but a useful approximation, when $|Y| \ge d$, is

$$\begin{aligned} \sum_{i=0}^{d} \binom{|Y|}{i} &= \sum_{i=0}^{d} \frac{|Y|!}{i!\,(|Y|-i)!} \le \sum_{i=0}^{d} \frac{|Y|^i}{i!} = \sum_{i=0}^{d} \frac{d^i \left(\frac{|Y|}{d}\right)^i}{i!} \\ &\le \left(\frac{|Y|}{d}\right)^d \cdot \sum_{i=0}^{\infty} \frac{d^i}{i!} = \left(\frac{|Y|}{d}\right)^d e^d, \end{aligned} \tag{4.11}$$

where the last step follows from the Taylor expansion of e^d. □

Given $\mathcal{F} = \{S_1, \ldots, S_m\}$ with VC-dimension at most d, the above proof shows that by repeatedly shifting $\mathcal{F}$ with elements of X, one can get a set system $\mathcal{F}' = \{S'_1, \ldots, S'_m\}$ such that $S'_i \subseteq S_i$ and $|S'_i| \le d$ for $i = 1, \ldots, m$. Thus one can think of the set S_i being 'fixed' or 'uniquely identified' by the set S'_i of size at most d. This justifies our earlier statement that the VC-dimension of a set system is intuitively analogous to the fact that three points 'fix' a disk in $\mathbb{R}^2$, or d points 'fix' a hyperplane in $\mathbb{R}^d$ and so on.

The proof of our second result also uses the shifting technique.

Lemma 4.9. *Given a set system $\mathcal{F} = \{S_1, \ldots, S_m\}$ on X, let $G_U(\mathcal{F}) = (\mathcal{F}, E_{\mathcal{F}})$ be its unit distance graph. If* VC-dim$(\mathcal{F}) \le d$, *then $|E_{\mathcal{F}}| \le d \cdot |\mathcal{F}|$.*

Proof. Repeatedly shift $\mathcal{F}$ with elements of X (in any order) and let $\mathcal{F}'$ be the resulting set system where shifting does not change any set. Let $E_{\mathcal{F}'}$ be the edges of the unit distance graph on $\mathcal{F}'$. We will need the following properties:

(1) $\mathcal{F}'$ is downward closed,
(2) $|\mathcal{F}| = |\mathcal{F}'|$,
(3) VC-dim$(\mathcal{F}') \le$ VC-dim$(\mathcal{F})$, and
(4) $|E_{\mathcal{F}'}| \ge |E_{\mathcal{F}}|$.

The first three properties have already been shown. Assuming the fourth one to be true as well, we first conclude the proof of Lemma 4.9.

Charge each edge $e \in E_{\mathcal{F}'}$ to the larger of the two sets of $\mathcal{F}'$ corresponding to the two vertices of e. As $\mathcal{F}'$ is downward closed, each $S' \in \mathcal{F}'$ has precisely $|S'|$ edges in $E_{\mathcal{F}'}$ charged to it—namely all the edges between S'

and each subset of S' of size $|S'|-1$. As VC-dim $(\mathcal{F}') \leq$ VC-dim $(\mathcal{F}) \leq d$, the largest set in $\mathcal{F}'$ has size at most d and so each set of $\mathcal{F}'$ gets charged at most d edges of $E_{\mathcal{F}'}$. Thus

$$|E_{\mathcal{F}}| \leq |E_{\mathcal{F}'}| \leq d\,|\mathcal{F}'| = d\,|\mathcal{F}|.$$

It remains to show that if $\mathcal{F}_a$ is the set system obtained by shifting $\mathcal{F}$ with any $a \in X$, then $|E_{\mathcal{F}_a}| \geq |E_{\mathcal{F}}|$.

Consider an edge $\{S, S_1\} \in E_{\mathcal{F}}$, where $S, S_1 \in \mathcal{F}$ and $S_1 = S \cup \{x\}$, for some $x \in X$. How can this edge *not* be present in the unit distance graph of $\mathcal{F}_a$? If $a \notin S$ and $x = a$, then S_1 will not be shifted and so the edge remains in $E_{\mathcal{F}_a}$. On the other hand, if $a \notin S$ and $x \neq a$, then $a \notin S_1$ and again both S, S_1 are unchanged by shifting with a and the edge remains in $E_{\mathcal{F}_a}$.

Thus it remains to consider the case when $a \in S$. Then $x \neq a$ and $a \in S_1$. The only way $\{S, S_1\}$ is not in $E_{\mathcal{F}_a}$ is if precisely one of S, S_1 gets shifted by a.

S gets shifted to S' by a, $S_1 = S_1'$ remains unchanged: This implies that the set $S_2 = S_1 \setminus \{a\}$ is already in $\mathcal{F}$. See the figure.

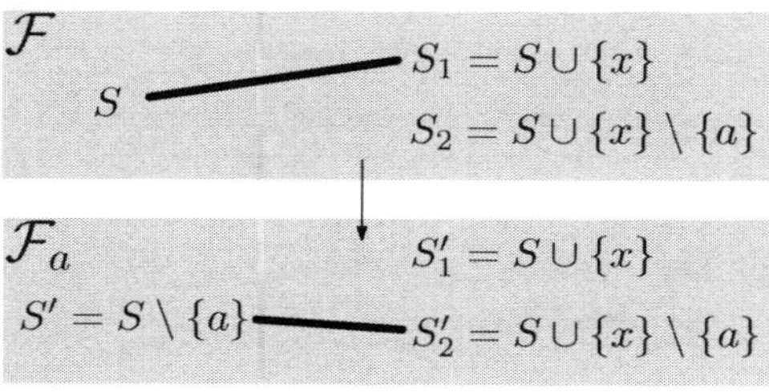

The edge between S' and S_1' is no longer present in $E_{\mathcal{F}_a}$; however there is now an edge between S' and S_2'. Since for each such pair $S \in \mathcal{F}$ and $x \in X$, the edge $\{S, S \cup \{x\}\}$ in $E_{\mathcal{F}}$ gets replaced with the edge $\{S \setminus \{a\}, S \cup \{x\} \setminus \{a\}\}$ in $E_{\mathcal{F}_a}$, the replacements are all distinct. That is, there is a one-to-one correspondence of replacements of edges of $E_{\mathcal{F}}$ with those of $E_{\mathcal{F}_a}$.

S is unchanged, $S \cup \{x\}$ gets shifted: This case is similar to the previous one. Then $S \setminus \{a\}$ already exists in $\mathcal{F}$ and thus in $\mathcal{F}_a$. Therefore the edge $\{S, S \cup \{x\}\}$ in $E_{\mathcal{F}}$ gets replaced by the edge $\{S \setminus \{a\}, S \cup \{x\} \setminus \{a\}\}$ in $E_{\mathcal{F}_a}$.

This concludes the proof. □

It is not hard to see that most geometric set systems have small VC-dimension. We conclude this section by presenting tight bounds on the VC-dimension of some basic geometric set systems.

LEMMA 4.12. *Let $\mathcal{H}$ be the family of all hyperplanes in $\mathbb{R}^d$. Then* VC-dim $(\mathcal{H}) = d$.

PROOF. As any set of d points in general position is shattered by $\mathcal{H}$, we have VC-dim $(\mathcal{H}) \geq d$. For the upper bound, the case $d = 2$ is easy to see and so inductively assume that the family of hyperplanes in $\mathbb{R}^{d-1}$ has VC-dimension at most $d - 1$.

Assume that there exists a set P of $d+1$ points in $\mathbb{R}^d$ that is shattered by $\mathcal{H}$. Then P must lie on some hyperplane H in $\mathbb{R}^d$; without loss of generality we can assume that $H = \mathbb{R}^{d-1}$. Let P' be any d points of P. For every $Q \subseteq P'$, let H_Q be a hyperplane in $\mathbb{R}^d$ containing precisely the points of Q. As H_Q does not contain the point in $P \setminus P'$, H_Q must be distinct from H and so $H_Q \cap H$ is a hyperplane in $\mathbb{R}^{d-1}$. But then P' (consisting of d points in $\mathbb{R}^{d-1}$) is shattered by the set of hyperplanes $\{H_Q \cap H : Q \subseteq P'\}$ in $\mathbb{R}^{d-1}$, a contradiction to the inductive hypothesis. □

LEMMA 4.13. *Let $\mathcal{H}^+$ be the family of all half-spaces in $\mathbb{R}^d$. Then* VC-dim $(\mathcal{H}^+) = d+1$.

PROOF. Clearly VC-dim $(\mathcal{H}^+) \geq d+1$, as any set of $d+1$ points in general position is shattered by $\mathcal{H}^+$. For the upper bound, the case $d = 1$ or $d = 2$ is easy to see and so inductively assume that the family of half-spaces in $\mathbb{R}^{d-1}$ has VC-dimension at most d.
Let P be any set of $d+2$ points in $\mathbb{R}^d$. If conv (P) contains a point of P, then clearly the interior point cannot be separated from the remaining points by a half-space. Otherwise P is in convex position, and let H be a hyperplane spanned by a set $P' \subset P$ of size d, with the two remaining points in $P \setminus P'$ lying on different sides of H[3]. Let q be the intersection point of the line segment spanned by $P \setminus P'$ with H. By induction on the dimension, the set $P' \cup \{q\}$ in $\mathbb{R}^{d-1}$ of size $d+1$ cannot be shattered by half-spaces; that is, P' can be partitioned into two disjoint subsets that cannot be separated by a hyperplane lying in H. Replacing q by $P \setminus P'$ in the appropriate set of this partition gives us two disjoint subsets of P that cannot be separated by a hyperplane in $\mathbb{R}^d$ and thus P cannot be shattered by half-spaces. □

LEMMA 4.14. *Let $\mathcal{B}$ be the family of all balls in $\mathbb{R}^d$. Then* VC-dim $(\mathcal{B}) = d+1$.

PROOF. Assume that a set of points P in $\mathbb{R}^d$ is shattered by the primal set system induced by balls. That is, for any $Q \subseteq P$ there exists a ball B with $Q = B \cap P$ and a ball B' with $P \setminus Q = B' \cap P$. Then the hyperplane passing through $\partial B \cap \partial B'$ (or simply separating B and B' if the two balls are disjoint) separates Q from $P \setminus Q$. Thus if a set of points is shattered by the primal set system induced by balls in $\mathbb{R}^d$, then it is shattered by the primal set system induced by half-spaces in $\mathbb{R}^d$. The proof now follows from Lemma 4.13. □

More generally, primal set systems induced by polynomial inequalities can be realized, using the so-called Veronese maps, by primal set systems induced by half-spaces in some higher dimensional space (see the discussion). Formally, each d-variate polynomial $f(x_1, \ldots, x_d)$ defines the set S_f on $\mathbb{R}^d$, where $S_f = \{p \in \mathbb{R}^d : f(p) \geq 0\}$. Veronese maps imply the following (stated without proof).

LEMMA 4.15. *Let $\mathcal{R}_{d,D}$ be a primal set system induced on $\mathbb{R}^d$ by the sets defined by all d-variable polynomials of degree at most D. Then* VC-dim $(\mathcal{R}_{d,D}) \leq \binom{d+D}{d}$.

[3]Such a hyperplane H always exists: let Δ be a simplex spanned by any $d+1$ points of P. The remaining point lies outside Δ, and so there is a hyperplane through a facet of Δ that separates this point from the other remaining vertex of Δ.

Bibliography and discussion. The notion of VC-dimension was introduced in [**VC71**]. The shifting proof of Lemma 4.9 is from [**Hau95**]; the statement was first proven in [**HLW94**] using induction. We refer the reader to [**Fra87**] for the use of the shifting technique to prove several other theorems. The more commonly presented proofs of Lemma 4.3 and Lemma 4.9 use induction on VC-dimension; see the texts [**Bol86, Mat99**]. A beautiful linear algebraic proof of Lemma 4.3 is given in [**FP83**]. For the bounds on VC-dimension derived from Veronese maps we refer to reader to [**Mat02**]. A thorough study of VC-dimension and related parameters for the set systems induced by half-spaces and related objects can be found in [**Ezr+20**]. A study of VC-dimension from the perspective of learning theory can be found in [**DGL96**, Chapter 12]. The structure of distance graphs are quite well-understood for certain set systems (e.g., see [**CKP20**]).

[Bol86] B. Bollobás, *Combinatorics*, Cambridge University Press, Cambridge, 1986. Set systems, hypergraphs, families of vectors and combinatorial probability. MR866142

[CKP20] V. Chepoi, K. Knauer, and M. Philibert, *Two-dimensional partial cubes*, Electron. J. Combin. **27** (2020), no. 3, Paper No. 3.29, 40, DOI 10.37236/8934. MR4245142

[DGL96] L. Devroye, Györfi and G. Lugosi. *A Probabilistic Theory of Pattern Recognition.* Springer, 1996.

[Ezr+20] E. Ezra, S. Har-Peled, H. Kaplan, and M. Sharir, *Decomposing arrangements of hyperplanes: VC-dimension, combinatorial dimension, and point location*, Discrete Comput. Geom. **64** (2020), no. 1, 109–173, DOI 10.1007/s00454-019-00141-7. MR4110531

[FP83] P. Frankl and J. Pach, *On the number of sets in a null t-design*, European J. Combin. **4** (1983), no. 1, 21–23, DOI 10.1016/S0195-6698(83)80004-3. MR694464

[Fra87] P. Frankl, *The shifting technique in extremal set theory*, Surveys in combinatorics 1987 (New Cross, 1987), London Math. Soc. Lecture Note Ser., vol. 123, Cambridge Univ. Press, Cambridge, 1987, pp. 81–110. MR905277

[HLW94] D. Haussler, N. Littlestone, and M. K. Warmuth, *Predicting $\{0,1\}$-functions on randomly drawn points*, Inform. and Comput. **115** (1994), no. 2, 248–292, DOI 10.1006/inco.1994.1097. MR1304811

[Hau95] D. Haussler, *Sphere packing numbers for subsets of the Boolean n-cube with bounded Vapnik-Chervonenkis dimension*, J. Combin. Theory Ser. A **69** (1995), no. 2, 217–232, DOI 10.1016/0097-3165(95)90052-7. MR1313896

[Mat02] J. Matoušek. *Lectures in Discrete Geometry.* Springer, 2002.

[Mat99] J. Matoušek. *Geometric Discrepancy: An Illustrated Guide.* Springer, 1999.

2. Shallow-cell Complexity

Indeed, probabilities are merely ratios of sizes of sets, and so probabilistic analysis is merely combinatorics. Yet, as is often the case in mathematics and science, using certain definitions and notations (rather than others) may simplify the analysis tremendously. Specifically, in many cases, the analysis is much simpler to carry out in terms of probabilities than in terms of set sizes.

Oded Goldreich

Recall the definition of the shallow-cell complexity of a set system.

DEFINITION 4.4. A set system $(X, \mathcal{F})$ has shallow-cell complexity $\varphi_{\mathcal{F}}(\cdot,\cdot)$ if for any positive integer k and any finite $Y \subseteq X$, the number of sets in $\mathcal{F}|_Y$ of size at most k is upper bounded by $|Y| \cdot \varphi_{\mathcal{F}}\big(|Y|, k\big)$.

For a family $\mathcal{R}$ of geometric objects in $\mathbb{R}^d$—e.g., the family of all half-spaces—the shallow-cell complexity of $\mathcal{R}$ is defined to be the shallow-cell complexity of the primal set system $\big(\mathbb{R}^d, \mathcal{R}\big)$.

Determining the shallow-cell complexity of a given set system $(X, \mathcal{F})$ seems difficult, as one is asking for a lot of information about the set system: to upper bound, for every integer $k \geq 1$ and every $Y \subseteq X$, the number of sets in $\mathcal{F}|_Y$ of size at most k (recall that these sets were called the ($\leq k$)-level-sets of $\mathcal{F}|_Y$ and denoted by $(\mathcal{F}|_Y)_{\leq k}$). However, the probabilistic averaging technique of Chapter 1 illustrates that for most geometric set systems one only needs an upper bound when k is a constant to derive an upper bound for every k. We have already seen two examples of this.

Primal set system induced by disks: $\varphi(n, k) = O\left(k^2\right)$ (Lemma 1.2).

Dual set system induced by disks : $\varphi(n, k) = O(k)$ (Lemma 1.9).

In this section we study the shallow-cell complexity of two geometric set systems, one primal and one dual.

Dual set systems. Let $\mathcal{O} = \{O_1, \ldots, O_n\}$ be a set of geometric objects in $\mathbb{R}^2$, each bounded by a Jordan curve in the plane. Recall that the dual set system induced on $\mathcal{O}$ by $\mathbb{R}^2$ is defined as

$$\mathcal{R}(\mathcal{O}) = \Big\{\mathcal{O}_p \,:\, p \in \mathbb{R}^2\Big\}, \qquad \text{where} \qquad \mathcal{O}_p = \Big\{O \in \mathcal{O} \colon O \ni p\Big\}.$$

We will assume that the boundaries of every pair of objects of $\mathcal{O}$ intersect in at most a constant number of points and that no point of $\mathbb{R}^2$ lies on the boundary of more than two objects of $\mathcal{O}$. For any $\mathcal{O}' \subseteq \mathcal{O}$, let $\mathcal{V}(\mathcal{O}')$ denote the finite set of points that lie on the intersection of the boundaries of any two objects of $\mathcal{O}'$.

Our goal is to upper bound $|\mathcal{R}_{\leq k}(\mathcal{O}')|$, for any $\mathcal{O}' \subseteq \mathcal{O}$ and any integer k. As before, this can be upper bounded, within constant factors, by $|\mathcal{V}_{\leq k}(\mathcal{O}')| + |\mathcal{O}'|$, where

$$\mathcal{V}_{\leq k}(\mathcal{O}') = \Big\{v \in \mathcal{V}(\mathcal{O}') : v \text{ is contained in the } \textit{interior} \text{ of } \leq k \text{ objects of } \mathcal{O}'\Big\}.$$

The case $k = 0$ is related to the *union complexity* of $\mathcal{O}$. Denote the union of $\mathcal{O}$ by

$$\mathcal{U}(\mathcal{O}) = \bigcup_{O \in \mathcal{O}} O.$$

That is, $p \in \mathcal{U}(\mathcal{O})$ if and only if p lies in at least one object of $\mathcal{O}$. Note that the set of vertices lying on the boundary of the union is precisely the set $\mathcal{V}_{\le 0}(\mathcal{O})$. The complexity of the union $\mathcal{U}(\mathcal{O})$ is the total number of faces on the boundary of $\mathcal{U}(\mathcal{O})$—in our case it consists of the vertices of $\mathcal{V}_{\le 0}(\mathcal{O})$ as well as the induced arcs between the vertices of $\mathcal{V}_{\le 0}(\mathcal{O})$. See the figure.

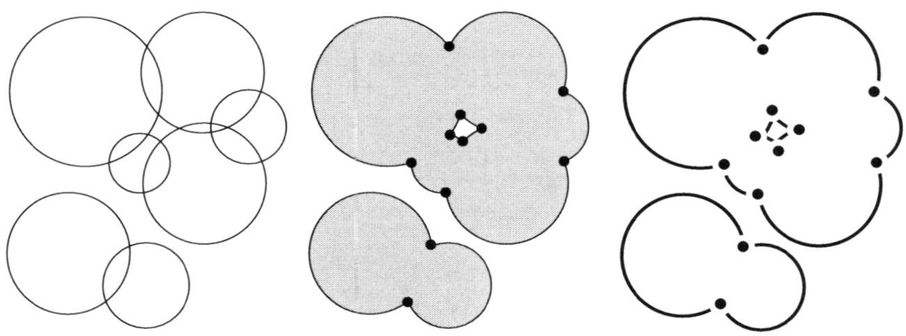

We say that $\mathcal{O}$ has union complexity $f\colon \mathbb{Z}^+ \to \mathbb{R}$ if the complexity of the union of any $\mathcal{O}' \subseteq \mathcal{O}$ is at most $f(|\mathcal{O}'|)$. The following lemma uses the probabilistic averaging technique of Chapter 1.

LEMMA 4.16. *Given a set $\mathcal{O}$ of n objects in the plane with union complexity $f(\cdot)$, where $f(m) \ge m$, and an integer $k \in \left[0, \frac{n}{2}\right]$,*

$$|\mathcal{V}_{\le k}(\mathcal{O})| = O\left((k+1)^2 \cdot f\left(\frac{n}{k+1}\right)\right).$$

This implies that the dual set system $\mathcal{R}(\mathcal{O})$ induced on $\mathcal{O}$ by $\mathbb{R}^2$ has shallow-cell complexity

$$\varphi_{\mathcal{R}(\mathcal{O})}(m, k) = O\left(\frac{(k+1)^2}{m} \cdot f\left(\frac{m}{k+1}\right)\right).$$

PROOF. Let S be a subset of $\mathcal{O}$ of size $t = \frac{n}{k+1}$ chosen uniformly at random from all t-sized subsets of $\mathcal{O}$ (for simplicity assume that n is a multiple of $k+1$).

> This is different from our usual sampling distribution which is to pick each point independently with probability p. The reason is that we will need to prove $\mathrm{E}[f(|S|)] \le f(np)$. There are two technical ways to calculate this: compute $\mathrm{E}[f(|S|)]$ by conditioning on $|S|$ and summing up an exponential series over all sizes of S or sampling a set of a fixed size. We have chosen the second way, but the first method also works.

We bound the expected size of $\mathcal{V}_{\le 0}(S)$ in two ways.

Upper bound: From the definition of union complexity,

$$|\mathcal{V}_{\le 0}(S)| \le f(|S|) = f(t).$$

Lower bound: A vertex $v \in \mathcal{V}_{\leq k}(\mathcal{O})$ ends up as a vertex of $\mathcal{V}_{\leq 0}(S)$ if and only if both of the two defining objects of v are present in S and none of the at most k objects containing v in their interior are present in S. The probability of this happening is at least $\frac{\binom{n-2-k}{t-2}}{\binom{n}{t}}$ and so by linearity of expectation,

$$\mathrm{E}\left[|\mathcal{V}_{\leq 0}(S)|\right] \geq |\mathcal{V}_{\leq k}(\mathcal{O})| \cdot \frac{\binom{n-2-k}{t-2}}{\binom{n}{t}}.$$

Combining the upper and lower bounds,

$$|\mathcal{V}_{\leq k}(\mathcal{O})| \cdot \frac{\binom{n-2-k}{t-2}}{\binom{n}{t}} \quad \leq \quad \mathrm{E}\left[|\mathcal{V}_{\leq 0}(S)|\right] \quad \leq \quad f(t).$$

Simplifying the above, we get

$$\begin{aligned}
|\mathcal{V}_{\leq k}(\mathcal{O})| &\leq f(t) \cdot \frac{\binom{n}{t}}{\binom{n-2-k}{t-2}} = f(t) \cdot \frac{n!}{(n-t)!\, t!} \frac{(t-2)!\,(n-k-t)!}{(n-k-2)!} \\
&= f(t) \cdot \frac{n(n-1)}{t(t-1)} \cdot \frac{(n-2)}{(n-k-2)} \frac{(n-3)}{(n-k-3)} \cdots \frac{(n-t+1)}{(n-k-t+1)} \\
&= f(t) \cdot \frac{n(n-1)}{t(t-1)} \cdot \frac{1}{1-\frac{k}{n-2}} \frac{1}{1-\frac{k}{n-3}} \cdots \frac{1}{1-\frac{k}{n-t+1}} \\
&= f(t) \cdot \frac{n(n-1)}{t(t-1)} \cdot \prod_{i=0}^{t-3} \frac{1}{1-\frac{k}{n-i-2}} \leq f(t) \cdot \frac{n(n-1)}{t(t-1)} \cdot \left(e^{\frac{2k}{n-t+1}}\right)^{t-2} \\
&= O\left(\frac{n^2}{t^2} \cdot f\left(\frac{n}{k+1}\right)\right) = O\left((k+1)^2 \cdot f\left(\frac{n}{k+1}\right)\right).
\end{aligned}$$

The third-to-last step used the fact that $1 - x \geq e^{-2x}$ for $x \in [0, 0.79]$; this holds in our case since $t = \frac{n}{k+1}$ and so $\frac{k}{n-t+1} = \frac{k(k+1)}{n(k+1)-n+(k+1)} = \frac{k(k+1)}{nk+k+1} \leq \frac{k+1}{n+1} \leq \frac{3}{4}$ for $k \in \left[0, \frac{n}{2}\right]$. The second-to-last step follows as $\frac{2k(t-2)}{n-t+1} \leq \frac{2kn}{n(k+1)-n+(k+1)} = \frac{2kn}{nk+k+1} \leq 2$. This completes the first part of the proof.
Let $\mathcal{O}' \subseteq \mathcal{O}$. By the definition of union complexity we have $|\mathcal{V}_{\leq 0}(\mathcal{O}')| \leq f(|\mathcal{O}'|)$ and thus the above calculation holds for any $\mathcal{O}'$. This implies that $|\mathcal{V}_{\leq k}(\mathcal{O}')| = O\left((k+1)^2 \cdot f\left(\frac{|\mathcal{O}'|}{k+1}\right)\right)$. Thus

$$\begin{aligned}
\left|\left(\mathcal{R}(\mathcal{O})|_{\mathcal{O}'}\right)_{\leq k}\right| &= |\mathcal{R}_{\leq k}(\mathcal{O}')| = O\left(|\mathcal{V}_{\leq k}(\mathcal{O}')| + |\mathcal{O}'|\right) \\
&= O\left((k+1)^2 \cdot f\left(\frac{|\mathcal{O}'|}{k+1}\right)\right),
\end{aligned}$$

and we get the stated bound for the shallow-cell complexity of $\mathcal{R}(\mathcal{O})$. □

For disks in the plane we have $f(t) = O(t)$ (see Claim 1.10) and we recover the $O(nk)$ bound of Lemma 1.9.
When $\mathcal{T}$ is a set of axis-aligned rectangles in the plane, we have $f(t) = O\left(t^2\right)$ and so Lemma 4.16 implies an upper bound of $O\left(n^2\right)$ for the size of $\mathcal{V}_{\leq k}(\mathcal{T})$. This is just the trivial bound, as we assumed that the boundaries of every pair of objects intersect in a constant number of points. Furthermore this is tight—even for $k = 0$,

a 'grid' of axis-aligned rectangles shows that there can be $\Omega\left(n^2\right)$ vertices on the boundary of the union of the rectangles. See the figure.

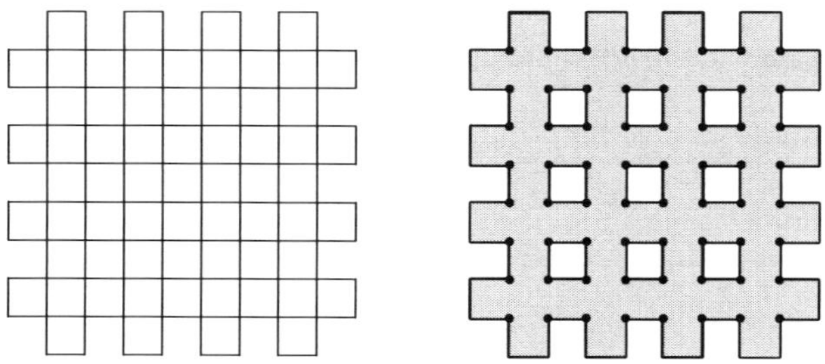

Primal set systems. In this section we consider an interesting example of a primal set system that is 'approximable' by a set system with smaller shallow-cell complexity. This is the primal set system $\mathcal{R}$ induced on a set P of n points in $\mathbb{R}^2$ by axis-aligned rectangles in the plane.

It is not hard to show that for any integer $k \geq 2$, the number of subsets of P of size at most k induced by axis-aligned rectangles is $O\left(n^2 k^2\right)$. That is, $\varphi_{\mathcal{R}}(m, k) = O\left(mk^2\right)$. Furthermore, the quadratic dependence on n cannot be improved. See the figure for an example, consisting of $2n$ points $\{p_1, \ldots, p_n, q_1, \ldots, q_n\}$, such that the number of subsets of size at most k induced on these points by axis-aligned rectangles is $\Omega\left(n^2 k\right)$. In particular, for each $i \geq j$, there is an axis-aligned rectangle containing precisely the set $\{p_i, q_j\}$, and so even for $k = 2$, there are $\Omega\left(n^2\right)$ sets induced on this set by rectangles.

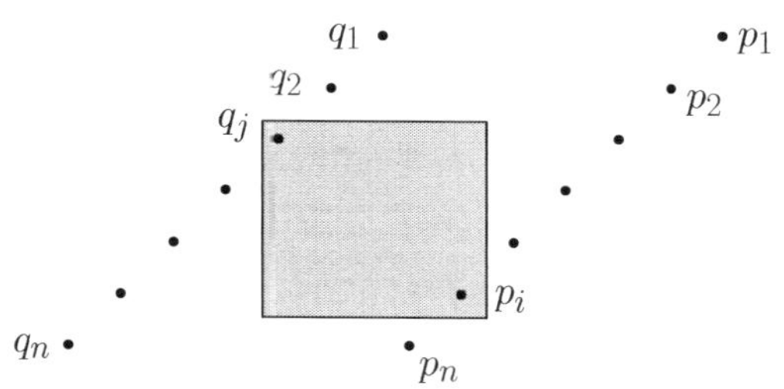

We now turn to the proof of the main theorem of this section.

THEOREM 4.17. *Let P be a set of n points in $\mathbb{R}^2$ and $\epsilon > 0$ a given parameter. Let $\mathcal{R}$ be the primal set system induced on P by axis-aligned rectangles in the plane. Then there exists a set system $\mathcal{R}'$ on P such that*

(1) *for any integer $k \geq 1$ and $Y \subseteq P$, $\left|\left(\mathcal{R}'|_Y\right)_{\leq k}\right| = O\left(|Y|\, k^2 \log \frac{1}{\epsilon}\right)$, implying that $\varphi_{\mathcal{R}'}(m, k) = O\left(k^2 \log \frac{1}{\epsilon}\right)$, and*
(2) *for any $R \in \mathcal{R}$ with $|R| \geq \epsilon n$, there exists a $R' \in \mathcal{R}'$ such $R' \subseteq R$ and $|R'| \geq \frac{|R|}{2}$.*

PROOF. For simplicity assume that n is a power of 2.

We construct the set system $\mathcal{R}'$ on P using a hierarchical spatial decomposition of P.

Set $P_0 = P$ and let l_0 be a vertical line, said to be at *level* 0, that divides P_0 into two sets P_1 and P_2 of size $\frac{n}{2}$ each. Let $\mathcal{R}_0$ be a set system on P_0 consisting of all subsets of P_0 induced by rectangles 'anchored' at l_0—that is, rectangles with one vertical edge lying on l_0. See the figure.

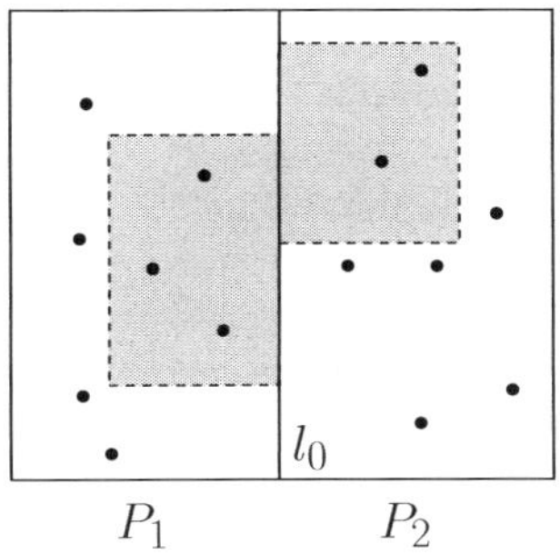

Now recurse independently on P_1 and P_2: divide P_1 into two equal-sized subsets of size $\frac{n}{4}$ each by a vertical line l_1 (of level 1) and let $\mathcal{R}_1$ consist of all sets induced on P_1 by rectangles anchored on l_1. Similarly divide P_2 with the line l_2 (of level 1) and add subsets of P_2 induced by rectangles anchored on l_2 to $\mathcal{R}_2$.

At the i-th level, equipartition the 2^i disjoint subsets $P_{2^i-1}, \ldots, P_{2^{i+1}-2}$ of P by vertical lines and for each $j = 2^i - 1, \ldots, 2^{i+1} - 2$, let $\mathcal{R}_j$ be the subsets of P_j induced by rectangles anchored on l_j. Finally set

$$\mathcal{R}' = \bigcup_{i=0}^{\lceil \log \frac{1}{\epsilon} \rceil} \bigcup_{j=2^i-1}^{2^{i+1}-2} \mathcal{R}_j.$$

Note that we only go until level $\lceil \log \frac{1}{\epsilon} \rceil$, since we only care about those rectangles containing at least ϵn points of P.

CLAIM 4.18. For any integer k and $Y \subseteq P$, $\left|(\mathcal{R}'|_Y)_{\leq k}\right| = O\left(|Y|\, k^2 \log \frac{1}{\epsilon}\right)$.

PROOF. The key fact is that for any P_j and the corresponding line l_j, the set system $\mathcal{R}_j$ induced on P_j by rectangles anchored on l_j has the property that for any $Y \subseteq P_j$, $\left|(\mathcal{R}_j|_Y)_{\leq k}\right| = O\left(|Y|\, k^2\right)$. We present the proof of this (Lemma 4.19) at the end.

Given any integer k and $Y \subseteq P$, we upper bound $\left|(\mathcal{R}'|_Y)_{\leq k}\right|$ level-by-level. As P is partitioned into disjoint subsets at each level,

$$\begin{aligned}\left|(\mathcal{R}'|_Y)_{\leq k}\right| &= \sum_{i=0}^{\lceil \log \frac{1}{\epsilon} \rceil} \sum_{j=2^i-1}^{2^{i+1}-2} O\left(|Y \cap P_j| \cdot k^2\right) = \sum_{i=0}^{\lceil \log \frac{1}{\epsilon} \rceil} O\left(|Y| \cdot k^2\right) \\ &= O\left(|Y|\, k^2 \log \frac{1}{\epsilon}\right).\end{aligned}$$

□

To see 2., consider any rectangle $R \in \mathcal{R}$ with $|R| \geq \epsilon n$. Then the line l intersected by R with the minimum level is unique and so R contains at least $\frac{|R|}{2}$ points from

one of the two sides of l and the intersection of R with this side is an anchored rectangle present in $\mathcal{R}'$.
We remark that the point here is to continue the recursive construction for $\log \frac{1}{\epsilon}$ steps instead of $\log n$ steps, as this suffices to prove the statement for rectangles containing at least ϵn points of P. □

LEMMA 4.19. *Let Q be a set of n points in $\mathbb{R}^2$ and let l be a vertical line in $\mathbb{R}^2$. Let $\mathcal{F}$ be the primal set system induced on Q by axis-aligned rectangles with a vertical edge lying on l. Then for any $Y \subseteq Q$, $\left|(\mathcal{F}|_Y)_{\leq k}\right| = O\left(|Y|\, k^2\right)$. That is, $\varphi_{\mathcal{F}}(m, k) = O\left(k^2\right)$.*

PROOF. See the figure for an illustration. It suffices to upper bound the number of sets due to anchored rectangles all lying on the same side of l.

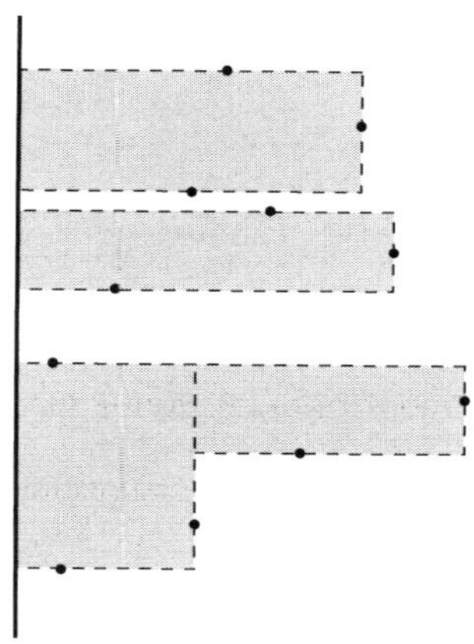

As before, within constant factors, it suffices to restrict ourselves to sets induced by *canonical rectangles* on Q—that is, axis-aligned rectangles containing a point of Q on each of their three bounding edges.
For each $q \in Q$, there can be only $O\left(k^2\right)$ canonical rectangles containing at most k points of Q in their interior and whose *right vertical edge* contains q: we can only move the top and bottom edges of each such rectangle and these can only induce $O\left(k^2\right)$ distinct subsets of Q of size at most k.
Thus the number of canonical rectangles containing at most k points of Q in their interior is $O\left(nk^2\right)$. The same bound applies to any $Y \subseteq Q$, implying that $\left|(\mathcal{F}|_Y)_{\leq k}\right| = O\left(|Y|\, k^2\right)$. □

We note that another proof of Lemma 4.19 follows using the averaging technique of Chapter 1.

Bibliography and discussion. The notion of shallow-cell complexity was formally introduced in [**Cha+12**]. Sharir [**Sha91**] gave other applications of bounding level sets of geometric set systems. For the shallow-cell complexity of dual set systems and its relation to union complexity, see the survey by Agarwal et al. [**APS08**]. The statement on axis-aligned rectangles is implicitly proved and used by Aronov et al. [**AES10**] to show the existence of asymptotically optimal ϵ-nets for the primal set system induced by axis-aligned rectangles in the plane.

[AES10] B. Aronov, E. Ezra, and M. Sharir, *Small-size ϵ-nets for axis-parallel rectangles and boxes*, SIAM J. Comput. **39** (2010), no. 7, 3248–3282, DOI 10.1137/090762968. MR2678074

[APS08] P. K. Agarwal, J. Pach, and M. Sharir, *State of the union (of geometric objects)*, Surveys on discrete and computational geometry, Contemp. Math., vol. 453, Amer. Math. Soc., Providence, RI, 2008, pp. 9–48, DOI 10.1090/conm/453/08794. MR2405676

[Cha+12] T. M. Chan, E. Grant, J. Könemann, and M. Sharpe, *Weighted capacitated, priority, and geometric set cover via improved quasi-uniform sampling*, Proceedings of the Twenty-Third Annual ACM-SIAM Symposium on Discrete Algorithms, ACM, New York, 2012, pp. 1576–1585. MR3205315

[Sha91] M. Sharir, *On k-sets in arrangements of curves and surfaces*, Discrete Comput. Geom. **6** (1991), no. 6, 593–613, DOI 10.1007/BF02574706. MR1121331

CHAPTER 5

Packings of Set Systems

The ϵ-net constructions so far have crucially used the fact that the given set systems are induced by configurations of objects in Euclidean space. For example, the constructions of ϵ-nets for the set system induced by disks in the plane—in Chapters 2 and 3—are based on the idea of maximal empty disks avoiding a random sample $S \subseteq P$ (the empty canonical disks spanned by S; see Claim 2.14). While this lead to optimal bounds for ϵ-nets for this and some other set systems, it also limited the use of these techniques to set systems induced by geometric objects. We now start to build the machinery required to generalize the earlier constructions to purely abstract settings.

In this chapter we focus on a specific aspect of Euclidean space: **packings**.

A basic instance of geometric packing is the following:

Let $\mathcal{C}$ be the cube $[0, n]^d$ in $\mathbb{R}^d$ and $P \subseteq \mathcal{C}$ be a set of points such that $\text{dist}(p_i, p_j) \geq \delta$ for every $p_i, p_j \in P$. What is an upper bound on $|P|$?

The observation is that the $|P|$ balls of radius $\frac{\delta}{2}$ centered at each $p \in P$, denoted by $\text{Ball}\left(p, \frac{\delta}{2}\right)$, must be pairwise disjoint. As these $|P|$ balls, each with volume at least $\left(\frac{\delta}{2}\right)^d$, are contained in the cube $\left[-\frac{\delta}{2}, n + \frac{\delta}{2}\right]^d$, we get

$$|P| \cdot \left(\frac{\delta}{2}\right)^d \leq \sum_{p \in P} \text{vol}\left(\text{Ball}\left(p, \frac{\delta}{2}\right)\right) \leq \text{vol}\left(\left[-\frac{\delta}{2}, n + \frac{\delta}{2}\right]^d\right) = (n + \delta)^d,$$

and thus $|P| = O\left(\left(\frac{n}{\delta}\right)^d\right)$.

The geometric notion of packing relies, among other things, on the notion of distance; the packing question then concerns the number of geometric objects that can exist while being pair-wise distant from each other.

In moving to abstract set systems the notion of distance between points is replaced, in a natural way, by the cardinality of the *symmetric difference* between sets.

Given two finite sets X and Y, the symmetric difference of X and Y is denoted by

$$\Delta(X, Y) = (X \setminus Y) \cup (Y \setminus X).$$

This chapter deals with the combinatorial analogs of packing statements where the underlying distance measure is the symmetric difference between sets. The main technical result will be the following important theorem which generalizes the packing properties of geometric objects to abstract set systems.

THEOREM 5.1. *Let $\mathcal{F} = \{S_1, \ldots, S_m\}$ be a set system on a set X of n elements and let $d \geq 1, \delta \in [n]$ be given integer parameters such that* $\text{VC-dim}(\mathcal{F}) \leq d$ *and for* *every* $1 \leq i < j \leq m$, *we have* $|\Delta(S_i, S_j)| \geq \delta$. *Then*

$$|\mathcal{F}| \leq 2 \cdot \mathrm{E}\left[|\mathcal{F}|_A|\right],$$

where A *is a subset of size* $s = \left\lceil \frac{8dn}{\delta} \right\rceil - 1$ *picked uniformly at random from* X *(without replacement).*

1. Packing Half-Spaces in $\mathbb{R}^d$

In nearly every theory [in Physics] there exist steps that are omitted in the theoretical papers and not treated in the textbooks. These steps are obviously designed to keep the experimental physicists in their place.

Hans Frauenfelder

The main theorem for this section is an immediate consequence of ϵ-net constructions of Chapter 3; it can be seen as a warm-up to the powerful packing statement of the next section.

THEOREM 5.2. *Let k, d and $\delta \geq 2$ be positive integers and P a set of n points in $\mathbb{R}^d$. Let $\mathcal{R}$ be a primal set system on P where each set of $\mathcal{R}$ is induced by a half-space in $\mathbb{R}^d$, such that*

(1) $|R| \leq k$ *for all $R \in \mathcal{R}$, and*
(2) $|\Delta(R, R')| \geq \delta$ *for all distinct $R, R' \in \mathcal{R}$.*

Then $|\mathcal{R}| = O\left(\frac{n^{\lfloor \frac{d}{2} \rfloor} k^{\lceil \frac{d}{2} \rceil}}{\delta^d}\right)$, where the constant in the asymptotic notation depends on d.

Overview of ideas. The main idea is captured in the proof of the following simpler problem in the dual setting (more later).

> Let $\mathcal{H}$ be a set of n hyperplanes in $\mathbb{R}^d$ and P a set of points in $\mathbb{R}^d$ in general position such that for any pair of points $p, q \in P$, the line segment $\overline{pq}$ intersects at least $\delta \geq 2$ hyperplanes of $\mathcal{H}$. What is the maximum cardinality of P?
>
> The answer is $|P| = O\left(\left(\frac{n}{\delta}\right)^d\right)$, derived as follows. Applying Theorem 3.18 to $\mathcal{H}$ with $r = \frac{n}{\delta-1}$, there exists a partition of $\mathbb{R}^d$ into a set Ξ of $O\left(r^d\right) = O\left(\left(\frac{n}{\delta}\right)^d\right)$ simplices such that the interior of each simplex of Ξ intersects at most $\frac{n}{r} = \delta - 1$ hyperplanes of $\mathcal{H}$. This implies that each simplex of Ξ can contain at most one point of P in its interior. Furthermore, since P is in general position, each simplex can contain at most $d\,(d+1)$ points of P on its boundary and thus there can be at most $O\left(\left(\frac{n}{\delta}\right)^d\right) \cdot d\,(d+1)$ such points.

Theorem 5.2 is a generalization of this statement.

PROOF OF THEOREM 5.2. Let $\mathcal{R} = \{R_1, \ldots, R_t\}$, where R_i is induced by the half-space h_i in $\mathbb{R}^d$ and let $\mathcal{H} = \{h_1, \ldots, h_t\}$ be these t half-spaces. We are given that each half-space in $\mathcal{H}$ contains at most k points of P and for any two indices $1 \leq i < j \leq t$, $\left|\Delta\left(h_i \cap P, h_j \cap P\right)\right| \geq \delta$.

To simplify the exposition, we will assume that P and $\mathcal{H}$ are in general position; in particular that no $d+2$ bounding hyperplanes of the half-spaces in $\mathcal{H}$ have a point in common and that no point of P lies on the bounding hyperplane of any half-space in $\mathcal{H}$. By partitioning $\mathcal{R}$ into two sets and bounding them separately, we can also assume without loss of generality that each half-space in $\mathcal{H}$ is downward-facing; that is, it contains the point $(0, \ldots, 0, -\lambda)$, for any sufficiently large $\lambda > 0$.

Using point-hyperplane duality, let $P^* = \{p^* \colon p \in P\}$ be the set of n hyperplanes dual to the points of P and let $\mathcal{H}^* = \{q_i^* \colon h_i \in \mathcal{H}\}$ be the set of n points dual to the bounding hyperplanes of the half-spaces in $\mathcal{H}$. That is,

$$\begin{aligned} p = (p_1, \ldots, p_d) \quad &\rightarrow \quad p^* : p_1 x_1 + \cdots + p_{d-1} x_{d-1} - x_d = -p_d, \\ \partial h_i : a_1 x_1 + \cdots + a_{d-1} x_{d-1} + x_d = b \quad &\rightarrow \quad q_i^* = (a_1, \ldots, a_{d-1}, b). \end{aligned}$$

See the figure for an example.

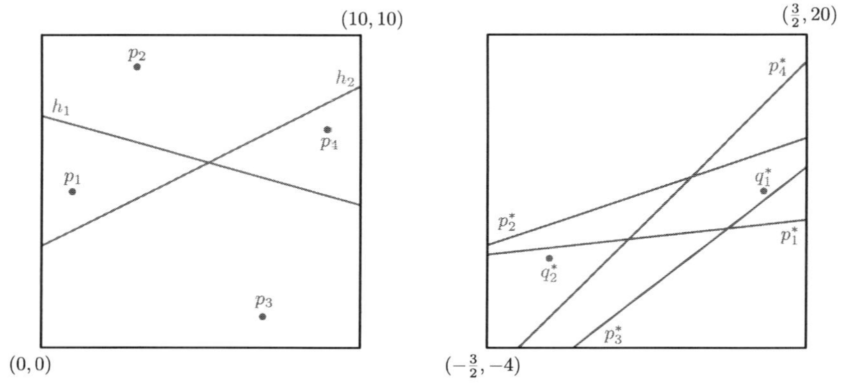

For each point $q^* \in \mathcal{H}^*$, let $l(q^*)$ be the set of hyperplanes of P^* lying below q^* with respect to the x_d axis; that is, $|l(q^*)|$ is the level of q^* with respect to P^* (see Definition 1.15). By the properties of point-hyperplane duality[1] and the assumption that no point of P lies on the bounding hyperplane of any half-space of $\mathcal{H}$, we get the following correspondence:

$$\begin{aligned} h_i \in \mathcal{H} \text{ contains at most } k \text{ points of } P \quad &\Longleftrightarrow \quad q_i^* \in \mathcal{H}^* \text{ has } |l(q_i^*)| \leq k, \\ |\Delta(h_i \cap P, h_j \cap P)| \geq \delta \quad &\Longleftrightarrow \quad |\Delta(l(q_i^*), l(q_j^*))| \geq \delta. \end{aligned}$$

Recall the existence of shallow-cuttings for hyperplanes.

DEFINITION 3.20. Given a set $\mathcal{H}$ of n hyperplanes in $\mathbb{R}^d$, a real parameter $r > 0$ and an integer $k \geq 0$, a $(\frac{1}{r}, k)$-shallow cutting is a set Ξ of interior-disjoint simplices (some of which are unbounded) such that

- the union of the simplices in Ξ covers all points in $\mathbb{R}^d$ of level at most k with respect to $\mathcal{H}$, and
- the interior of each simplex in Ξ intersects at most $\frac{n}{r}$ hyperplanes of $\mathcal{H}$.

THEOREM 3.21. *Let $\mathcal{H}$ be a set of n hyperplanes in $\mathbb{R}^d$, $r > 0$ a real and $k \geq 0$ an integer. Then there exists a $(\frac{1}{r}, k)$-shallow cutting of $\mathcal{H}$ of size*

$$O\left(r^{\lfloor \frac{d}{2} \rfloor} \left(\frac{kr}{n} + 1\right)^{\lceil \frac{d}{2} \rceil}\right).$$

[1]In particular, that it preserves sidedness—a point $p \in \mathbb{R}^d$ lies below a hyperplane h if and only if the point dual to h lies above the hyperplane dual to p.

Construct a $\left(\frac{\delta-1}{n}, k\right)$-shallow cutting Ξ for the hyperplanes in P^*. By Theorem 3.21 and the fact that $k \geq \left\lceil \frac{\delta}{2} \right\rceil$, we have

$$|\Xi| = O\left(\left(\frac{n}{\delta-1}\right)^{\lfloor \frac{d}{2} \rfloor}\left(\frac{k\left(\frac{n}{\delta-1}\right)}{n}+1\right)^{\lceil \frac{d}{2} \rceil}\right) = O\left(\frac{n^{\lfloor \frac{d}{2} \rfloor} k^{\lceil \frac{d}{2} \rceil}}{\delta^d}\right). \tag{5.3}$$

As each point in $\mathcal{H}^*$ has level at most k with respect to P^*, $\mathcal{H}^*$ lies in the union of the simplices of Ξ. Furthermore, no two points of $\mathcal{H}^*$ lie in the interior of the same simplex:

> Assume that $q_i^*, q_j^* \in \mathcal{H}^*$ lie in the interior of $\Delta \in \Xi$. Then all hyperplanes of P^* in the symmetric difference $\Delta\left(l\left(q_i^*\right), l\left(q_j^*\right)\right)$ must intersect the interior of Δ and so
>
> $$\left|\Delta\left(l\left(q_i^*\right), l\left(q_j^*\right)\right)\right| \leq \frac{|P^*|}{n/(\delta-1)} \leq \delta - 1,$$
>
> a contradiction.

By general position assumption, at most $d(d+1)$ points of $\mathcal{H}^*$ can lie on the boundary of any simplex of Ξ and thus there can be only $d(d+1) \cdot |\Xi|$ such points. Equation (5.3) now implies that

$$|\mathcal{R}| = |\mathcal{H}| = |\mathcal{H}^*| = O\left(d(d+1)\,|\Xi|\right) = O\left(\frac{n^{\lfloor \frac{d}{2} \rfloor} k^{\lceil \frac{d}{2} \rceil}}{\delta^d}\right).$$

□

Bibliography and discussion. The use of cuttings to prove the bound for hyperplanes is from [**CW89**], and the use of shallow-cuttings is from [**DEG16, MR17**].

[CW89] B. Chazelle and E. Welzl, *Quasi-optimal range searching in spaces of finite VC-dimension*, Discrete Comput. Geom. **4** (1989), no. 5, 467–489, DOI 10.1007/BF02187743. MR1014739

[DEG16] K. Dutta, E. Ezra, and A. Ghosh, *Two proofs for shallow packings*, Discrete Comput. Geom. **56** (2016), no. 4, 910–939, DOI 10.1007/s00454-016-9824-0. MR3561795

[MR17] N. H. Mustafa and S. Ray, *ε-Mnets: hitting geometric set systems with subsets*, Discrete Comput. Geom. **57** (2017), no. 3, 625–640, DOI 10.1007/s00454-016-9845-8. MR3614776

2. A Packing Theorem for Combinatorial Set Systems

When presented with the probabilistic method, the probabilist J. Doob remarked: 'Well, this is very nice, but it's really just a counting argument'.

Joel Spencer

The main result of this section is the following beautiful statement.

THEOREM 5.1. *Let $\mathcal{F} = \{S_1, \ldots, S_m\}$ be a set system on a set X of n elements and let $d \geq 1, \delta \in [n]$ be given integer parameters such that* VC-dim$(\mathcal{F}) \leq d$ *and for every $1 \leq i < j \leq m$, we have $|\Delta(S_i, S_j)| \geq \delta$. Then*

$$|\mathcal{F}| \leq 2 \cdot \mathrm{E}\left[|\mathcal{F}|_A|\right],$$

where A is a subset of size $s = \left\lceil \frac{8dn}{\delta} \right\rceil - 1$ picked uniformly at random from X (without replacement).

Overview of ideas. Let $R \subseteq X$ be a uniform random sample of size at least $\frac{n}{\delta}$ (without replacement) and consider the set system $\mathcal{F}|_R$. We would like to argue that, on average, each set of $\mathcal{F}$ maps to a distinct set of $\mathcal{F}|_R$ and so $|\mathcal{F}| = \mathrm{E}\left[|\mathcal{F}|_R|\right]$. However, it could be that $S, S' \in \mathcal{F}$ map to the same set of $\mathcal{F}|_R$. This happens precisely when no element in the symmetric difference $\Delta(S, S')$ is picked in R. As each element of X is picked into R with probability at least $\frac{1}{\delta}$ and the symmetric difference of every pair of sets is at least δ, in expectation we will pick at least one element from the symmetric difference of these two sets and thus S and S' will not be 'merged' in $\mathcal{F}|_R$.

This intuition will be implemented in the proof using the averaging technique of Chapter 1, as follows. Recall first the unit distance graph $G_U(\mathcal{F}|_R)$ on the vertex set $\mathcal{F}|_R$: two sets of $\mathcal{F}|_R$ are connected by an edge if and only if their symmetric difference has size precisely one.

For a random sample R of size $s = \Theta\left(\frac{n}{\delta}\right)$, we will essentially (but not quite!) double-count the expected number of pairs of sets in $\mathcal{F}$ that end up at unit symmetric difference in $\mathcal{F}|_R$.

Upper bound: This follows from the fact (Lemma 4.9) that there are at most $d|\mathcal{F}|_R|$ pairs of sets at unit symmetric difference in $\mathcal{F}|_R$, for *any* choice of $R \subseteq X$.

Lower bound: For every pair $S_i, S_j \in \mathcal{F}$, we will lower bound the probability that the symmetric difference of the two sets $\{S_i \cap R, S_j \cap R\}$ is exactly one. This, together with linearity of expectation, implies a lower bound on the expected number of pairs of sets of $\mathcal{F}$ that end up at unit symmetric difference in $\mathcal{F}|_R$.

Putting these bounds together will give an upper bound on $|\mathcal{F}|$!

However, there are two technical issues that must be overcome (the formulation below is intentionally imprecise).

(1) We will count the expected number of pairs of sets of $\mathcal{F}$ that are at unit symmetric difference in $G_U(\mathcal{F}|_R)$. However Lemma 4.9 only gives an upper bound on the pairs of sets of $\mathcal{F}|_R$ at unit distance in $G_U(\mathcal{F}|_R)$.

These two are not the same, as each set of $\mathcal{F}|_R$ potentially corresponds to *many* sets of $\mathcal{F}$. Therefore one has to look at the *weighted* unit distance graph on the sets of $\mathcal{F}|_R$. Towards that, for each $S' \in \mathcal{F}|_R$,

define the weight $w(S')$ to be the number of sets of $\mathcal{F}$ mapping to S'.
In other words, $w(S')$ is the cardinality of the preimage of S' under the projection $\mathcal{F}|_R$.

Thus instead of counting the number of edges in the unit distance graph $G_U(\mathcal{F}|_R)$, we will sum up the weight of the edges in $G_U(\mathcal{F}|_R)$, where an edge $\{S'_i, S'_j\}$ is assigned the weight $w(S'_i) \cdot w(S'_j)$. In fact, to simplify certain calculations, we will instead define the weight as

$$w(\{S'_i, S'_j\}) = \min\{w(S'_i), w(S'_j)\}.$$

Note that for any $a, b > 0$[2],

$$\min\{a,b\} \cdot \frac{(a+b)}{2} \leq ab \leq \min\{a,b\} \cdot (a+b). \tag{5.4}$$

Thus if $a+b$ is a fixed constant, $\min\{w(S'_i), w(S'_j)\}$ can be substituted for $w(S'_i) \cdot w(S'_j)$.[3]

(2) We need to lower bound, for any two sets $S_i, S_j \in \mathcal{F}$, the probability that $\{S_i \cap R, S_j \cap R\}$ ends up as an edge in $G_U(\mathcal{F}|_R)$. This occurs if and only if *exactly* one element of $\Delta(S_i, S_j)$ is picked into R; the probability of this occurring depends only on $|\Delta(S_i, S_j)|$.

Given that we are only interested in the sum of these $\binom{|\mathcal{F}|}{2}$ probabilities (that is, the expected number of pairs at unit distance in $\mathcal{F}|_R$), a clever idea that simplifies calculations is to use double-counting to count this sum in a uniform way. Rather than summation over pairs of sets in $\mathcal{F}$, we count, for the i-th element of R, the expected number of pairs at unit distance due to *that* element. By symmetry, this value is the same for all elements of R. So conditioned on the set A of the first $|R|-1$ elements of R, we only have to compute the expected number of pairs of sets that are put at unit distance by the $|R|$-th random element.

Putting these ideas together will give the formal proof, to which we turn to next.

PROOF OF THEOREM 5.1. Let R be a uniform random sample of X of size $s = \left\lceil \frac{8dn}{\delta} \right\rceil$ (that is, without replacement). We can think of constructing R by first picking a uniform random sample A of size $s-1$ from X, choosing an element a uniformly at random from $X \setminus A$, and setting $R = A \cup \{a\}$[4].

---

[2] Assume $a \leq b$. Then it is equivalent to the fact that $\frac{a}{2} + \frac{b}{2} \leq b \leq a+b$.

[3] Essentially, for each edge $\{S'_i, S'_j\}$, we are counting the *normalized* number of pairs at unit distance; that is, the quantity $\frac{w(S'_i) \cdot w(S'_j)}{w(S'_i) + w(S'_j)}$. This is equal, within a factor of two, to $\min\{w(S'_i), w(S'_j)\}$, which is the weight function we will use.

[4] These two distributions are equivalent: the probability that a fixed set of size t is chosen is precisely $t \cdot \left(1/\binom{n}{t-1}\right) \cdot 1/(n-(t-1)) = 1/\binom{n}{t}$.

Let $G_U(\mathcal{F}|_R) = (\mathcal{F}|_R, E_R)$ be the unit distance graph on $\mathcal{F}|_R$. For each $S' \in \mathcal{F}|_R$, define $w(S')$ to be the number of sets of $\mathcal{F}$ mapping to S':

$$w(S') = \big| \{S \in \mathcal{F} \,:\, S \cap R = S'\} \big|.$$

Define the weight of an edge $\{S'_i, S'_j\} \in E_R$ as

$$w\left(\{S'_i, S'_j\}\right) = \min\left\{w(S'_i), w(S'_j)\right\}.$$

Let $W = \sum_{e \in E_R} w(e)$ be the total weight of the edges of $G_U(\mathcal{F}|_R)$.

We will count $\mathrm{E}[W]$ **in two ways.**

First we present an upper bound on W and hence on $\mathrm{E}[W]$ (recall that $m = |\mathcal{F}|$ and VC-dim$(\mathcal{F}) \leq d$).

CLAIM 5.5.

$$W \leq 2d \cdot m.$$

PROOF. By Lemma 4.9, $|E_R| \leq d\,|\mathcal{F}|_R|$. Thus

$$\sum_{S' \in \mathcal{F}|_R} \deg(S') = 2\,|E_R| \leq 2d\,|\mathcal{F}|_R|,$$

and so there exists a set $S' \in \mathcal{F}|_R$ with $\deg(S') \leq \frac{2d\,|\mathcal{F}|_R|}{|\mathcal{F}|_R|} = 2d$. By the definition of edge weights, the weight of each edge incident to S' is at most $w(S')$. Thus the sum of the weight of all the edges incident to S' is at most $2d \cdot w(S')$.

Remove S' from $G_U(\mathcal{F}|_R)$. The remaining graph is still a unit distance graph on $|\mathcal{F}|_R| - 1$ vertices and the weight of its edges can be bounded inductively. Thus we get

$$W \leq 2d \sum_{S' \in \mathcal{F}|_R} w(S') = 2d \cdot m.$$

□

Next a lower bound on $\mathrm{E}[W]$.

CLAIM 5.6.

$$\mathrm{E}[W] \geq 4dm - 4d\,\mathrm{E}\left[|\mathcal{F}|_A|\right].$$

PROOF. Let W_1 be the weight of the edges in $G_U(\mathcal{F}|_R)$ where the element a is the symmetric difference. By symmetry, we have

$$\mathrm{E}[W] = s \cdot \mathrm{E}[W_1],$$

and so it suffices to give a lower bound on $\mathrm{E}[W_1]$.

We first give a lower bound on $\mathrm{E}[W_1]$ conditioned on the choice of the first $s-1$ elements A.

CLAIM 5.7. For any fixed $Y \subseteq X$ of size $s-1$,

$$\mathrm{E}\left[W_1 \mid A = Y\right] \geq \frac{\delta}{2n}\left(m - |\mathcal{F}|_Y|\right).$$

PROOF. The expectation here is only over the choice of a. Note that $R = Y \cup \{a\}$.

Consider a set $Q \in \mathcal{F}|_Y$ and let $\mathcal{F}_Q$ be the sets of $\mathcal{F}$ mapping to Q—namely the sets whose projection onto Y is Q. Let $b = |\mathcal{F}_Q|$.

Once a has been chosen, $\mathcal{F}_Q$ will be split into two sets: those sets of $\mathcal{F}_Q$ that contain a, say there are b_1 of these, and those that do not contain a, say $b_2 = b - b_1$ in number. These two sets are now at unit distance in $\mathcal{F}|_R$, with the element a being their symmetric difference; the weight of the edge between these two sets is $\min\{b_1, b_2\}$. Then W_1 is the total weight of such edges for all $Q \in \mathcal{F}|_Y$.

For each pair of sets in $\mathcal{F}_Q$, the probability that the randomly chosen last element a will cause their symmetric difference in R to be 1 is at least $\frac{\delta}{n-(s-1)} \geq \frac{\delta}{n}$. Therefore the expected contribution of each pair of sets in $\mathcal{F}_Q$ to the term $b_1 b_2$ is at least $\frac{\delta}{n}$. Noting that $b = b_1 + b_2$ is independent of the choice of a, summation over all pairs of sets in $\mathcal{F}_Q$ yields the following lower bound on the expected contribution of the sets in $\mathcal{F}_Q$ to W_1:

$$\begin{aligned}
\mathrm{E}[\min\{b_1, b_2\}] &\geq \mathrm{E}\left[\frac{b_1 b_2}{b_1 + b_2}\right] \quad \Big(\text{by Equation (5.4)}\Big)\\
&= \frac{\mathrm{E}[b_1 b_2]}{b_1 + b_2} \quad \Big(\text{as } |\mathcal{F}_Q| = b = b_1 + b_2 \text{ is independent of choice of } a\Big)\\
&= \frac{\sum_{S,S' \in \mathcal{F}_Q} \Pr\left[a \text{ lies in the symmetric difference of } S \text{ and } S'\right]}{b}\\
&\geq \frac{\sum_{S,S' \in \mathcal{F}_Q} \delta/n}{b} = \frac{|\mathcal{F}_Q|\left(|\mathcal{F}_Q| - 1\right) \cdot \delta/n}{2\,|\mathcal{F}_Q|} = \frac{\delta}{2n} \cdot \left(|\mathcal{F}_Q| - 1\right).
\end{aligned}$$

Summing up over all sets of $\mathcal{F}|_Y$,

$$\begin{aligned}
\mathrm{E}\left[W_1 \mid A = Y\right] &\geq \sum_{Q \in \mathcal{F}|_Y} \frac{\delta}{2n}\Big(|\mathcal{F}_Q| - 1\Big)\\
&= \frac{\delta}{2n}\left(\sum_{Q \in \mathcal{F}|_Y} |\mathcal{F}_Q| - \sum_{Q \in \mathcal{F}|_Y} 1\right) = \frac{\delta}{2n}\Big(m - |\mathcal{F}|_Y|\Big).
\end{aligned}$$

□

Finally we compute a lower bound for $\mathrm{E}[W]$:

$$\begin{aligned}
\mathrm{E}[W] &= s \cdot \mathrm{E}[W_1] = s \cdot \sum_{\substack{Y \subseteq X\\ |Y| = s-1}} \mathrm{E}\left[W_1 | A = Y\right] \cdot \Pr[A = Y]\\
&\geq s \cdot \sum_{\substack{Y \subseteq X\\ |Y| = s-1}} \frac{\delta}{2n}\Big(m - |\mathcal{F}|_Y|\Big) \cdot \Pr\left[A = Y\right]\\
&= \frac{s\delta}{2n}\left(m \sum_{\substack{Y \subseteq X\\ |Y| = s-1}} \Pr\left[A = Y\right] - \sum_{\substack{Y \subseteq X\\ |Y| = s-1}} |\mathcal{F}|_Y| \cdot \Pr\left[A = Y\right]\right)\\
&= \frac{s\delta}{2n}\left(m - \mathrm{E}\left[|\mathcal{F}|_A|\right]\right) \geq 4dm - 4d\,\mathrm{E}\left[|\mathcal{F}|_A|\right],
\end{aligned}$$

where the last step follows from substituting $s = \left\lceil \frac{8dn}{\delta} \right\rceil$. □

Combining the upper and lower bounds on $\mathrm{E}[W]$,

$$4dm - 4d\,\mathrm{E}\left[|\mathcal{F}|_A|\right] \;\leq\; \mathrm{E}[W] \;\leq\; 2dm,$$

$$\text{implying that} \quad m \leq 2\,\mathrm{E}\left[|\mathcal{F}|_A|\right].$$

This completes the proof of Theorem 5.1. □

Bibliography and discussion. The original packing lemma, with the stunning idea presented here, is from Haussler [**Hau95**], who gave the proof for the specific case of set systems with bounded VC-dimension. The more general bound in terms of the expected projection size presented here was proven in [**Mus16**]. Our exposition mostly follows that of Matoušek [**Mat99**, Section 5.3].

[Hau95] D. Haussler, *Sphere packing numbers for subsets of the Boolean n-cube with bounded Vapnik-Chervonenkis dimension*, J. Combin. Theory Ser. A **69** (1995), no. 2, 217–232, DOI 10.1016/0097-3165(95)90052-7. MR1313896

[Mat99] J. Matoušek. *Geometric Discrepancy: An Illustrated Guide*. Springer, 1999.

[Mus16] N. H. Mustafa, *A simple proof of the shallow packing lemma*, Discrete Comput. Geom. **55** (2016), no. 3, 739–743, DOI 10.1007/s00454-016-9767-5. MR3473678

3. Applications of the Packing Theorem

Different mathematicians study papers in different ways, but when I read a mathematical paper in a field in which I'm conversant, I concentrate on the thoughts that are between the lines. I might look over several paragraphs or strings of equations and think to myself, 'Oh yeah, they're putting in enough rigamarole to carry such-and-such an idea.' When the idea is clear, the formal setup is usually unnecessary and redundant. I often feel that I could write it out myself more easily than figuring out what the authors actually wrote. It's like a new toaster that comes with a 16-page manual. If you already understand toasters and if the new toaster looks like previous toasters you've encountered, you might just plug it in and see if it works, rather than first reading all the details in the manual.

William Thurston

Given a set P of points in $[0,n]^d$, if the distance between every pair of points of P is at least δ, then $|P| = O\left(\left(\frac{n}{\delta}\right)^d\right)$. Our first application of Theorem 5.1 is an analogous statement for set systems of bounded VC-dimension.

THEOREM 5.8. *Let $\mathcal{F} = \{S_1, \ldots, S_m\}$ be a set system on a set X of n elements, with* VC-dim$(\mathcal{F}) \leq d$. *Let $\delta \in [n]$ be an integer such that for every $1 \leq i < j \leq m$, we have $|\Delta(S_i, S_j)| \geq \delta$ (that is, the size of the symmetric difference between every pair of sets of $\mathcal{F}$ is at least δ). Then $|\mathcal{F}| \leq 2\left(\frac{8en}{\delta}\right)^d$.*

We remark that by applying Theorem 5.8 with $\delta = 1$, one recovers the statement of Lemma 4.3 with a slightly weaker bound.

Our second application is a packing lemma where we further assume that each set of $\mathcal{F}$ has size at most k.

DEFINITION 5.9. For positive integers k and δ, a set system $(X, \mathcal{F})$ is called a (k, δ)-packing if

(1) $|S| \leq k$ for all $S \in \mathcal{F}$, and
(2) $|\Delta(S, S')| \geq \delta$ for all $S, S' \in \mathcal{F}$.

Recall the notion of shallow-cell complexity of a set system.

DEFINITION 4.4. A set system $(X, \mathcal{F})$ has shallow-cell complexity $\varphi_{\mathcal{F}}(\cdot,\cdot)$ if for any positive integer k and any finite $Y \subseteq X$, the number of sets in $\mathcal{F}|_Y$ of size at most k is upper bounded by $|Y| \cdot \varphi_{\mathcal{F}}(|Y|, k)$.

For a family $\mathcal{R}$ of geometric objects in $\mathbb{R}^d$—e.g., the family of all half-spaces—the shallow-cell complexity of $\mathcal{R}$ is defined to be the shallow-cell complexity of the primal set system $(\mathbb{R}^d, \mathcal{R})$.

THEOREM 5.10. *Given positive integers k, d and $\delta \in [k]$, let $\mathcal{F} = \{S_1, \ldots, S_m\}$ be a set system on a set X of n elements. Further assume that $\mathcal{F}$ is a (k, δ)-packing, $\mathcal{F}$ has shallow-cell complexity $\varphi_{\mathcal{F}}(\cdot,\cdot)$ and* VC-dim$(\mathcal{F}) \leq d$. *Then*

$$|\mathcal{F}| \leq \frac{48dn}{\delta} \cdot \varphi_{\mathcal{F}}\left(\frac{8dn}{\delta}, \frac{24dk}{\delta}\right).$$

Overview of ideas. We present the intuition behind the proof of Theorem 5.10. Pick each element of X independently with probability $p = \frac{1}{\delta}$ to get a random sample R. We have $\mathrm{E}[|R|] = \frac{n}{\delta}$ and for each $S_i \in \mathcal{F}$,

$$\mathrm{E}[|S_i \cap R|] = \frac{|S_i|}{\delta} \leq \frac{k}{\delta}.$$

Now consider any two sets $S_i \cap R, S_j \cap R$ in $\mathcal{F}|_R$. Their symmetric difference consists of the elements of $\Delta(S_i, S_j) \cap R$, and so

$$\mathrm{E}[|\Delta(S_i \cap R, S_j \cap R)|] = \frac{|\Delta(S_i, S_j)|}{\delta} \geq 1.$$

Thus one can hope that the sets of $\mathcal{F}$ remain distinct in $\mathcal{F}|_R$, and that each set will have size at most $\frac{k}{\delta}$ in $\mathcal{F}|_R$. Now using the shallow-cell complexity of $\mathcal{F}$,

$$|\mathcal{F}| = |\mathcal{F}|_R| \leq \frac{n}{\delta} \cdot \varphi_{\mathcal{F}}\left(\frac{n}{\delta}, \frac{k}{\delta}\right).$$

Converting this intuition into a proof will be done using Theorem 5.1.

PROOF OF THEOREM 5.8. From Theorem 5.1, we have

$$|\mathcal{F}| \leq 2 \cdot \mathrm{E}\left[\,|\mathcal{F}|_A|\,\right], \text{ where } A \subseteq X \text{ is a uniform random sample of size } \left\lceil \frac{8dn}{\delta} \right\rceil - 1.$$

For the case of set systems with VC-dimension at most d, $|\mathcal{F}|_A|$ can be upper bounded independent of the specific choice of A. Thus using Lemma 4.3 to upper bound $|\mathcal{F}|_A|$, we get

$$|\mathcal{F}| \leq 2 \max_{\substack{A \subseteq X,\\ |A| = \lceil \frac{8dn}{\delta} \rceil - 1}} |\mathcal{F}|_A| \leq 2\left(\frac{e|A|}{d}\right)^d \leq 2\left(\frac{8en}{\delta}\right)^d.$$

□

PROOF OF THEOREM 5.10. Let $A \subseteq X$ be a uniform random sample of size $\frac{8dn}{\delta} - 1$. Note that for any $S \in \mathcal{F}$, $\mathrm{E}[|S \cap A|] \leq \frac{8dk}{\delta}$ as $|S| \leq k$. Define

$$\mathcal{F}^A = \left\{ S \in \mathcal{F} \colon |S \cap A| > 3 \cdot \frac{8dk}{\delta} \right\}.$$

Markov's inequality (Equation (1 26)) implies that for any $S \in \mathcal{F}$,

$$\Pr\left[S \in \mathcal{F}^A\right] = \Pr\left[|S \cap A| > 3 \cdot \frac{8dk}{\delta}\right] \leq \frac{1}{3}.$$

The sets of $\mathcal{F}|_A$ can be partitioned into two collections: each $S \cap A \in \mathcal{F}|_A$ with $|S \cap A| > 3 \cdot \frac{8dk}{\delta}$ belongs to the first collection, otherwise the second collection. The number of sets in the first collection can be upper bounded by $|\mathcal{F}^A|$ while each set in the second collection has size at most $3 \cdot \frac{8dk}{\delta}$. Thus we have

$$\begin{aligned}
\mathrm{E}\left[|\mathcal{F}|_A|\right] &\le \mathrm{E}\left[|\mathcal{F}^A|\right] + \mathrm{E}\left[|(\mathcal{F}\setminus\mathcal{F}^A)|_A|\right] \\
&\le \sum_{S\in\mathcal{F}} \Pr\left[S\in\mathcal{F}^A\right] + |A|\cdot\varphi_{\mathcal{F}}\left(|A|, 3\cdot\frac{8dk}{\delta}\right) \\
&\le \frac{|\mathcal{F}|}{3} + \frac{8dn}{\delta}\cdot\varphi_{\mathcal{F}}\left(\frac{8dn}{\delta}, \frac{24dk}{\delta}\right),
\end{aligned}$$

where the number of sets in the projection of $\mathcal{F}\setminus\mathcal{F}^A$ onto A is bounded using the shallow-cell complexity $\varphi_{\mathcal{F}}(\cdot,\cdot)$. Now the bound follows from applying Theorem 5.1:

$$\begin{aligned}
|\mathcal{F}| \le 2\cdot\mathrm{E}\left[|\mathcal{F}|_A|\right] &\le 2\left(\frac{|\mathcal{F}|}{3} + \frac{8dn}{\delta}\cdot\varphi_{\mathcal{F}}\left(\frac{8dn}{\delta}, \frac{24dk}{\delta}\right)\right) \\
\implies |\mathcal{F}| &\le 6\cdot\left(\frac{8dn}{\delta}\cdot\varphi_{\mathcal{F}}\left(\frac{8dn}{\delta}, \frac{24dk}{\delta}\right)\right).
\end{aligned}$$

□

The computation of $\mathrm{E}\left[|\mathcal{F}|_A|\right]$ can be rephrased as follows. First note that for any $Y\subseteq X$ and *any* $\mathcal{F}'\subseteq\mathcal{F}$, we have

$$|\mathcal{F}|_Y| \le |\mathcal{F}'| + |(\mathcal{F}\setminus\mathcal{F}')|_Y|.$$

By the definition of expectation,

$$\mathrm{E}\left[|\mathcal{F}|_A|\right] = \sum_{\substack{Y\subseteq X\\ |Y|=\frac{8dn}{\delta}-1}} |\mathcal{F}|_Y|\cdot\Pr[A=Y]$$

Letting $\mathcal{F}^Y$ be the sets of $\mathcal{F}$ with projected size at least $3\cdot\frac{8dk}{\delta}$ in $\mathcal{F}|_Y$,

$$\begin{aligned}
&\le \sum_{\substack{Y\subseteq X\\ |Y|=\frac{8dn}{\delta}-1}} \left(\left|\mathcal{F}^Y\right| + \left|\left(\mathcal{F}\setminus\mathcal{F}^Y\right)|_Y\right|\right)\cdot\Pr[A=Y] \\
&= \mathrm{E}\left[\left|\mathcal{F}^A\right|\right] + \sum_{\substack{Y\subseteq X\\ |Y|=\frac{8dn}{\delta}-1}} \left|\left(\mathcal{F}\setminus\mathcal{F}^Y\right)|_Y\right|\cdot\Pr[A=Y] \\
&\le \frac{|\mathcal{F}|}{3} + \frac{8dn}{\delta}\cdot\varphi_{\mathcal{F}}\left(\frac{8dn}{\delta}, \frac{24dk}{\delta}\right).
\end{aligned}$$

We remark here that the packing lemma for half-spaces proven earlier, Theorem 5.2, follows immediately by Theorem 5.10 and Theorem 1.16. This proof—without the use of any spatial properties—illustrates the usefulness of abstraction and not surprisingly, this will carry over to the study of ϵ-nets.

Bibliography and discussion. The shallow packing lemma for some geometric set systems was first shown in [**DEG165c**]. The statement was then generalized and the proof simplified in [**Mus16**], whose presentation we have essentially followed here. The technical trick in the proof is to upper bound the desired quantity $|\mathcal{F}|$ by a function that involves $|\mathcal{F}|$ itself, and is a technique to shorten an iterative/inductive/recursive argument.

[DEG165c] K. Dutta, E. Ezra, and A. Ghosh, *Two proofs for shallow packings*, Discrete Comput. Geom. **56** (2016), no. 4, 910–939, DOI 10.1007/s00454-016-9824-0. MR3561795

[Mus16] N. H. Mustafa, *A simple proof of the shallow packing lemma*, Discrete Comput. Geom. **55** (2016), no. 3, 739–743, DOI 10.1007/s00454-016-9767-5. MR3473678

CHAPTER 6

Epsilon-Nets: Combinatorial Bounds

The initial study of ϵ-nets in the field of computational geometry started in the 1980s with the work of Clarkson who showed the existence of ϵ-nets of size $O\left(\frac{1}{\epsilon}\log\frac{1}{\epsilon}\right)$ for specific geometric set systems. He was mainly interested in their algorithmic applications, such as nearest-neighbor queries for a set of points in Euclidean space. Chapter 2 is largely based on his work.

Independently, Haussler and Welzl showed similar bounds in a purely abstract setting, needing just that the given set system has bounded VC-dimension. In fact, what was needed, given a set system $(X, \mathcal{F})$, was the property that there exist an absolute constant d such that

$$\text{for all } Y \subseteq X,\ |\mathcal{F}|_Y| = O\left(|Y|^d\right) \quad \text{(see Lemma 4.3).}$$

They showed, surprisingly, that this is already a sufficient condition for the existence of small ϵ-nets; the following will be the first theorem of this chapter.

THEOREM 6.1. *Let $(X, \mathcal{F})$ be a finite set system, $d \geq 1$ an integer such that* VC-dim$(\mathcal{F}) \leq d$, *and $\epsilon \in \left(0, \frac{1}{2}\right)$ a given parameter. Let N be a uniform random sample of X of size $t = \left\lceil \frac{56\,d}{\epsilon} \ln \frac{1}{\epsilon} \right\rceil$. Then N is an ϵ-net of $\mathcal{F}$ with probability at least $\frac{1}{2}$.*

It was later shown that the above bound is optimal within constant factors; that is, for every positive integer n, integer $d \geq 2$ and small-enough parameter $\epsilon > 0$, there is a set system $(X, \mathcal{F})$ with $|X| = n$ and VC-dim$(\mathcal{F}) \leq d$, such that any ϵ-net of $\mathcal{F}$ has size $\Omega\left(\frac{d}{\epsilon}\log\frac{1}{\epsilon}\right)$. This lower bound will be presented in Chapters 10 and 11.

Over the past thirty years, it has been observed that improvements to the Clarkson and Haussler–Welzl bounds are possible for a variety of geometric set systems. We have already seen an example in Chapter 3: $O\left(\frac{1}{\epsilon}\right)$-sized ϵ-nets exist for set systems induced by disks in the plane. Early work towards $o\left(\frac{1}{\epsilon}\log\frac{1}{\epsilon}\right)$ upper bounds was fundamentally geometric, involving spatial partitioning along the ideas seen in Chapter 3.

Over the next twenty years, through the work of Aronov, Chan, Clarkson, Ezra, Ray, Sharir, Varadarajan and others, it was realized that geometry is not really needed. In fact, somewhat surprisingly, an entire suite of optimal bounds can be obtained entirely combinatorially, with the shallow-cell complexity being a key parameter of a set system that dictates the size of ϵ-nets.

The reason that the shallow-cell complexity of a set system comes into play is the following. Given a set system $(X, \mathcal{F})$, the probability that a set $F \in \mathcal{F}$ is not hit by a uniform random sample $N \subseteq X$ decreases exponentially with the size of F. On the other hand, the number of sets of $\mathcal{F}$ of size at most k is an increasing function of k whose growth is upper bounded by the shallow-cell complexity of the set system. It turns out that the interplay between these two dictates the sizes of ϵ-nets.

The second main theorem of this chapter will be the following.

THEOREM 6.2. *Let $(X, \mathcal{F})$ be a finite set system with shallow-cell complexity $\varphi_{\mathcal{F}}(\cdot, \cdot)$ and with* VC-dim $(\mathcal{F}) \leq d$. *Then for any $\epsilon \in \left(0, \frac{1}{2}\right)$ there exists an ϵ-net of $\mathcal{F}$ of size*

$$O\left(\frac{d}{\epsilon} + \frac{1}{\epsilon} \log \varphi_{\mathcal{F}}\left(\frac{16d}{\epsilon}, 48d\right)\right).$$

Consider the primal set system $\mathcal{R}$ induced by disks in the plane: as $\varphi_{\mathcal{R}}(m, k) = O\left(k^2\right)$ (Lemma 1.2) and VC-dim $(\mathcal{R}) \leq 3$, Theorem 6.2 implies the existence of ϵ-nets of size $O\left(\frac{1}{\epsilon}\right)$. Thus we recover Theorem 3.3 from just the shallow-cell complexity of $\mathcal{R}$!

1. A First Bound using Ghost Sampling

Mathematics is one of a few fields in which one can do top-level work without a lot of life experience, something that might be key in the arts or humanities. One does not have to have experience raising children through school, dealing with family tragedies, and so forth, to be able to find three numbers whose fourth powers add up to another one.

Noam Elkies

We prove the following.

THEOREM 6.1. *Let $(X, \mathcal{F})$ be a finite set system, $d \geq 1$ an integer such that* VC-dim$(\mathcal{F}) \leq d$*, and $\epsilon \in \left(0, \frac{1}{2}\right)$ a given parameter. Let N be a uniform random sample of X of size $t = \left\lceil \frac{56\,d}{\epsilon} \ln \frac{1}{\epsilon} \right\rceil$. Then N is an ϵ-net of $\mathcal{F}$ with probability at least $\frac{1}{2}$.*

Set $n = |X|$. We can assume that each set in $\mathcal{F}$ has size at least ϵn.

For ease of calculations, we will allow an element to be picked into N multiple times. That is, N will be a *sequence* of size $t = \left\lceil \frac{56d}{\epsilon} \ln \frac{1}{\epsilon} \right\rceil$, where each element in this sequence is chosen uniformly at random from X. In a natural way, a sequence Y is an ϵ-net of $\mathcal{F}$ if Y contains at least one element from each set of $\mathcal{F}$.

Throughout this section we will work with sequences instead of sets. Moreover, the *size* of the intersection of any set $R \in \mathcal{F}$ with a sequence will count multiplicities.

Overview of ideas. Let $\mathcal{T} = X^t$ denote the set of all t-sized sequences of elements of X, and let $\mathcal{T}_b \subseteq \mathcal{T}$ be the sequences which are *not* an ϵ-net of $\mathcal{F}$. Our goal is to upper bound $|\mathcal{T}_b|$. In particular, we will show that $|\mathcal{T}_b| < \frac{|\mathcal{T}|}{2}$, implying that N—constructed by picking t elements uniformly at random, with replacement—is an ϵ-net of $\mathcal{F}$ with probability at least $\frac{1}{2}$. This implies the same property for a uniform random sample of X of size t, proving Theorem 6.1.

Fix an ordering of the sets of $\mathcal{F}$ and

for each $Y \in \mathcal{T}_b$, let $R_Y \in \mathcal{F}$ be the *first* set in the ordering for which Y fails—that is, for which $R_Y \cap Y = \emptyset$.

We will use the probabilistic averaging technique from Chapter 1. That is, we take a uniform random sequence S of X of a certain size s—with replacement, so $S \in X^s$—and examine the relationship between $\mathcal{T}_b$ and S. Specifically,

we count the expected number of sequences $Y \in \mathcal{T}_b$ s.t. $|R_Y \cap S| \geq \frac{\epsilon s}{2}$.

Lower bound: On one hand, for any $Y \in \mathcal{T}_b$,

$$\mathrm{E}\left[\,|R_Y \cap S|\,\right] = \sum_{i=1}^{s} \frac{|R_Y|}{n} \geq \sum_{i=1}^{s} \epsilon = \epsilon s, \tag{6.3}$$

keeping in mind that each element of R_Y is counted with multiplicity in $|R_Y \cap S|$. A tail bound will then imply that the expected number of sets $Y \in \mathcal{T}_b$ for which $|R_Y \cap S| \geq \frac{\epsilon s}{2}$ is at least $\frac{|\mathcal{T}_b|}{2}$.

Upper bound: On the other hand, this expectation can be calculated exactly:

$$\frac{1}{n^s}\sum_{Z\in X^s}\left|\left\{Y\in\mathcal{T}_b\,:\,|R_Y\cap Z|\,\geq\frac{\epsilon s}{2}\right\}\right|$$

$$=\frac{1}{n^s}\left|\left\{(Z,Y):\,Z\in X^s,\,Y\in\mathcal{T}_b\subseteq X^t,\,|R_Y\cap Z|\geq\frac{\epsilon s}{2}\right\}\right|. \tag{6.4}$$

Call each of the above (Z, Y) a *satisfying pair*. That is,

$$Z\in X^s,\quad Y\in\mathcal{T}_b\subseteq X^t,\qquad |R_Y\cap Y|=0\quad\text{while}\quad |R_Y\cap Z|\geq\tfrac{\epsilon s}{2}.$$

We will show that these constraints together force an upper bound on the total number of satisfying pairs.

Combining the lower and upper bounds will then give the desired bound on $|\mathcal{T}_b|$.

PROOF OF THEOREM 6.1. For $Z \in X^s$, define

$$\mathcal{T}_Z=\left\{Y\in\mathcal{T}_b\,:\,|R_Y\cap Z|\,\geq\frac{\epsilon s}{2}\right\}.$$

Set $s = t$ and let S be an element chosen uniformly at random from X^s.
We count the expected size of $\mathcal{T}_S$ in two ways.

LEMMA 6.5 (Lower bound).

$$\mathrm{E}\left[\,|\mathcal{T}_S|\,\right]=\sum_{Y\in\mathcal{T}_b}\Pr\left[|R_Y\cap S|\geq\frac{\epsilon s}{2}\right]\geq\frac{1}{2}\cdot|\mathcal{T}_b|.$$

PROOF. The proof follows from linearity of expectation and the next claim.

CLAIM 6.6. For any $R \in \mathcal{F}$, $\Pr\left[|R\cap S|\leq\frac{\epsilon s}{2}\right]<\frac{1}{2}$.

PROOF. For $i = 1, \ldots, s$, let Y_i be an indicator random variable that is 1 if and only if the i-th element of S is in R. Then

$$|R\cap S|=\sum_{i=1}^{s}Y_i,\quad\text{where}\quad\Pr\left[Y_i=1\right]=\frac{|R|}{n}\geq\epsilon.$$

Then Chernoff's bound (Theorem 1.20) applied to the s variables $\{Y_1, \ldots, Y_s\}$ with $\delta = \frac{1}{2}$ implies that

$$\Pr\left[|R\cap S|\leq\left(1-\frac{1}{2}\right)\epsilon s\right]\leq\exp\left(-\frac{\epsilon s}{8}\right)\leq\exp\left(-\frac{56d\ln\frac{1}{\epsilon}}{8}\right)<\frac{1}{2},$$

recalling that $s=t=\left\lceil\frac{56d}{\epsilon}\ln\frac{1}{\epsilon}\right\rceil$. □

The upper bound of $\frac{1}{2}$ in Claim 6.6 can be replaced by $o(1)$ but it doesn't matter for us as this only changes the multiplicative constant factor in the final bound (see discussion). □

The upper bound on $\mathrm{E}\left[|\mathcal{T}_S|\right]$ is implied by the following combinatorial statement.

LEMMA 6.7 (Upper bound).

$$\sum_{Z\in X^s}|\mathcal{T}_Z|=\left|\left\{(Z,Y):\;\;Z\in X^s,\,Y\in\mathcal{T}_b\subseteq X^t,\,|R_Y\cap Z|\geq\frac{\epsilon s}{2}\right\}\right|<\frac{n^{2t}}{4}. \tag{6.8}$$

PROOF. The trick to showing Equation (6.8) is to use averaging again.

For each $U \in X^{s+t}$ and $R \in \mathcal{F}$, let $\Psi(U,R)$ be the number of ways of partitioning U into two subsequences $Z \in X^s$ and $Y \in X^t$ such that (Z,Y) is a satisfying pair, with $R_Y = R$.

As each satisfying pair (Z,Y) can be combined into a sequence of size $s+t$ in $\binom{s+t}{t}$ ways, we have

$$\binom{s+t}{t} \sum_{Z \in X^s} |\mathcal{T}_Z| = \sum_{U \in X^{s+t}} \sum_{R \in \mathcal{F}} \Psi(U,R). \tag{6.9}$$

We make two observations:

(1) For each fixed $U \in X^{s+t}$, let $\mathcal{F}' \subseteq \mathcal{F}$ be such that all the sets in $\mathcal{F}'$ have the same intersection with U. Then all $R \in \mathcal{F}'$ have the same intersection with any (Z,Y) derived from U, implying that $\Psi(U,R)$ is (possibly) non-zero only for the first set, according to our initial ordering, of $\mathcal{F}'$. As there are $|\mathcal{F}|_U|$ distinct intersections of sets of $\mathcal{F}$ with U, there are at most $|\mathcal{F}|_U|$ sets $R \in \mathcal{F}$ for which $\Psi(U,R)$ is non-zero.

(2) For a fixed $R \in \mathcal{F}$ with $|R \cap U| \geq \frac{\epsilon s}{2}$, there are at most $\binom{s+t-\frac{\epsilon s}{2}}{t}$ ways to select Y from U such that Y does not contain any element of R. This is an upper bound on $\Psi(U,R)$ for any fixed U and R.

The above two observations together with Equation (6.9) imply that $\sum_{Z \in X^s} |\mathcal{T}_Z|$ can be upper bounded by

$$\frac{1}{\binom{s+t}{t}} \cdot \sum_{U \in X^{s+t}} |\mathcal{F}|_U| \cdot \binom{s+t-\frac{\epsilon s}{2}}{t} \leq \sum_{U \in X^{s+t}} \cdot \left(\frac{e\,(s+t)}{d}\right)^d \cdot \frac{\binom{s+t-\frac{\epsilon s}{2}}{t}}{\binom{s+t}{t}},$$

where the second step follows from Lemma 4.3. It remains to simplify this upper bound:

$$\begin{aligned}
&= n^{2t} \cdot \left(\frac{2et}{d}\right)^d \cdot \frac{2t - \frac{\epsilon t}{2}}{2t} \, \frac{2t - \frac{\epsilon t}{2} - 1}{2t-1} \cdots \frac{t - \frac{\epsilon t}{2} + 1}{t+1} \quad \left(\text{recalling that } s = t\right) \\
&= n^{2t} \cdot \left(\frac{2et}{d}\right)^d \cdot \left(1 - \frac{\frac{\epsilon t}{2}}{2t}\right) \cdots \left(1 - \frac{\frac{\epsilon t}{2}}{t+1}\right) \leq n^{2t} \cdot \left(\frac{2et}{d}\right)^d \cdot \left(1 - \frac{\frac{\epsilon t}{2}}{2t}\right)^t \\
&\leq n^{2t} \cdot \left(\frac{2et}{d}\right)^d \cdot e^{-\frac{\epsilon t}{4}} \leq n^{2t} \cdot \left(\frac{112\,e}{\epsilon}\ \ln\frac{1}{\epsilon}\right)^d \cdot e^{-14d \ln\frac{1}{\epsilon}} \\
&= n^{2t} \cdot \left(\frac{112\,e}{\epsilon}\ \ln\frac{1}{\epsilon}\right)^d \cdot \epsilon^{14d} < n^{2t} \cdot \left(\frac{112\,e}{\epsilon^2}\right)^d \cdot \epsilon^{14d} \\
&< n^{2t} \cdot (112\,e)^d \cdot \epsilon^{12d} < \frac{n^{2t}}{4},
\end{aligned}$$

as $(112e)^d \cdot \epsilon^{12d} \leq (112e)^d \cdot \left(\frac{1}{2}\right)^{12d} < \frac{1}{4}$. □

Combining the upper and lower bounds,

$$\frac{1}{2} \cdot |\mathcal{T}_b| \quad \leq \quad \mathrm{E}\left[|\mathcal{T}_S|\right] \quad = \quad \frac{1}{n^s} \sum_{Z \in X^s} |\mathcal{T}_Z| \quad < \quad \frac{n^t}{4},$$

gives $|\mathcal{T}_b| < \frac{n^t}{2} = \frac{|\mathcal{T}|}{2}$, as required. □

A remark: the upper bound used in the proof, that

$$|\mathcal{F}|_U| \le \left(\frac{e|U|}{d}\right)^d,$$

is a consequence of the fact that VC-dim $(\mathcal{F}) \le d$. If instead we only used $|\mathcal{F}|_U| = O(|U|)^d$, the final bound would come out to be $O\left(\frac{d}{\epsilon}\log\frac{d}{\epsilon}\right)$.

We conclude with two remarks.

Ghost sampling: A clever double-counting trick in the proof is to upper bound the number of satisfying pairs (Z, Y)—where $Z \in X^s$ and $Y \in X^t$—by enumerating over all $(s+t)$-sized subsets of X. This trick was also used in the proof of Theorem 5.1, though in that case we had $t = 1$.
In statistics and learning theory literature, this instance of double-counting is called *ghost sampling*, and it is a useful technique to avoid discretization when the base set X is infinite. For the case when the set system $(X, \mathcal{F})$ is finite, there are simpler proofs (see Chapter 12).
In the proof that we presented, we only wanted the probability that the random sample N succeeds to be an ϵ-net of $\mathcal{F}$ to be non-zero, which is sufficient to guarantee the existence of an ϵ-net of the required size. By introducing this probability as a parameter in the sample size and re-working the above proof, we arrive at the following statement (a proof is presented in Chapter 12).

THEOREM 6.10. *Let $(X, \mathcal{F})$ be a finite set system, $d \in \mathbb{N}$ a positive integer such that* VC-dim$(\mathcal{F}) \le d$, *and $\epsilon \in \left(0, \frac{1}{2}\right)$ a given parameter. Then there exists an absolute constant $C_6 > 0$ such that a random sample N constructed by picking each point of X independently with probability $\frac{C_6}{\epsilon|X|}\ln\frac{1}{\epsilon^d\gamma}$ is an ϵ-net of $\mathcal{F}$ with probability at least $1-\gamma$.*

Iterative View: As is often the case with double-counting proofs, one can 'unroll' the proof of Theorem 6.1 to an iterative version, as follows[1]. Let $(X_0, \mathcal{F}_0)$ be a set system with $|X_0| = n$ and $\epsilon > 0$ a parameter such that each set of $\mathcal{F}_0$ has size at least ϵn. A straightforward application of Chernoff's bound implies that there exists a $X_1 \subseteq X_0$ such that each $S \in \mathcal{F}_0$ contains at least $\frac{\epsilon n}{2}$ elements of X_1 and further

$$|X_1| \le |X_0| \cdot \left(\frac{1}{2} + \sqrt{\frac{10\log|\mathcal{F}_0|}{\epsilon n}}\right).$$

Now one can repeat this step for the set system $(X_1, \mathcal{F}_1 = \mathcal{F}|_{X_1})$ to get a set $X_2 \subseteq X_1$ and so on. After the i-th iteration we have a set X_i such that each $S \in \mathcal{F}_0$ contains at least $\frac{\epsilon n}{2^i}$ elements of X_i and furthermore,

$$|X_i| \le |X_{i-1}| \cdot \left(\frac{1}{2} + \sqrt{\frac{10\log|\mathcal{F}_{i-1}|}{\frac{\epsilon n}{2^{i-1}}}}\right) \le \quad \cdots \quad \le n \cdot \prod_{j=0}^{i-1}\left(\frac{1}{2} + \sqrt{\frac{10\log|\mathcal{F}_j|}{\frac{\epsilon n}{2^j}}}\right).$$

[1] Also this can be done by an inductive argument, though that somewhat obscures the ideas.

We continue the iterations as long as iteration i satisfies $\sqrt{\frac{10\log|\mathcal{F}_i|}{\frac{\epsilon n}{2^i}}} \leq \frac{1}{2}$. Say the above procedure runs for t iterations. Now a calculation shows that for all $i \leq t$,

$$|X_i| \leq n \cdot \prod_{j=0}^{i-1}\left(\frac{1}{2} + \sqrt{\frac{10\log|\mathcal{F}_j|}{\frac{\epsilon n}{2^j}}}\right) \leq c_1 \cdot \frac{n}{2^i}, \tag{6.11}$$

where c_1 is a sufficiently large constant. We set $t = \left\lfloor \log \frac{\epsilon n}{c'\, d\, \log \frac{1}{\epsilon}} \right\rfloor$ for a sufficiently large constant c' depending only on c_1. Then it can be verified that for all $i \leq t$,

$$\sqrt{\frac{10\log|\mathcal{F}_i|}{\frac{\epsilon n}{2^i}}} \leq \sqrt{\frac{10\log|\mathcal{F}_t|}{\frac{\epsilon n}{2^t}}} \leq \sqrt{\frac{10\log\left(\frac{ec_1 n}{2^t d}\right)^d}{\frac{\epsilon n}{2^t}}} \leq \sqrt{\frac{10\log\left(\frac{ec_1 c' \log\frac{1}{\epsilon}}{\epsilon}\right)}{c'\log\frac{1}{\epsilon}}} \leq \frac{1}{2},$$

where the second step follows from Equation (6.11) and Lemma 4.3.
Finally, each $S \in \mathcal{F}_0$ contains at least $\frac{\epsilon n}{2^t} \geq c' d \log \frac{1}{\epsilon} \geq 1$ points of X_t. Thus X_t is an ϵ-net of $\mathcal{F}_0$, with

$$|X_t| \leq c_1 \cdot \frac{n}{2^t} = c_1 \cdot \frac{n}{\frac{\epsilon n}{c'\, d\, \log\frac{1}{\epsilon}}} = O\left(\frac{d}{\epsilon}\log\frac{1}{\epsilon}\right).$$

An elegant strengthened form of this idea is the basis of Chapter 8.

Bibliography and discussion. A slightly weaker bound than the main theorem was first shown in [**HW876a**], and built upon the work in [**VC716a**]. The bound of this section is from [**Blu+89**]. The proof is usually stated entirely in probabilistic language while we have chosen a more combinatorial exposition that makes clear its basis in the probabilistic averaging technique of Chapter 1. The proof also follows immediately from the more general notion of ϵ-approximations (see Chapter 13).
The constant '56' in Theorem 6.1 can be replaced, with more precise calculations, by $1 + o(1)$! In particular, it was shown in [**KPW926a**] that a uniform random sample obtained by $\frac{d}{\epsilon}\left(\log\frac{1}{\epsilon} + 2\log\log\frac{1}{\epsilon} + 3\right)$ independent draws is an ϵ-net with probability at least $1 - e^{-d}$.

[Blu+89] A. Blumer, A. Ehrenfeucht, D. Haussler, and M. K. Warmuth, *Learnability and the Vapnik-Chervonenkis dimension*, J. Assoc. Comput. Mach. **36** (1989), no. 4, 929–965, DOI 10.1145/76359.76371. MR1072253

[HW876a] D. Haussler and E. Welzl, *ϵ-nets and simplex range queries*, Discrete Comput. Geom. **2** (1987), no. 2, 127–151, DOI 10.1007/BF02187876. MR884223

[KPW926a] J. Komlós, J. Pach, and G. Woeginger, *Almost tight bounds for ϵ-nets*, Discrete Comput. Geom. **7** (1992), no. 2, 163–173, DOI 10.1007/BF02187833. MR1139078

[VC716a] V. N. Vapnik and A. Ja. Červonenkis, *The uniform convergence of frequencies of the appearance of events to their probabilities* (Russian, with English summary), Teor. Verojatnost. i Primenen. **16** (1971), 264–279. MR0288823

2. Optimal ϵ-Nets using Packings

The question you raise, 'how can such a formulation lead to computations?' doesn't bother me in the least! Throughout my whole life as a mathematician, the possibility of making explicit, elegant computations has always come out by itself, as a byproduct of a thorough conceptual understanding of what was going on. Thus I never bothered about whether what would come out would be suitable for this or that, but just tried to understand—and it always turned out that understanding was all that mattered.

Alexandre Grothendieck

Given a set system $(X, \mathcal{F})$, we now show the existence of small ϵ-nets of $\mathcal{F}$ as a function of its shallow-cell complexity, which we first recall.

DEFINITION 4.4. A set system $(X, \mathcal{F})$ has shallow-cell complexity $\varphi_{\mathcal{F}}(\cdot, \cdot)$ if for any positive integer k and any finite $Y \subseteq X$, the number of sets in $\mathcal{F}|_Y$ of size at most k is upper bounded by $|Y| \cdot \varphi_{\mathcal{F}}(|Y|, k)$.

For a family $\mathcal{R}$ of geometric objects in $\mathbb{R}^d$—e.g., the family of all half-spaces—the shallow-cell complexity of $\mathcal{R}$ is defined to be the shallow-cell complexity of the primal set system $(\mathbb{R}^d, \mathcal{R})$.

The main theorem we will prove in this section is the following.

THEOREM 6.2. *Let $(X, \mathcal{F})$ be a finite set system with shallow-cell complexity $\varphi_{\mathcal{F}}(\cdot, \cdot)$ and with* VC-dim$(\mathcal{F}) \leq d$. *Then for any $\epsilon \in \left(0, \frac{1}{2}\right)$ there exists an ϵ-net of $\mathcal{F}$ of size*

$$O\left(\frac{d}{\epsilon} + \frac{1}{\epsilon} \log \varphi_{\mathcal{F}}\left(\frac{16d}{\epsilon}, 48d\right)\right).$$

Overview of ideas. The proof requires three ideas. For the moment assume that each set in $\mathcal{F}$ has size exactly ϵn.

(1) Fix any *maximal* subset $\mathcal{P} \subseteq \mathcal{F}$ such that every pair of sets of $\mathcal{P}$ have symmetric difference at least $\frac{\epsilon n}{2}$. In other words, $\mathcal{P}$ is a maximal $\left(\epsilon n, \frac{\epsilon n}{2}\right)$-packing of $\mathcal{F}$ (see Definition 5.9). Setting $\delta = \frac{\epsilon n}{2}$, $k = \epsilon n$ and applying Theorem 5.10, we have

$$|\mathcal{P}| \leq \frac{48dn}{\delta} \cdot \varphi_{\mathcal{F}}\left(\frac{8dn}{\delta}, \frac{24dk}{\delta}\right) = O\left(\frac{d}{\epsilon} \cdot \varphi_{\mathcal{F}}\left(O\left(\frac{d}{\epsilon}\right), O(d)\right)\right).$$

The key point here is that for each $F \in \mathcal{F} \setminus \mathcal{P}$, the maximality of $\mathcal{P}$ implies that there exists a set $S \in \mathcal{P}$ such that the size of the symmetric difference between F and S is at most $\frac{\epsilon n}{2}$. In other words, F contains at least $\frac{\epsilon n}{2}$ points from S, where $|S| = \epsilon n$ by assumption. Thus a $\frac{1}{2}$-net N_S for the set system $(S, \mathcal{F}|_S)$ must hit F and consequently $\bigcup_{S \in \mathcal{P}} N_S$ is an ϵ-net of $\mathcal{F}$. Here the sets of $\mathcal{P}$ play the role of canonical objects of Chapter 2.

Simply picking a $\frac{1}{2}$-net N_S *separately* for each $(S, \mathcal{F}|_S)$, $S \in \mathcal{P}$, where each N_S is of constant size by Theorem 6.10, will give an ϵ-net of $\mathcal{F}$ of total size

$$\sum_{S \in \mathcal{P}} |N_S| = O(|\mathcal{P}|) = O\left(\frac{d}{\epsilon} \cdot \varphi_{\mathcal{F}}\left(O\left(\frac{d}{\epsilon}\right), O(d)\right)\right).$$

This is too big.

(2) The next idea is to 'amortize' the size of the $\frac{1}{2}$-nets by first picking a random sample $R \subseteq X$.

For a fixed $S \in \mathcal{P}$ and $(S, \mathcal{F}|_S)$, Theorem 6.10 states that a random sample of S constructed by picking each point of S independently with probability

$$p = \Theta\left(\frac{1}{(1/2)\,|S|}\log\frac{1}{\gamma} + \frac{d}{(1/2)\,|S|}\log\frac{1}{(1/2)}\right) = \Theta\left(\frac{1}{\epsilon n}\log\frac{1}{\gamma} + \frac{d}{\epsilon n}\right).$$

is a $\frac{1}{2}$-net of $\mathcal{F}|_S$ with probability at least $1-\gamma$.

Instead of sampling points separately from each $S \in \mathcal{P}$, we will construct a 'global' random sample R by picking each point of X independently with the above probability p.

For a fixed $S \in \mathcal{P}$, R fails to be a $\frac{1}{2}$-net of $\mathcal{F}|_S$ with probability at most γ. By the union bound, the probability that there exists a $S \in \mathcal{P}$ such that R fails to be a $\frac{1}{2}$-net of $\mathcal{F}|_S$ is at most $|\mathcal{P}| \cdot \gamma$. Setting $\gamma = \frac{1}{|\mathcal{P}|+1}$ implies that, with non-zero probability, R is a $\frac{1}{2}$-net for all $\mathcal{F}|_S$, $S \in \mathcal{P}$. Then

$$\begin{aligned}
\mathrm{E}\left[|R|\right] = np &= O\left(\frac{1}{\epsilon}\log\frac{1}{\gamma} + \frac{d}{\epsilon}\right) \\
&= O\left(\frac{1}{\epsilon}\log\left(\frac{d}{\epsilon}\cdot\varphi_{\mathcal{F}}\left(O\left(\frac{d}{\epsilon}\right), O(d)\right)\right) + \frac{d}{\epsilon}\right) \\
&= O\left(\frac{1}{\epsilon}\log\frac{d}{\epsilon}\right) + O\left(\frac{1}{\epsilon}\log\varphi_{\mathcal{F}}\left(O\left(\frac{d}{\epsilon}\right), O(d)\right)\right) + O\left(\frac{d}{\epsilon}\right).
\end{aligned}$$

The large additional term $O\left(\frac{1}{\epsilon}\log\frac{d}{\epsilon}\right)$ still remains.

(3) Since the expected number of sets in $\mathcal{P}$ for which R fails is at most $|\mathcal{P}|\cdot\gamma$, we had set $\gamma < \frac{1}{|\mathcal{P}|}$ to ensure the existence of a set R with no failures. Instead, we will set γ such that the expected number of $\mathcal{F}|_S$, $S \in \mathcal{P}$, for which R fails to be a $\frac{1}{2}$-net is $O\left(\frac{1}{\epsilon}\right)$. The key point is that for each of these failed sets of $\mathcal{P}$, we can afford to separately add a $O(d)$-sized $\frac{1}{2}$-net for a total of $O\left(\frac{d}{\epsilon}\right)$ additional points. In other words, we set γ so that $|\mathcal{P}| \cdot \gamma = \Theta\left(\frac{1}{\epsilon}\right)$. Then the size of the initial random sample R is

$$\begin{aligned}
O\left(\frac{1}{\epsilon}\log\frac{1}{\gamma} + \frac{d}{\epsilon}\right) &= O\left(\frac{1}{\epsilon}\log\left(\epsilon\,|\mathcal{P}|\right) + \frac{d}{\epsilon}\right) \\
&= O\left(\frac{1}{\epsilon}\log\varphi_{\mathcal{F}}\left(O\left(\frac{d}{\epsilon}\right), O(d)\right) + \frac{d}{\epsilon}\right),
\end{aligned}$$

as desired.

We now turn to the formal proof with complete calculations.

Proof of Theorem 6.2. For an integer $j \geq 0$, set

$$\epsilon_j = 2^j \cdot \epsilon \qquad \text{and} \qquad \delta_j = \frac{\epsilon_j n}{2}.$$

For each $j = 1, \ldots, \lceil \log \frac{1}{\epsilon} \rceil$, define

$$\mathcal{F}_j = \{ S \in \mathcal{F} \ : \ \epsilon_{j-1} n \ \leq \ |S| \ < \ \epsilon_j n \}.$$

Further let $\mathcal{P}_j$ be a *maximal* subset of $\mathcal{F}_j$ satisfying the property that

$$\text{for all } S, S' \in \mathcal{P}_j, \qquad |\Delta(S, S')| \geq \delta_j.$$

As each set in $\mathcal{F}_j$ has size less than $\epsilon_j n$, Theorem 5.10 implies that

$$|\mathcal{P}_j| \leq \frac{48\, dn}{\delta_j} \cdot \varphi_{\mathcal{F}} \left(\frac{8\, dn}{\delta_j}, \frac{24\, d\epsilon_j n}{\delta_j} \right) = \frac{96\, dn}{\epsilon_j n} \cdot \varphi_{\mathcal{F}} \left(\frac{16\, dn}{\epsilon_j n}, \frac{48\, d\epsilon_j n}{\epsilon_j n} \right)$$

$$= \frac{96\, d}{\epsilon_j} \cdot \varphi_{\mathcal{F}} \left(\frac{16\, d}{\epsilon_j}, 48\, d \right). \tag{6.12}$$

CLAIM 6.13. Let $j \in \{1, \ldots, \lceil \log \frac{1}{\epsilon} \rceil\}$. Suppose the set $N_j \subseteq X$ is a $\frac{1}{2}$-net *simultaneously* for all these set systems:

$$\left\{ (S, \mathcal{F}_j|_S) \ : \ S \in \mathcal{P}_j \right\}.$$

Then N_j hits all the sets of $\mathcal{F}_j$.

PROOF. Let $F \in \mathcal{F}_j$. If $F \in \mathcal{P}_j$, then clearly it is hit by the $\frac{1}{2}$-net of $\mathcal{F}_j|_F$. Otherwise by the maximality of $\mathcal{P}_j$, there exists a $S \in \mathcal{P}_j$ such that

$$|\Delta(F, S)| = |F \setminus S| + |S \setminus F| < \delta_j \qquad \text{and hence} \qquad |F \setminus S| < \delta_j - |S \setminus F|.$$

As $|F| \geq 2^{j-1} \epsilon n = \delta_j$, it follows that

$$|F \cap S| = |F| - |F \setminus S| \geq \delta_j - (\delta_j - |S \setminus F|) = |S \setminus F|.$$

This implies that $|F \cap S| \geq \frac{|S|}{2}$ and as $F \cap S \in \mathcal{F}_j|_S$, F is hit by the $\frac{1}{2}$-net of $\mathcal{F}_j|_S$. $\square$

Thus it suffices to compute, for each $j = 1, \ldots, \lceil \log \frac{1}{\epsilon} \rceil$, a set N_j such that

$$N_j \ \text{ is a } \tfrac{1}{2}\text{-net of all the } |\mathcal{P}_j| \text{ set systems } (S, \mathcal{F}_j|_S),\ S \in \mathcal{P}_j.$$

We can then return $\bigcup_j N_j$ as an ϵ-net of $\mathcal{F}$. We will construct each N_j separately for each index $j \in \{1, \ldots, \lceil \log \frac{1}{\epsilon} \rceil\}$ by computing the following two sets R_j and M_j, and setting $N_j = R_j \cup M_j$.

Constructing R_j: Let R_j be a sample constructed by picking each point of X independently with probability

$$C_6 \cdot \left(\frac{d}{(1/2) \cdot \epsilon_{j-1} n} \ln \frac{1}{(1/2)} + \frac{1}{(1/2) \cdot \epsilon_{j-1} n} \ln \left(d \cdot \varphi_{\mathcal{F}} \left(\frac{16d}{\epsilon_j}, 48d \right) \right) \right).$$

where C_6 is the constant from Theorem 6.10. For each $S \in \mathcal{P}_j$, we have $|S| \in [\epsilon_{j-1} n, \epsilon_j n)$ and so Theorem 6.10 applied to

$$\text{the set system } (S, \mathcal{F}|_S) \quad \text{with} \quad \epsilon = \frac{1}{2}, \quad \gamma = \frac{1}{d \cdot \varphi_{\mathcal{F}} \left(\frac{16d}{\epsilon_j}, 48d \right)},$$

implies that R_j is a $\frac{1}{2}$-net of $\mathcal{F}|_S$ with probability at least $1 - \gamma$. Furthermore,

$$\mathrm{E}\left[|R_j| \right] = O \left(\frac{d}{\epsilon_j} + \frac{1}{\epsilon_j} \log \left(d \cdot \varphi_{\mathcal{F}} \left(\frac{16d}{\epsilon_j}, 48d \right) \right) \right).$$

Constructing M_j: Initialize $M_j = \emptyset$. For each $S \in \mathcal{P}_j$ for which R_j *fails* to be a $\frac{1}{2}$-net, construct a $\frac{1}{2}$-net of $(S, \mathcal{F}_j|_S)$ of size $O(d)$ (by Theorem 6.10) and add it to M_j. Then

$$\begin{aligned}
\mathrm{E}\left[|M_j|\right] &= \sum_{S\in\mathcal{P}_j} \Pr\left[R_j \text{ is not a } \frac{1}{2}\text{-net of } \mathcal{F}_j|_S\right] \cdot \left(\text{size of } \frac{1}{2}\text{-net of } \mathcal{F}_j|_S\right) \\
&\le \sum_{S\in\mathcal{P}_j} \frac{1}{d \cdot \varphi_{\mathcal{F}}\left(\frac{16d}{\epsilon_j}, 48d\right)} \cdot O(d) = |\mathcal{P}_j| \cdot \frac{1}{d \cdot \varphi_{\mathcal{F}}\left(\frac{16d}{\epsilon_j}, 48d\right)} \cdot O(d) \\
&\le \frac{96d}{\epsilon_j} \cdot \varphi_{\mathcal{F}}\left(\frac{16d}{\epsilon_j}, 48d\right) \cdot \frac{1}{d \cdot \varphi_{\mathcal{F}}\left(\frac{16d}{\epsilon_j}, 48d\right)} \cdot O(d) = O\left(\frac{d}{\epsilon_j}\right),
\end{aligned}$$

where the second-to-last step uses the upper bound on $|\mathcal{P}_j|$ given in Equation (6.12).

Thus we can conclude with the expected size of the final ϵ-net N of $\mathcal{F}$:

$$\begin{aligned}
\mathrm{E}[|N|] &= \sum_{j=1}^{\lceil \log \frac{1}{\epsilon} \rceil} \mathrm{E}\left[|R_j| + |M_j|\right] = \sum_{j=1}^{\lceil \log \frac{1}{\epsilon} \rceil} \mathrm{E}\left[|R_j|\right] + \sum_{j=1}^{\lceil \log \frac{1}{\epsilon} \rceil} \mathrm{E}\left[|M_j|\right] \\
&= \sum_{j=1}^{\lceil \log \frac{1}{\epsilon} \rceil} O\left(\frac{d}{\epsilon_j} + \frac{1}{\epsilon_j} \log\left(d \cdot \varphi_{\mathcal{F}}\left(\frac{16d}{\epsilon_j}, 48d\right)\right)\right) + \sum_{j=1}^{\lceil \log \frac{1}{\epsilon} \rceil} O\left(\frac{d}{\epsilon_j}\right) \\
&= \sum_{j=1}^{\lceil \log \frac{1}{\epsilon} \rceil} O\left(\frac{d}{2^j\epsilon} + \frac{1}{2^j\epsilon}\log d + \frac{1}{2^j\epsilon} \log \varphi_{\mathcal{F}}\left(\frac{16d}{2^j\epsilon}, 48d\right)\right) + \sum_{j=1}^{\lceil \log \frac{1}{\epsilon} \rceil} O\left(\frac{d}{2^j\epsilon}\right) \\
&= O\left(\frac{d}{\epsilon}\right) + O\left(\frac{1}{\epsilon}\right) \sum_{j=1}^{\lceil \log \frac{1}{\epsilon} \rceil} \frac{1}{2^j} \cdot \log \varphi_{\mathcal{F}}\left(\frac{16d}{2^j\epsilon}, 48d\right) \\
&\le O\left(\frac{d}{\epsilon}\right) + O\left(\frac{1}{\epsilon}\right) \sum_{j=1}^{\lceil \log \frac{1}{\epsilon} \rceil} \frac{1}{2^j} \cdot \log \varphi_{\mathcal{F}}\left(\frac{16d}{\epsilon}, 48d\right) \\
&= O\left(\frac{d}{\epsilon} + \frac{1}{\epsilon} \log \varphi_{\mathcal{F}}\left(\frac{16d}{\epsilon}, 48d\right)\right).
\end{aligned}$$

□

Bibliography and discussion. The proof in this chapter is from [**MDG18**], building on ideas from several earlier works [**AES10, CF90, CV07, PR08**]. In particular, [**PR08, AES10**] are important papers that made progress on ϵ-nets for several basic geometric set systems. It was shown in [**KMP17**] that the bound on sizes of ϵ-nets as a function of their shallow-cell complexity is tight.

[AES10] B. Aronov, E. Ezra, and M. Sharir, *Small-size ϵ-nets for axis-parallel rectangles and boxes*, SIAM J. Comput. **39** (2010), no. 7, 3248–3282, DOI 10.1137/090762968. MR2678074

[CF90] B. Chazelle and J. Friedman, *A deterministic view of random sampling and its use in geometry*, Combinatorica **10** (1990), no. 3, 229–249, DOI 10.1007/BF02122778. MR1092541

[CV07] K. L. Clarkson and K. Varadarajan, *Improved approximation algorithms for geometric set cover*, Discrete Comput. Geom. **37** (2007), no. 1, 43–58, DOI 10.1007/s00454-006-1273-8. MR2279863

[KMP17] A. Kupavskii, N. H. Mustafa, and J. Pach, *Near-optimal lower bounds for ϵ-nets for half-spaces and low complexity set systems*, A journey through discrete mathematics, Springer, Cham, 2017, pp. 527–541. MR3726612

[MDG18] N. H. Mustafa, K. Dutta, and A. Ghosh, *A simple proof of optimal epsilon nets*, Combinatorica **38** (2018), no. 5, 1269–1277, DOI 10.1007/s00493-017-3564-5. MR3884789

[PR08] E. Pyrga and S. Ray, *New existence proofs for ϵ-nets*, Computational geometry (SCG'08), ACM, New York, 2008, pp. 199–207, DOI 10.1145/1377676.1377708. MR2504287

CHAPTER 7

Epsilon-Nets: An Algorithm

The goal of this chapter is to study the following algorithm to compute an ϵ-net of a given set system $(X, \mathcal{F})$. The precise constants will be specified later.

General Net-Finder Algorithm $\left((X, \mathcal{F}), d, \varphi_{\mathcal{F}}, \epsilon\right)$
Specification: VC-dim $(\mathcal{F}) \leq d, \epsilon > 0$.

N_0: a uniform random sample of X of size $\Theta\left(\frac{1}{\epsilon}\log \varphi_{\mathcal{F}}\left(\Theta\left(\frac{d}{\epsilon}\right), \Theta\left(d\right)\right)\right)$.
$N = N_0$.
while *there exists a set* $S \in \mathcal{F}$, $|S| \geq \epsilon |X|$, *not hit by* N **do**
  add $\Theta(d)$ uniformly chosen random elements of S to N.
return N.

The main theorem of this chapter shows that the General Net-Finder Algorithm computes ϵ-nets of expected size equal, within constant factors, to those guaranteed by Theorem 6.2. This algorithm has two main advantages over earlier constructions.

Simplicity: The algorithm is oblivious to the intricate combinatorial constructs that we have seen earlier—packings, spatial partitioning, empty canonical objects and so on. In fact some of these ingredients from earlier work are present but *only* in the proof—the 'complexity' is shifted from the algorithm to its analysis.

Adaptivity: The choice of the $\Theta(d)$ elements added to N in each iteration is guided not only by the choice of the initial sample N_0, but also takes into account all elements added so far to N. This is not the case for Theorem 6.2, where after the initial random sample, the additional points for each 'failed event' were added and analysed independently. The new algorithm adds points to N only when it is required. While this additional 'adaptivity' does not improve the theoretical worst-case bound, it can give better bounds in practice.

We remark that the unhit set S at each iteration need not be random—it can be *any* unhit set. The analysis of this algorithm will only use the randomness of the initial sample as well as the $\Theta(d)$ elements picked in each previous iteration. The specific choice of the set S is irrelevant.

As we will see in Chapter 16, most known bounds on ϵ-net sizes follow from Theorem 6.2. Thus General Net-Finder Algorithm computes ϵ-nets whose sizes are equal, within constant factors, to the best known bounds for commonly studied geometric set systems.

For the case of set systems $\mathcal{F}$ with $\varphi_{\mathcal{F}}(n, k) = O(k^c)$ for an absolute constant $c \geq 1$, it is not even necessary to take an initial random sample. The algorithm simplifies to the following.

Linear Net-Finder Algorithm $\left((\boldsymbol{X}, \boldsymbol{\mathcal{F}}), \boldsymbol{d}, \boldsymbol{\varphi_{\mathcal{F}}}, \boldsymbol{\epsilon}\right)$
Specification: VC-dim $(\mathcal{F}) \leq d, \varphi_{\mathcal{F}}(n, k) = O(k^c)$ for a constant $c \geq 1$.

$N = \emptyset$.
while *there exists a set* $S \in \mathcal{F}$, $|S| \geq \epsilon |X|$, *not hit by* N **do**
 add $\Theta(d)$ uniformly chosen random elements of S to N.
return N.

By Theorem 6.2, such a $\mathcal{F}$ has an ϵ-net of size $O\left(\frac{1}{\epsilon}\right)$. The second theorem of this chapter shows that Linear Net-Finder Algorithm constructs ϵ-nets of expected size $O\left(\frac{1}{\epsilon}\right)$ in these cases as well.

1. Special Case: Linear-sized Nets

Every lecture should make only one main point. The German philosopher G. W. F. Hegel wrote that any philosopher who uses the word 'and' too often cannot be a good philosopher. I think he was right, at least insofar as lecturing goes. Every lecture should state one main point and repeat it over and over, like a theme with variations.

Gian-Carlo Rota

Let $(X, \mathcal{F})$ be a set system with shallow-cell complexity $\varphi_{\mathcal{F}}(\cdot, \cdot)$ and VC-dim$(\mathcal{F}) \leq d$. We will study the following algorithm for computing an ϵ-net N of $\mathcal{F}$.

Linear Net-Finder Algorithm $\left((X, \mathcal{F}), d, \varphi_{\mathcal{F}}, \epsilon\right)$
Specification: VC-dim$(\mathcal{F}) \leq d, \varphi_{\mathcal{F}}(n, k) = O(k^c)$ for a constant $c \geq 1$.

$N = \emptyset$.
while *there exists a set* $S \in \mathcal{F}$, $|S| \geq \epsilon |X|$, *not hit by* N **do**
  N_S: pick each $x \in S$ independently with prob. $C_6 \cdot \left(\frac{1}{\epsilon' |S|} \ln 2 + \frac{d}{\epsilon' |S|} \ln \frac{1}{\epsilon'}\right)$,
    where $\epsilon' = \frac{1}{100}$ and C_6 is the constant from Theorem 6.10.
  $N = N \cup N_S$.
return N.

In order to avoid dealing with implementation-specific details of the above algorithm, we will assume the existence of an 'oracle' that can return an unhit set in our set system with respect to the current set N. While the algorithm itself is oblivious to the structure of $\mathcal{F}$, efficient implementation of the oracle depends on the specific geometric properties of $\mathcal{F}$. This is outside the scope of this text; see discussion.

Here is the main theorem of this section.

THEOREM 7.1. *Let $(X, \mathcal{F})$ be a finite set system, and let c, d be two absolute constants such that the shallow-cell complexity of $\mathcal{F}$ is $\varphi_{\mathcal{F}}(n, k) = O(k^c)$ and* VC-dim$(\mathcal{F}) \leq d$. *Then for any $\epsilon > 0$, Linear Net-Finder Algorithm returns an ϵ-net of expected size $O\left(\frac{1}{\epsilon}\right)$, where the constant in the asymptotic notation depends on c and d.*
Furthermore it runs for an expected $O\left(\frac{1}{\epsilon}\right)$ iterations.

In particular, the Linear Net-Finder Algorithm computes an ϵ-net of expected size $O\left(\frac{1}{\epsilon}\right)$ for the primal set systems induced by half-spaces in $\mathbb{R}^2$, half-spaces in $\mathbb{R}^3$, pseudo-disks and disks in $\mathbb{R}^2$ as well as for the dual set systems induced on objects in the plane with linear union complexity.

Overview of ideas. Recall that the key idea in the proof of Theorem 6.2 is to construct a maximal packing $\mathcal{P} \subseteq \mathcal{F}$; that is, a maximal $\mathcal{P} \subseteq \mathcal{F}$ such that every pair of sets of $\mathcal{P}$ have symmetric difference at least $\frac{\epsilon n}{2}$.

Using Theorem 5.10, we get

$$|\mathcal{P}| = O\left(\frac{d}{\epsilon} \varphi_{\mathcal{F}}\left(O\left(\frac{d}{\epsilon}\right), O(d)\right)\right) = O\left(\frac{1}{\epsilon}\right),$$

using our assumption on $\varphi_{\mathcal{F}}$ and that c, d are considered absolute constants. Instead of constructing $\mathcal{P}$ as earlier, the proof now proceeds as follows.

Fix a maximal packing $\mathcal{P} \subseteq \mathcal{F}$. At the i-th iteration, say the oracle returns an unhit set $S_i \in \mathcal{F}$. Linear Net-Finder Algorithm then adds a uniform random subset $N_{S_i} \subseteq S_i$ of expected size $\Theta(1)$ to N. By the maximality of $\mathcal{P}$, S_i shares a large fraction of its elements with some $Y \in \mathcal{P}$ (call it the closest set of S_i).

Now let $j > i$ be a later iteration and S_j an unhit set returned by the oracle at iteration j such that the closest set of S_j in $\mathcal{P}$ is also Y. Thus S_j also shares a large fraction of its elements with Y, yet is unhit by the random sample N_{S_i}. This will be used to show that expected symmetric difference between $S_j \cap Y$ and $S_i \cap Y$ is large—in fact a constant fraction of $|Y|$.

For each $Y \in \mathcal{P}$, let $\mathcal{F}_Y \subseteq \mathcal{F}$ be the sets, considered over all iterations, whose closest set in $\mathcal{P}$ is Y. Then by the above reasoning, the sets in $\{S \cap Y : S \in \mathcal{F}_Y\}$ have, on average, pairwise large symmetric difference. Applying Theorem 5.10 to these sets will allow us to derive the upper bound $\mathrm{E}[|\mathcal{F}_Y|] = O(1)$.

Together with the upper bound on $|\mathcal{P}|$, we get

$$\mathrm{E}[|N|] = \sum_{Y \in \mathcal{P}} \mathrm{E}[|\mathcal{F}_Y|] \cdot \Theta(1) = O\left(\frac{1}{\epsilon}\right).$$

In the proof below, we will assume that each set $S \in \mathcal{F}$ considered by the algorithm has size $[\epsilon n, 2\epsilon n)$. Then we will show that an expected $O\left(\frac{1}{\epsilon}\right)$ elements are added to N.

The general case follows directly by grouping the sets considered by the algorithm by their sizes—all sets of size $[2^i \epsilon n, 2^{i+1} \epsilon n)$ go into group i, for $i = 0, \ldots, \log \frac{1}{\epsilon}$. The algorithm can be seen as constructing simultaneously a $(2^i \epsilon)$-net for group i. The proof below implies that for each group i, the expected number of elements of X added to N is $O\left(\frac{1}{2^i \epsilon}\right)$. Then the expected total number of elements added to N over all groups can be upper bounded by

$$\sum_{i=0}^{\log \frac{1}{\epsilon}} O\left(\frac{1}{2^i \epsilon}\right) = O\left(\frac{1}{\epsilon}\right).$$

Let $\beta \in [4\epsilon', 1]$ be a constant to be specified later.

Let $\mathcal{P}$ be a maximal $(2\epsilon n, (1-\beta)\, 2\epsilon n)$-packing of $\mathcal{F}$, and let $\mathcal{P} = \{P^1, \ldots, P^m\}$. By Theorem 5.10,

$$\begin{aligned} m = |\mathcal{P}| &\leq \frac{48dn}{(1-\beta)\, 2\epsilon n} \varphi_{\mathcal{F}}\left(\frac{8dn}{(1-\beta)\, 2\epsilon n}, \frac{24d \cdot 2\epsilon n}{(1-\beta)\, 2\epsilon n}\right) \\ &= O\left(\frac{d}{(1-\beta)\,\epsilon} \cdot \varphi_{\mathcal{F}}\left(\frac{4d}{(1-\beta)\,\epsilon}, \frac{24d}{(1-\beta)}\right)\right). \end{aligned} \tag{7.2}$$

Say the Linear Net-Finder Algorithm continues for t steps.
Let S_i be the unhit set returned by the oracle at the i-th step and let N_{S_i} be the random sample of S_i added to N.

As $\mathcal{P}$ is a *maximal* $(2\epsilon n, (1-\beta)\, 2\epsilon n)$-packing of $\mathcal{F}$, for each S_i there exists an index $j \in [m]$ with $|\Delta\left(S_i, P^j\right)| < (1-\beta)\, 2\epsilon n$ (it is possible that $P^j = S_i$). Assign S_i to the set P^j. For each $j \in [m]$, let n_j be the number of sets of $\{S_1, \ldots, S_t\}$ assigned to P^j and denote them by

$$\mathcal{S}^j = \left\{S_1^j, \ldots, S_{n_j}^j\right\},$$

listed *in the order* considered by the Linear Net-Finder Algorithm.

LEMMA 7.3. *For each $j \in [m]$ and $i \in [n_j]$,*

$$|S_i^j \cap P^j| > \frac{|P^j| + |S_i^j| - (1-\beta) 2\epsilon n}{2} \geq \beta \epsilon n.$$

PROOF. Note that

$$\begin{aligned} |S_i^j| + |P^j| &= |P^j \setminus S_i^j| + |S_i^j \setminus P^j| + 2\,|S_i^j \cap P^j| \\ &< (1-\beta)\, 2\epsilon n + 2\,|S_i^j \cap P^j|. \end{aligned}$$

Re-arranging the terms gives the first inequality. The second follows from the fact that $|P^j|, |S_i^j| \geq \epsilon n$. □

For each $j \in [m]$, define

$$\mathcal{S}'^j = \left\{S \in \mathcal{S}^j \colon N_S \text{ is an } \epsilon'\text{-net of the set system } (S, \mathcal{F}|_S)\right\}.$$

The core of the argument is the following lemma.

LEMMA 7.4. *For any $j \in [m]$,*

$$|\mathcal{S}'^j| = O\left(\frac{d}{(\beta - 2\epsilon')} \cdot \varphi_{\mathcal{F}}\left(\frac{8d}{(\beta - 2\epsilon')}, \frac{24d}{(\beta - 2\epsilon')}\right)\right).$$

PROOF. Let $n'_j = |\mathcal{S}'^j|$. By re-labeling the sets of $\mathcal{S}^j$, we can assume that $\mathcal{S}'^j = \{S_1^j, \ldots, S_{n'_j}^j\}$, again listed here in the order that they were considered by the Linear Net-Finder Algorithm. Consider their projection onto P^j:

$$\mathcal{T}'^j = \left\{T_1^j, \ldots, T_{n'_j}^j\right\}, \quad \text{where } T_i^j = S_i^j \cap P^j.$$

Then observe the following:

- For each $i \in [n'_j]$, Lemma 7.3 implies that
$$|T_i^j| \geq \beta \epsilon n.$$
- For each $1 \leq k < l \leq n'_j$, the set $N_{S_k^j}$ was added to N in the algorithm *before* the set S_l^j was considered. As $N_{S_k^j}$ is an ϵ'-net of $\mathcal{F}|_{S_k^j}$ and the set

S_l^j is not hit by $N_{S_k^j}$, it must be that S_l^j contains less than ϵ'-th fraction of the points of S_k^j, implying that

$$|T_k^j \cap T_l^j| = |S_k^j \cap S_l^j \cap P^j| \le |S_k^j \cap S_l^j| < \epsilon' \cdot |S_k^j| < \epsilon' \cdot 2\epsilon n.$$

The above two facts imply that for any T_l^j and T_k^j,

$$\left|\Delta\left(T_l^j, T_k^j\right)\right| = \left|T_k^j \setminus T_l^j\right| + \left|T_l^j \setminus T_k^j\right| \ge 2\left(\beta\epsilon n - \epsilon' \cdot 2\epsilon n\right) = \left(\beta - 2\epsilon'\right) 2\epsilon n.$$

Thus the sets of $\mathcal{T}'^j$ form a $\left(|P^j|, (\beta - 2\epsilon')\, 2\epsilon n\right)$-packing over the elements of P^j. By Theorem 5.10, we have

$$|\mathcal{S}'^j| = |\mathcal{T}'^j| = O\left(\frac{d}{(\beta - 2\epsilon')} \cdot \varphi_{\mathcal{F}}\left(\frac{8d}{(\beta - 2\epsilon')}, \frac{24d}{(\beta - 2\epsilon')}\right)\right).$$

□

LEMMA 7.5. *For any $j \in [m]$,*

$$\mathrm{E}\left[|\mathcal{S}^j|\right] = O\left(\frac{d}{(\beta - 2\epsilon')} \cdot \varphi_{\mathcal{F}}\left(\frac{8d}{(\beta - 2\epsilon')}, \frac{24d}{(\beta - 2\epsilon')}\right)\right).$$

PROOF. For each $S_i^j \in \mathcal{S}^j$, $N_{S_i^j}$ is constructed by picking each element of S_i^j independently with probability

$$C_6 \cdot \left(\frac{1}{\epsilon' |S_i^j|} \ln 2 \; + \; \frac{d}{\epsilon' |S_i^j|} \ln \frac{1}{\epsilon'}\right).$$

Let Y_i be an indicator random variable that is 1 if and only if $N_{S_i^j}$ is an ϵ'-net of $\mathcal{F}|_{S_i^j}$. Then Theorem 6.10 implies that

$$\Pr\left[Y_i = 1\right] = \Pr\left[N_{S_i^j} \text{ is a } \epsilon'\text{-net of } \mathcal{F}|_{S_i^j}\right] \ge \frac{1}{2}.$$

Note that the value of Y_i is independent of the values of Y_j, $j < i$.
Now consider a random experiment consisting of a sequence of independent Bernoulli trials, where the i-th trial is a 'success' if $Y_i = 1$. Then for any integer r, the expected number of trials to get r successes is at most $\frac{r}{(1/2)}$ (it follows a negative binomial distribution). On the other hand, the number of successes is upper bounded by Lemma 7.4, and the proof follows. □

We now return to the proof of the main theorem.

THEOREM 7.1. *Let $(X, \mathcal{F})$ be a finite set system, and let c, d be two absolute constants such that the shallow-cell complexity of $\mathcal{F}$ is $\varphi_{\mathcal{F}}(n, k) = O\left(k^c\right)$ and $\text{VC-dim}\left(\mathcal{F}\right) \le d$. Then for any $\epsilon > 0$, Linear Net-Finder Algorithm returns an ϵ-net of expected size $O\left(\frac{1}{\epsilon}\right)$, where the constant in the asymptotic notation depends on c and d.*
Furthermore it runs for an expected $O\left(\frac{1}{\epsilon}\right)$ iterations.

PROOF. Clearly the algorithm only stops when N is an ϵ-net. Thus it remains to bound the expected size of N.

At each iteration, from the set $S_i \in \mathcal{F}$ returned by the oracle, we add at most

$$\mathrm{E}\left[|N_{S_i}|\right] = C_6 \cdot \left(\frac{1}{\epsilon'} \ln 2 \; + \; \frac{d}{\epsilon'} \ln \frac{1}{\epsilon'}\right).$$

new elements to N. Furthermore, the expected number of iterations is

$$\begin{aligned}
\mathrm{E}\left[\sum_{j=1}^{m} |\mathcal{S}^j|\right] &= \sum_{j=1}^{m} \mathrm{E}\left[|\mathcal{S}^j|\right] \\
&= m \cdot O\left(\frac{d}{(\beta - 2\epsilon')} \cdot \varphi_{\mathcal{F}}\left(\frac{8d}{(\beta - 2\epsilon')}, \frac{24d}{(\beta - 2\epsilon')}\right)\right) \quad \Big(\text{Lemma 7.5}\Big) \\
&= O\left(\frac{d}{(1-\beta)\,\epsilon} \cdot \varphi_{\mathcal{F}}\left(\frac{4d}{(1-\beta)\,\epsilon}, \frac{24d}{(1-\beta)}\right)\right) \\
&\quad \cdot O\left(\frac{d}{(\beta - 2\epsilon')} \cdot \varphi_{\mathcal{F}}\left(\frac{8d}{(\beta - 2\epsilon')}, \frac{24d}{(\beta - 2\epsilon')}\right)\right) \quad \Big(\text{Equation (7.2)}\Big).
\end{aligned}$$

Setting β to be any small constant, say $\beta = \frac{1}{25}$ and $\epsilon' = \frac{\beta}{4} = \frac{1}{100}$, we get

$$\mathrm{E}\left[|N|\right] = O\left(\frac{1}{\epsilon}\right),$$

where the constant in the asymptotic notation depends on c and d. □

Bibliography and discussion. The algorithm presented is taken from [**Mus19**]. For geometric set systems, the existence of efficient implementation of such oracles follow from the extensive work on range searching, reporting and emptiness data-structures (see [**Aga18**]).

[Aga18] P. K. Agarwal, *Range searching*, Handbook of discrete and computational geometry, CRC Press Ser. Discrete Math. Appl., CRC, Boca Raton, FL, 1997, pp. 575–598. MR1730187

[Mus19] N. H. Mustafa, *Computing optimal epsilon-nets is as easy as finding an unhit set*, 46th International Colloquium on Automata, Languages, and Programming, LIPIcs. Leibniz Int. Proc. Inform., vol. 132, Schloss Dagstuhl. Leibniz-Zent. Inform., Wadern, 2019, pp. Art. No. 87, 12. MR3984904

2. General Case

Thought is only a flash in the middle of a long night, but the flash that means everything.

Henri Poincaré

Let $(X, \mathcal{F})$ be a set system with VC-dim $(\mathcal{F}) \leq d$, and shallow-cell complexity $\varphi_{\mathcal{F}}(\cdot,\cdot)$. In this section we extend the algorithm of the previous section to compute ϵ-nets of set systems as a function of their shallow cell complexity $\varphi_{\mathcal{F}}(\cdot,\cdot)$. The only difference in the algorithm is that instead of starting with an empty set N, we start with a random sample N_0 taken from X.

General Net-Finder Algorithm $\left((X,\mathcal{F}), d, \varphi_{\mathcal{F}}, \epsilon\right)$
Specification: VC-dim $(\mathcal{F}) \leq d, \epsilon > 0$.

N_0: pick each $x \in X$ i.i.d. with probability
$C_6 \cdot \left(\frac{2}{\beta\epsilon|X|} \ln\left(d^2\varphi_{\mathcal{F}}\left(\frac{8d}{\beta(1-\beta)\epsilon}, \frac{24d}{\beta(1-\beta)}\right)^2\right) + \frac{2\,d}{\beta\epsilon|X|}\ln\frac{2}{\beta}\right)$,
where $\beta = \frac{1}{25}$ and C_6 is the constant from Theorem 6.10.

$N = N_0$
while *there exists a set* $S \in \mathcal{F}$, $|S| \geq \epsilon\,|X|$, *not hit by* N **do**
  N_S: pick each $x \in S$ i.i.d. with probability $C_6 \cdot \left(\frac{1}{\epsilon'\,|S|}\ln 2 + \frac{d}{\epsilon'\,|S|}\ln\frac{1}{\epsilon'}\right)$,
  where $\epsilon' = \frac{1}{100}$.
  $N = N \cup N_S$.
return N.

We will again assume the existence of an oracle as a black-box that can return an unhit set in our set system with respect to the current candidate net N. Each iteration of the above algorithm makes one call to the oracle.

Our main theorem is the following

THEOREM 7.6. *Let $(X, \mathcal{F})$ be a finite set system with shallow-cell complexity $\varphi_{\mathcal{F}}(\cdot,\cdot)$ and VC-dim $(\mathcal{F}) \leq d$. Then for any $\epsilon \in \left(0, \frac{1}{2}\right]$, General Net-Finder Algorithm returns an ϵ-net of expected size*

$$O\left(\frac{1}{\epsilon}\log\left(\varphi_{\mathcal{F}}\left(\frac{210d}{\epsilon}, 630\,d\right)\right) + \frac{d}{\epsilon}\right).$$

Furthermore, it runs for an expected $O\left(\frac{1}{\epsilon}\right)$ iterations.

Overview of ideas. For the case $\varphi_{\mathcal{F}}(n,k) = O\left(k^c\right)$, the maximal packing that we constructed had size $O\left(\frac{1}{\epsilon}\right)$ and we showed that the iterative algorithm added an expected $O(1)$ new elements to N for each set of the packing (Theorem 7.1). For the general case that we consider now, a maximal packing will be bigger, with size $O\left(\frac{1}{\epsilon}\varphi_{\mathcal{F}}\left(O\left(\frac{d}{\epsilon}\right), O(d)\right)\right)$. So we cannot afford to add new points for each set of the packing.

To avoid that, we first take a global sample. Then the expected number of sets of the packing for which we have to add additional points becomes much smaller and the analysis proceeds as earlier.

As before, we assume that each set S considered by the algorithm has size $[\epsilon n, 2\epsilon n)$ and show that, in expectation, $O\left(\frac{1}{\epsilon}\right)$ additional points are added after the initial random sample N_0.

> In general, let group i consist of the sets of $\mathcal{F}$ of size $\left[2^i \epsilon n, 2^{i+1}\epsilon n\right)$. The algorithm can be seen as constructing simultaneously a $\left(2^i\epsilon\right)$-net for group i, for $i = 0, \ldots, \left\lceil \log \frac{1}{\epsilon} \right\rceil$. The proof below implies that for an initial random sample of expected size $O\left(\frac{1}{2^i\epsilon}\log\left(\varphi_{\mathcal{F}}\left(\frac{210d}{2^i\epsilon}, 630d\right)\right)\right)$, the algorithm runs for an expected $O\left(\frac{1}{2^i\epsilon}\right)$ iterations, each of which adds $\Theta(d)$ additional points, to compute a $\left(2^i\epsilon\right)$-net for the sets of group i. Then the expected size of the initial random sample N_0 over all groups can be upper bounded by
>
> $$\sum_i O\left(\frac{1}{2^i\epsilon}\log\left(\varphi_{\mathcal{F}}\left(\frac{210d}{2^i\epsilon}, 630d\right)\right)\right) = O\left(\frac{1}{\epsilon}\log\left(\varphi_{\mathcal{F}}\left(\frac{210d}{\epsilon}, 630d\right)\right)\right)$$
>
> and the expected total number of iterations is upper bounded by $\sum_i O\left(\frac{1}{2^i\epsilon}\right) = O\left(\frac{1}{\epsilon}\right)$.

The analysis continues that of the previous section. Recall that $\mathcal{F}$ consists of sets of size $[\epsilon n, 2\epsilon n)$, and let $\mathcal{P}$ be a $(2\epsilon n, (1-\beta)\, 2\epsilon n)$-packing of $\mathcal{F}$. We have $\mathcal{P} = \left\{P^1, \ldots, P^m\right\}$, where

$$m = O\left(\frac{d}{(1-\beta)\,\epsilon}\varphi_{\mathcal{F}}\left(\frac{4d}{(1-\beta)\,\epsilon}, \frac{24d}{(1-\beta)}\right)\right).$$

> Say the algorithm continues for t iterations and let $S_1, \ldots, S_t$ be the sets considered in these t iterations. Denote by N_{S_i} the random sample of S_i added to N at iteration i. For each $j \in [m]$, n_j is the number of sets of $\{S_1, \ldots, S_t\}$ assigned to $P^j \in \mathcal{P}$, and denote these sets by
>
> $$\mathcal{S}^j = \left\{S_1^j, \ldots, S_{n_j}^j\right\},$$
>
> listed in the order considered by the General Net-Finder Algorithm.

We restate the key lemmas.

> LEMMA 7.3. *For each $j \in [m]$ and $i \in [n_j]$,*
>
> $$|S_i^j \cap P^j| > \frac{|P^j| + |S_i^j| - (1-\beta)2\epsilon n}{2} \geq \beta\epsilon n.$$
>
> LEMMA 7.5. *For any $j \in [m]$,*
>
> $$\mathrm{E}\left[|\mathcal{S}^j|\right] = O\left(\frac{d}{(\beta - 2\epsilon')}\cdot\varphi_{\mathcal{F}}\left(\frac{8d}{(\beta-2\epsilon')}, \frac{24d}{(\beta-2\epsilon')}\right)\right).$$

We remark here that the expectation in Lemma 7.5 is over the choice of random points added to N during the t iterations. It is independent of the initial sample N_0 and holds for *any* choice of N_0.

PROOF OF THEOREM 7.6. Clearly the algorithm only stops when N is an ϵ-net. Thus it remains to bound its expected size.

Consider an index $j \in [m]$. By Lemma 7.3, for any $i \in [n_j]$,

$$|S_i^j \cap P^j| > \beta \epsilon n \geq \frac{\beta}{2} \cdot |P^j|. \tag{7.7}$$

If N_0 is a $\frac{\beta}{2}$-net of $(P^j, \mathcal{F}|_{P^j})$, then Equation (7.7) implies that any $S \in \mathcal{S}^j$ would be hit by N_0 and thus $\mathcal{S}^j = \emptyset$. By Theorem 6.10, for a fixed index j,

$$\Pr\left[N_0 \text{ is not a } \frac{\beta}{2}\text{-net of } \left(P^j, \mathcal{F}|_{P^j}\right)\right] \leq \frac{1}{d^2 \varphi_{\mathcal{F}}\left(\frac{8d}{\beta(1-\beta)\epsilon}, \frac{24d}{\beta(1-\beta)}\right)^2}. \tag{7.8}$$

From each S_i we add at most

$$\mathrm{E}\left[|N_{S_i}|\right] = |S_i| \cdot C_6 \cdot \left(\frac{1}{\epsilon'|S_i|} \log 2 \;+\; \frac{d}{\epsilon'|S_i|} \log \frac{1}{\epsilon'}\right) = O\left(\frac{d}{\epsilon'} \log \frac{1}{\epsilon'}\right) = O\left(d\right)$$

additional points to N. Furthermore, the expected number of iterations is

$$\begin{aligned}
\mathrm{E}\left[\sum_{j=1}^{m} |\mathcal{S}^j|\right] &= \sum_{j=1}^{m} \mathrm{E}\left[|\mathcal{S}^j|\right] \\
&= \sum_{j=1}^{m} \mathrm{E}\left[\; |\mathcal{S}^j| \;\middle|\; N_0 \text{ is not a } \frac{\beta}{2}\text{-net of } \left(P^j, \mathcal{F}|_{P^j}\right)\right] \\
&\cdot \Pr\left[N_0 \text{ is not a } \frac{\beta}{2}\text{-net of } \left(P^j, \mathcal{F}|_{P^j}\right)\right] \\
&= O\left(\frac{d}{(1-\beta)\,\epsilon} \varphi_{\mathcal{F}}\left(\frac{4d}{(1-\beta)\,\epsilon}, \frac{24d}{(1-\beta)}\right)\right) \cdot \frac{1}{d^2 \varphi_{\mathcal{F}}\left(\frac{8d}{\beta(1-\beta)\epsilon}, \frac{24d}{\beta(1-\beta)}\right)^2} \\
&\cdot O\left(\frac{d}{(\beta - 2\epsilon')} \cdot \varphi_{\mathcal{F}}\left(\frac{8d}{(\beta - 2\epsilon')}, \frac{24d}{(\beta - 2\epsilon')}\right)\right),
\end{aligned}$$

where the last step follows from substituting the value of m, Equation (7.8) and Lemma 7.5. By the assumption on the monotonicity of $\varphi_{\mathcal{F}}(\cdot, \cdot)$ in both arguments, we get

$$= O\left(\frac{1}{(1-\beta)\,(\beta - 2\epsilon')\,\epsilon}\right).$$

Putting everything together,

$$\begin{aligned}
\mathrm{E}\left[|N|\right] &= \mathrm{E}\left[|N_0|\right] + \mathrm{E}\left[\sum_{i=1}^{t} |N_{S_i}|\right] = \mathrm{E}\left[|N_0|\right] + O\left(d\right) \cdot \mathrm{E}\left[\sum_{j=1}^{m} |\mathcal{S}^j|\right] \\
&= C_6 \cdot \left(\frac{2}{\beta\epsilon} \ln\left(d^2 \varphi_{\mathcal{F}}\left(\frac{8d}{\beta\,(1-\beta)\,\epsilon}, \frac{24d}{\beta\,(1-\beta)}\right)^2\right) + \frac{2\,d}{\beta\epsilon} \ln \frac{2}{\beta}\right) \\
&+ O\left(\frac{d}{(1-\beta)\,(\beta - 2\epsilon')\,\epsilon}\right).
\end{aligned}$$

Setting β to be any small constant, say $\beta = \frac{1}{25}$, and $\epsilon' = \frac{\beta}{4} = \frac{1}{100}$, we get the desired bound. $\square$

Bibliography and discussion. The algorithm presented is from [**Mus19**].

[Mus19] N. H. Mustafa, *Computing optimal epsilon-nets is as easy as finding an unhit set*, 46th International Colloquium on Automata, Languages, and Programming, LIPIcs. Leibniz Int. Proc. Inform., vol. 132, Schloss Dagstuhl. Leibniz-Zent. Inform., Wadern, 2019, pp. Art. No. 87, 12. MR3984904

CHAPTER 8

Epsilon-Nets: Weighted Case

Let $(X, \mathcal{F})$ be a set system with shallow-cell complexity $\varphi_{\mathcal{F}}(m,k) = O(k^c)$ for an absolute constant c. Then for any $\epsilon > 0$, Theorem 6.2 states the existence of an ϵ-net N of $\mathcal{F}$ with

$$|N| = O\left(\frac{1}{\epsilon}\right).$$

We now consider a generalization where one is also given a weight function $w\colon X \to \mathbb{R}^+$. Then instead of minimizing the *cardinality* of the ϵ-net of $\mathcal{F}$, one would like to minimize its *weight*. Note that we still want N to hit sets of cardinality at least $\epsilon \cdot |X|$.

> We have already seen an ϵ-net problem involving a weight function $w(\cdot)$ where the goal was to pick an ϵ-net of minimum cardinality but hitting all sets of weight at least ϵ-th fraction of the total weight. By duplicating elements according to their weights it was shown to be equivalent to the unweighted cardinality-based ϵ-net problem.
>
> This, however, is not the case for the problem we address in this chapter.

Let $W = \sum_{x \in X} w(x)$ be the total weight of the elements of X. Then our aim is to construct an ϵ-net N of $\mathcal{F}$ where the average weight of an element of N is $O\left(\frac{W}{|X|}\right)$. The first theorem of this chapter shows that indeed that is possible: there exists an ϵ-net of $\mathcal{F}$, assuming that $\varphi_{\mathcal{F}}(m,k) = O(k^c)$ for an absolute constant c, of total weight

$$O\left(\frac{W}{|X|} \cdot \frac{1}{\epsilon}\right). \tag{8.1}$$

In fact, the first main result of this chapter will prove the following statement.

THEOREM 8.2. *Let $(X, \mathcal{F})$ be a set system on n elements and let $\epsilon > 0$ be a given parameter. Further assume that $\mathcal{F}$ has shallow-cell complexity $\varphi_{\mathcal{F}}(m,k) = O(k^2)$. Then there exists a probabilistic procedure to compute an ϵ-net N of $\mathcal{F}$ such that each $p \in X$ is present in N with probability $O\left(\frac{1}{\epsilon n}\right)$.*

Linearity of expectation then implies that the expected weight of N is the one given by Equation (8.1). Further, with the weights $w(x) = 1$ for all $x \in X$, one gets the unweighted problem, and so the above bound cannot be improved.

> At first glance, Theorem 8.2 is surprising: consider the set system formed by a set X of n points in $\mathbb{R}$ where the subsets are induced by intervals. Given any $\epsilon > 0$, consider the partition of X into $\lfloor \frac{1}{\epsilon} \rfloor$ disjoint sets $X_1, \ldots, X_{\lfloor \frac{1}{\epsilon} \rfloor}$, each containing at least ϵn contiguous points. Then if one picks each $p \in X$ into a uniform random sample R independently with probability $\Theta\left(\frac{1}{\epsilon n}\right)$,

the probability of *not* hitting a fixed X_i is $\left(1 - \Theta\left(\frac{1}{\epsilon n}\right)\right)^{\epsilon n} = \Theta(1)$. So, in expectation there will be $\Theta\left(\frac{1}{\epsilon}\right)$ sets that will *not* be hit by N—that is, picking points of X independently at random will leave a constant fraction of the sets unhit.

The algorithms of Chapter 3 constructed an ϵ-net via a 2-step process where the points of X were not picked independently: the first step was to pick an independent uniform random sample $R \subseteq X$ of size $\Theta\left(\frac{1}{\epsilon}\right)$ and the second step consisted of adding additional points from each of sets not hit by R, to get the final ϵ-net N. However, that introduces non-uniformity—not all points of X end up in N with asymptotically the same probability and it is not clear how to control the probability of each element of X being added to N.

The key new idea is to design a procedure that 'slows down' the construction of N. We will still construct an ϵ-net using random sampling—but now over *logarithmically* many rounds. This allows the points picked in each round to depend on the choices made in earlier rounds and enables us to carefully design a probability distribution such that each point is picked with roughly the same probability.

The second main result of this chapter generalizes Theorem 8.2 to give a bound on the size of weighted ϵ-nets as a function of the shallow-cell complexity of the set system.

1. Special Case: Linear-sized Nets

Mathematics is amazingly compressible: you may struggle a long time, step by step, to work through some process or idea from several approaches. But once you really understand it and have the mental perspective to see it as a whole, there is a tremendous mental compression. You can file it away, recall it quickly and completely when you need it, and use it as just one step in some other mental process. The insight that goes with this compression is one of the real joys of mathematics.

William Thurston

This section considers set systems with small shallow-cell complexity. The case for set systems with arbitrary shallow-cell complexity will be treated in the next section. Recall the notion of shallow-cell complexity.

> DEFINITION 4.4. A set system $(X, \mathcal{F})$ has shallow-cell complexity $\varphi_{\mathcal{F}}(\cdot,\cdot)$ if for any positive integer k and any finite $Y \subseteq X$, the number of sets in $\mathcal{F}|_Y$ of size at most k is upper bounded by $|Y| \cdot \varphi_{\mathcal{F}}\big(|Y|, k\big)$.
>
> For a family $\mathcal{R}$ of geometric objects in $\mathbb{R}^d$—e.g., the family of all half-spaces—the shallow-cell complexity of $\mathcal{R}$ is defined to be the shallow-cell complexity of the primal set system $\big(\mathbb{R}^d, \mathcal{R}\big)$.

The main theorem of this section is the following.

THEOREM 8.2. *Let $(X, \mathcal{F})$ be a set system on n elements and let $\epsilon > 0$ be a given parameter. Further assume that $\mathcal{F}$ has shallow-cell complexity $\varphi_{\mathcal{F}}(m,k) = O\left(k^2\right)$. Then there exists a probabilistic procedure to compute an ϵ-net N of $\mathcal{F}$ such that each $p \in X$ is present in N with probability $O\left(\frac{1}{\epsilon n}\right)$.*

> The proof of Theorem 8.2 works more generally for set systems $\mathcal{F}$ with $\varphi_{\mathcal{F}}(m,k) = O\left(k^c\right)$ for some absolute constant $c \geq 1$. However for expository purposes we state and prove it for the case $c = 2$.

The primal set system induced on a set of points in $\mathbb{R}^2$ by disks satisfies the conditions of Theorem 8.2—see Lemma 1.2—and thus Theorem 8.2 implies Theorem 3.3.

Overview of ideas. Assume that each set of $\mathcal{F}$ has size at least ϵn.

We will present a probabilistic iterative procedure that constructs an ϵ-net N over $\log\left(\Theta\left(\epsilon n\right)\right)$ rounds, where each round adds some elements to N. The elements added in round i will depend on the choices made in earlier rounds and thus there is a complicated dependency between the events $\{p \in N\colon p \in X\}$. Nevertheless we will prove that each element of X ends up in N with probability $O\left(\frac{1}{\epsilon n}\right)$.

The following is the key procedure applied at each round.

LEMMA 8.3. *Let $(X, \mathcal{F})$ be a set system with shallow-cell complexity $\varphi_{\mathcal{F}}(m,k) = O\left(k^2\right)$. Let $k \geq 400$ be a positive integer such that $|S| \geq k$ for all $S \in \mathcal{F}$. Construct a set Y by choosing each element of X independently with probability*

$$\frac{1}{2} + \sqrt{\frac{10 \ln k}{k}}.$$

Then one can construct a set $Q \subseteq X \setminus Y$ such that

(1) *for each $S \in \mathcal{F}$, either $|S \cap Y| \geq \frac{k}{2}$ or $S \cap Q \neq \emptyset$, and*
(2) *each $p \in X$ belongs to Q with probability $O\left(\frac{1}{k^2}\right)$, where the probability is over the choice of Y. The constant in this probability depends on the multiplicative constant in $\varphi_{\mathcal{F}}(m, k)$.*

To see the idea behind the proof of Lemma 8.3, note that as each $p \in X$ is added to Y with probability $\frac{1}{2} + \sqrt{\frac{10 \ln k}{k}}$, each set $S \in \mathcal{F}$ will contain greater than $\frac{|S|}{2} \geq \frac{k}{2}$ elements of Y *in expectation*. However there will be some failures—that is, sets of $\mathcal{F}$ from which less than $\frac{k}{2}$ elements are picked into Y. For each such $S \in \mathcal{F}$, we will add a carefully chosen element $p_S \in S$ to Q. The new insight here is that

> *it is possible to assign each $S \in \mathcal{F}$ to an element $p_S \in S$ such that each element $p \in P$ is assigned $O\left(k^3\right)$ sets of $\mathcal{F}$.*
>
> This follows from the fact that if $\varphi_{\mathcal{F}}(m, k) = O\left(k^2\right)$, then on average, a point of P is contained in only $O\left(k^3\right)$ sets of $\mathcal{F}$. To see this, assume for simplicity that all sets of $\mathcal{F}$ have size $\Theta(k)$; then this average is at most
>
> $$\frac{|\mathcal{F}| \cdot \Theta(k)}{n} = \frac{O\left(nk^2\right) \cdot \Theta(k)}{n} = O\left(k^3\right).$$

Thus each $p \in X$ can potentially be added to Q because of a failure of some $S \in \mathcal{F}$ from one of the $O\left(k^3\right)$ sets assigned to p. On the other hand, a calculation using Chernoff's bound will show that $\Pr\left[|S \cap Y| < \frac{k}{2}\right] = O\left(\frac{1}{k^5}\right)$ for each $S \in \mathcal{F}$. Together with the union bound, this implies that each $p \in X$ is added to Q with probability $O\left(\frac{1}{k^2}\right)$, as desired.

Before giving the proof of Lemma 8.3, we first see how to apply it iteratively to get the main result.

PROOF OF THEOREM 8.2. The algorithm works over t iterations, for an integer parameter t to be fixed later. We can assume that each set of $\mathcal{F}$ has size at least ϵn.

Weighted Linear Net-Finder Algorithm $\Big((X, \mathcal{F}), \epsilon > 0\Big)$.
Specification: $\varphi_{\mathcal{F}}(m, k) = O\left(k^2\right)$.

$X_0 = X$, $Q_0 = \emptyset$ and $\mathcal{F}_0 = \mathcal{F}$.
$i = 0$.

while $i < t$ **do**
- apply Lemma 8.3 to $(X_i, \mathcal{F}_i)$ with $k_i = \frac{\epsilon n}{2^i}$ to get the sets Y and Q.
- set $X_{i+1} = Y$ and $Q_{i+1} = Q$.
- let $\mathcal{F}_{i+1}$ be the set system obtained from $\mathcal{F}_i$ by first removing all sets of $\mathcal{F}_i$ hit by Q and then projecting $\mathcal{F}_i$ onto Y. That is,
$$\mathcal{F}_{i+1} = \left\{S \cap Y \ :\ S \in \mathcal{F}_i \ \text{ and } \ S \cap Q = \emptyset\right\}.$$
- $i = i + 1$.

return $N = X_t \cup Q_1 \cup \cdots \cup Q_t$.

Note that by induction and Lemma 8.3, each set in $\mathcal{F}_i$ has size at least $\frac{\epsilon n}{2^i}$. Let p_i be the probability of adding an element from X_i into X_{i+1}. That is,

$$p_i = \frac{1}{2} + \sqrt{\frac{10 \ln \frac{\epsilon n}{2^i}}{\frac{\epsilon n}{2^i}}}.$$

First observe that p_i is an increasing function of i.

CLAIM 8.4. For any $i \leq \log \frac{\epsilon n}{8}$,

$$\sqrt{\frac{10 \ln \frac{\epsilon n}{2^{i+1}}}{\frac{\epsilon n}{2^{i+1}}}} \geq 1.1 \cdot \sqrt{\frac{10 \ln \frac{\epsilon n}{2^i}}{\frac{\epsilon n}{2^i}}}.$$

PROOF. Setting $2^l = \frac{\epsilon n}{2^i}$,

$$\frac{\ln 2^{l-1}}{2^{l-1}} \geq (1.1)^2 \cdot \frac{\ln 2^l}{2^l} \iff 2(l-1) \geq (1.1)^2 \cdot l \iff l \geq \frac{2}{2 - 1.1^2} = 2.53\ldots$$

Thus the claim holds for $l \geq 3$—or equivalently, for $i \leq \log \frac{\epsilon n}{8}$. □

> We want to continue the iterative procedure as long as the probability of each element being picked is at most 1. Towards this, set
>
> (8.5) $$t = \left\lfloor \log \left(\frac{\epsilon n}{400} \right) \right\rfloor.$$
>
> Then as p_i is increasing with i (Claim 8.4), we have
>
> (8.6) $$\sqrt{\frac{10 \ln \frac{\epsilon n}{2^i}}{\frac{\epsilon n}{2^i}}} \leq \sqrt{\frac{10 \ln \frac{\epsilon n}{2^t}}{\frac{\epsilon n}{2^t}}} \leq \sqrt{\frac{10 \ln 400}{400}} \leq \frac{1}{2},$$
>
> and thus $p_i \leq 1$. Also $k_i \geq k_t = \frac{\epsilon n}{2^t} \geq 400$.

It remains to prove the following two statements.

1. **$N = X_t \cup Q_1 \cup \cdots \cup Q_t$ is an ϵ-net:** Each $S \in \mathcal{F}$ of size at least ϵn is either hit by some Q_i for $i \leq t$ or it contains at least $\frac{\epsilon n}{2^t} \geq 400$ elements of X_t.

2. **Each $p \in X$ is present in N with probability $O\left(\frac{1}{\epsilon n}\right)$:** The key is the following.

   CLAIM 8.7. For any $i \leq t$, $\Pr\left[p \in X_i\right] = O\left(\frac{1}{2^i}\right)$.

   PROOF.

$$\Pr\left[p \in X_i\right] \leq \prod_{j=0}^{i-1} \left(\frac{1}{2} + \sqrt{\frac{10 \ln \frac{\epsilon n}{2^j}}{\frac{\epsilon n}{2^j}}} \right) = \frac{1}{2^i} \prod_{j=0}^{i-1} \left(1 + 2\sqrt{\frac{10 \ln \frac{\epsilon n}{2^j}}{\frac{\epsilon n}{2^j}}} \right)$$

$$\leq \frac{1}{2^i} \prod_{j=0}^{i-1} \exp\left(2\sqrt{\frac{10 \ln \frac{\epsilon n}{2^j}}{\frac{\epsilon n}{2^j}}} \right) = \frac{1}{2^i} \exp\left(2\sqrt{10} \sum_{j=0}^{i-1} \sqrt{\frac{\ln \frac{\epsilon n}{2^j}}{\frac{\epsilon n}{2^j}}} \right).$$

   The summation in the above expression is upper bounded by a geometric series due to Claim 8.4 and thus

$$\Pr\left[p \in X_i\right] \leq \frac{1}{2^i} \exp\left(O\left(\sqrt{\frac{\ln \frac{\epsilon n}{2^i}}{\frac{\epsilon n}{2^i}}} \right) \right) = O\left(\frac{1}{2^i} \right),$$

   where the last step follows from Equation (8.6). □

Recalling that $\Pr[p \in Q_j] = O\left(\frac{1}{(k_{j-1})^2}\right) = O\left(\frac{1}{\left(\frac{\epsilon n}{2^{j-1}}\right)^2}\right)$ by Lemma 8.3,

$$\begin{aligned}
\Pr\left[p \in Q_1 \cup \cdots \cup Q_t\right] &= \sum_{j=1}^{t} \Pr[p \in X_{j-1}] \cdot \Pr[p \in Q_j] \\
&= \sum_{j=1}^{t} O\left(\frac{1}{2^{j-1}}\right) \cdot O\left(\frac{1}{\left(\frac{\epsilon n}{2^{j-1}}\right)^2}\right) \\
&= O\left(\frac{1}{(\epsilon n)^2}\right) \sum_{j=1}^{t} 2^j = O\left(\frac{1}{(\epsilon n)^2}\right) \cdot O(2^t) = O\left(\frac{1}{\epsilon n}\right).
\end{aligned}$$

Finally, by Claim 8.7,

$$\Pr\left[p \in X_t\right] = O\left(\frac{1}{2^t}\right) = O\left(\frac{1}{\epsilon n}\right).$$

This concludes the proof of Theorem 8.2. □

It remains to prove the key lemma.

LEMMA 8.3. *Let $(X, \mathcal{F})$ be a set system with shallow-cell complexity $\varphi_{\mathcal{F}}(m, k) = O(k^2)$. Let $k \geq 400$ be a positive integer such that $|S| \geq k$ for all $S \in \mathcal{F}$. Construct a set Y by choosing each element of X independently with probability*

$$\frac{1}{2} + \sqrt{\frac{10 \ln k}{k}}.$$

Then one can construct a set $Q \subseteq X \setminus Y$ such that

1. *for each $S \in \mathcal{F}$, either $|S \cap Y| \geq \frac{k}{2}$ or $S \cap Q \neq \emptyset$, and*
2. *each $p \in X$ belongs to Q with probability $O\left(\frac{1}{k^2}\right)$, where the probability is over the choice of Y. The constant in this probability depends on the multiplicative constant in $\varphi_{\mathcal{F}}(m, k)$.*

PROOF. A technical detail in calculating probabilities is that the sets of $\mathcal{F}$ can have different sizes. However it is not hard to see that the 'worst case' is when each set of $\mathcal{F}$ has size $\Theta(k)$, since the probability of choosing less than $\frac{k}{2}$ elements from $S \in \mathcal{F}$ decreases exponentially with $|S|$. Thus we restrict ourselves to $\mathcal{F}_{\leq 2k} \subseteq \mathcal{F}$ and show that any fixed element of X belongs to Q with probability $O\left(\frac{1}{k^2}\right)$.

The same proof shows that any fixed element of X is added to Q because of a set in $\mathcal{F}_{\leq 2^i k} \setminus \mathcal{F}_{\leq 2^{i-1} k}$ with probability $O\left(\frac{1}{(2^i k)^2}\right)$. The general bound then follows by summing up over all set sizes to get

$$\Pr\left[p \in Q\right] = \sum_{i=1}^{\lceil \log \frac{n}{k} \rceil} O\left(\frac{1}{(2^i k)^2}\right) = O\left(\frac{1}{k^2}\right).$$

LEMMA 8.8. *There is a way to assign each $S \in \mathcal{F}_{\leq 2k}$ to an element $p_S \in S$ such that each $p \in X$ is assigned $O(k^3)$ sets of $\mathcal{F}_{\leq 2k}$.*

PROOF. By assumption on the shallow-cell complexity of $\mathcal{F}$, $|\mathcal{F}_{\leq 2k}| = O\left(nk^2\right)$ and thus

$$\sum_{S \in \mathcal{F}_{\leq 2k}} |S| = O\left(nk^3\right).$$

By the pigeonhole principle there exists an element $p \in X$ belonging to $O\left(k^3\right)$ sets of $\mathcal{F}_{\leq 2k}$—all these sets are assigned to p. Remove p from X, remove the sets containing p from $\mathcal{F}_{\leq 2k}$ and reiterate for the remaining sets.
After the j-th step, there are $O\left((n-j)\ k^2\right)$ remaining sets of size at most $2k$ over the remaining $(n-j)$ elements of X. One of these $n-j$ elements will be contained in $O\left(k^3\right)$ remaining sets, which are then assigned to it. One can continue this process until all sets have been assigned to some element of X. □

The next statement gives an upper bound on the probability that a set $S \in \mathcal{F}_{\leq 2k}$ causes the element $p_S \in X$ to be added to Q. For notational ease, set

$$f(k) = \sqrt{\frac{10 \ln k}{k}}. \tag{8.9}$$

CLAIM 8.10. For any $S \in \mathcal{F}_{\leq 2k}$,

$$\Pr\left[|S \cap Y| < \frac{k}{2}\right] \leq \exp\left(-\frac{kf(k)^2}{2}\right).$$

PROOF. As $|S| \geq k$,

$$\mathrm{E}\left[|S \cap Y|\right] = |S| \cdot \left(\frac{1}{2} + f\left(k\right)\right) \geq \frac{k}{2} + kf(k),$$

Chernoff's bound (Corollary 1.23) implies that

$$\Pr\left[|S \cap Y| < \frac{k}{2}\right] = \Pr\left[|S \cap Y| < \left(1 - \frac{2f(k)}{1+2f(k)}\right) \cdot \left(\frac{k}{2} + kf(k)\right)\right]$$
$$\leq \exp\left(-\frac{1}{2}\left(\frac{2f(k)}{1+2f(k)}\right)^2 \cdot \left(\frac{k}{2} + kf(k)\right)\right) = \exp\left(-\frac{kf(k)^2}{1+2f(k)}\right) \leq e^{-\frac{kf(k)^2}{2}},$$

where the last step follows from the fact that $f(k) \leq \frac{1}{2}$ by Equation (8.6). □

Claim 8.10 is the reason for setting the value of $f\left(k\right)$ as we did. Lemma 8.8 states that any element in X can be added to Q due to one of $O\left(k^3\right)$ sets of $\mathcal{F}_{\leq 2k}$. Thus to be able to upper bound, using the union bound, the probability that any fixed element of X is added to Q due to a set of $\mathcal{F}_{\leq 2k}$, we must have

$$e^{-\frac{kf(k)^2}{2}} \cdot \Theta\left(k^3\right) \leq \frac{1}{k^2},$$

which is what dictates the value of $f\left(k\right)$ in Equation (8.9).

Now if Y has less than $\frac{k}{2}$ elements from S, the algorithm adds the designated element $p_S \in S$ to Q. Thus by Claim 8.10 and Lemma 8.8, for any $p \in X$,

$$\Pr\left[p \in Q\right] \leq \sum_{\substack{S \in \mathcal{F}_{\leq 2k} \\ p_S = p}} \Pr\left[|S \cap Y| < \frac{k}{2}\right] \leq \sum_{\substack{S \in \mathcal{F}_{\leq 2k} \\ p_S = p}} e^{-\frac{k}{2}\frac{10\ln k}{k}} = \sum_{\substack{S \in \mathcal{F}_{\leq 2k} \\ p_S = p}} e^{-5\ln k}$$
$$= O\left(k^3\right) \cdot \frac{1}{k^5} = O\left(\frac{1}{k^2}\right).$$

This finishes the proof of the lemma. □

Bibliography and discussion. A slightly non-optimal bound containing the key insight was discovered by Varadarajan [**Var10**], who applied it for the specific case of dual set systems induced by geometric objects in $\mathbb{R}^2$. His charging scheme was implicitly encoded in a permutation of the elements of X. The bound was later generalized and improved to the optimal bound by Chan et al. [**Cha+12**], whose proof and charging scheme we have essentially followed here.

[Cha+12] T. M. Chan, E. Grant, J. Könemann, and M. Sharpe, *Weighted capacitated, priority, and geometric set cover via improved quasi-uniform sampling*, Proceedings of the Twenty-Third Annual ACM-SIAM Symposium on Discrete Algorithms, ACM, New York, 2012, pp. 1576–1585. MR3205315

[Var10] K. Varadarajan, *Weighted geometric set cover via quasi-uniform sampling*, STOC'10—Proceedings of the 2010 ACM International Symposium on Theory of Computing, ACM, New York, 2010, pp. 641–647. MR2743313

2. General Case

Fejér's papers are particularly well written, they are very easy to read. This is due to his style of work: When he found an idea, he tended it with loving care; he tried to perfect it, simplify it, free it from unessentials; he worked on it carefully and minutely until the idea became transparently clear. He eventually produced a work of art, not of too large dimensions, but highly finished ... He attracted many people to mathematics by the success of his own work and by his personal charm. He sat in a coffee house with young people who could not help loving him and trying to imitate him as he wrote formulas on the menus and alternately spoke about mathematics and told stories about mathematicians. In fact, almost all Hungarian mathematicians who were his contemporaries or somewhat younger were personally influenced by him, and several started their mathematical career by working on his problems.

George Pólya

The main theorem of this section proves the general form of Theorem 8.2. Recall the definition of a well-behaved function.

DEFINITION 4.5. A function $\varphi(\cdot,\cdot)$ is (a,b)-well-behaved, for $a \in (1,2)$ and $b \in \mathbb{R}^+$, if it is non-decreasing in both arguments and for all positive integers $m \geq k \geq b \geq 2$,

$$\varphi(m,k) \leq \left(\varphi\left(\frac{m}{2},\frac{k}{2}\right)\right)^a.$$

THEOREM 8.11. *Let $(X,\mathcal{F})$ be a set system on n elements and $a \in (1,2)$, $b \in \mathbb{R}^+$ be parameters such that the shallow-cell complexity of $\mathcal{F}$, $\varphi_{\mathcal{F}}(\cdot,\cdot)$, is (a,b)-well-behaved. Let $\epsilon > 0$ be a given parameter and set $c_1 = \exp\left(\frac{\sqrt{2}}{\sqrt{2}-\sqrt{a}}\right)$ and $c_2 = b + 2^{\frac{a}{a-1}}$. Note that $c_1, c_2 \geq 1$. Then there exists a procedure to compute an ϵ-net N of $\mathcal{F}$ such that each $p \in X$ is present in N with probability*

$$O\left(\frac{c_1 c_2}{\epsilon n} + \frac{c_1}{\epsilon n}\ln \varphi_{\mathcal{F}}\left(\frac{c_1 n}{2^t},\frac{\epsilon n}{2^t}\right)\right),$$

where t is the largest integer in $\left[0, \log\frac{\epsilon n}{c_2}\right]$ such that for all $i < t$,

$$\sqrt{\frac{8\ln\left(\frac{\epsilon n}{2^i}\cdot\varphi_{\mathcal{F}}\left(\frac{c_1 n}{2^i},\frac{\epsilon n}{2^i}\right)\right)}{\frac{\epsilon n}{2^i}}} \leq \frac{1}{2}. \tag{8.12}$$

Furthermore, let $w: X \to \mathbb{R}$ be weights on the elements of X, with $W = \sum_{p\in X} w(p)$. Then there exists an ϵ-net of $\mathcal{F}$ of total weight

$$O\left(W\left(\frac{c_1 c_2}{\epsilon n} + \frac{c_1}{\epsilon n}\ln \varphi_{\mathcal{F}}\left(\frac{c_1 n}{2^t},\frac{\epsilon n}{2^t}\right)\right)\right).$$

The somewhat odd condition given by Equation (8.12) is due to the fact that we do not assume anything about the function $\varphi_{\mathcal{F}}(\cdot,\cdot)$, other than that it is well-behaved. For the case of shallow-cell complexity functions of natural geometric set systems, typically $t = \log(\Omega(\epsilon n))$ and so Theorem 8.11 gives a bound matching the one of Theorem 6.2.

The proof is similar to that of Theorem 8.2, the key lemma being the following generalization of Lemma 8.3.

LEMMA 8.13. *Let $(X, \mathcal{F})$ be a set system and k a positive integer such that $|S| \geq k$ for all $S \in \mathcal{F}$, and where $\sqrt{\frac{8\ln(k\cdot\varphi_{\mathcal{F}}(|X|,k))}{k}} \leq \frac{1}{2}$. Let Y be a uniform random sample of X of size*

$$|X| \cdot \left(\frac{1}{2} - \sqrt{\frac{8\ln\left(k \cdot \varphi_{\mathcal{F}}\left(|X|, k\right)\right)}{k}} \right).$$

Then one can construct a set $Q \subseteq X \setminus Y$ such that

(1) *for each $S \in \mathcal{F}$, either $|S \cap Y| \geq \frac{k}{2}$ or $S \cap Q \neq \emptyset$, and*
(2) *each $p \in X$ belongs to Q with probability $O\left(\frac{1}{k^2}\right)$, where the probability is over the choice of Y.*

Compared to the proof in the previous section, the calculations using Lemma 8.13 become technically more involved due to the fact that the probability of picking an element into Y involves the function $\varphi_{\mathcal{F}}(\cdot,\cdot)$, which additionally has $|X|$ as an argument.

To simplify the technical details, Lemma 8.13 is stated with a slightly different probability distribution from that of Lemma 8.3, with Y being a random sample of a *fixed* size. This simplifies later calculations, though the entire proof can also be made to work where each element of X is added to Y independently with probability $\frac{1}{2} + \sqrt{\frac{8\ln(k\cdot\varphi_{\mathcal{F}}(|X|,k))}{k}}$.

The algorithm is the same as earlier, except we will use Lemma 8.13.

Weighted Net-Finder Algorithm $\left((X, \mathcal{R}), \epsilon > 0\right)$.

$X_0 = X$, $Q_0 = \emptyset$ and $\mathcal{F}_0 = \mathcal{F}$.
$i = 0$.

while $i < t$ **do**
  apply Lemma 8.13 to $(X_i, \mathcal{F}_i)$ with $k_i = \frac{\epsilon n}{2^i}$ to get the sets Y and Q.
  set $X_{i+1} = Y$ and $Q_{i+1} = Q$.
  let $\mathcal{F}_{i+1}$ be the set system obtained from $\mathcal{F}_i$ by first removing all sets of $\mathcal{F}_i$ hit by Q and then projecting $\mathcal{F}_i$ onto Y. That is,

$$\mathcal{F}_{i+1} = \left\{ S \cap Y \;:\; S \in \mathcal{F}_i \text{ and } S \cap Q = \emptyset \right\}.$$

  $i = i + 1$.
return $N = X_t \cup Q_1 \cup \cdots \cup Q_t$.

Number of iterations t. Let t be the largest integer in $\left[0, \log \frac{\epsilon n}{c_2}\right]$ such that for all $i < t$,

$$\sqrt{\frac{8\ln\left(\frac{\epsilon n}{2^i} \cdot \varphi_{\mathcal{F}}\left(\frac{c_1 n}{2^i}, \frac{\epsilon n}{2^i}\right)\right)}{\frac{\epsilon n}{2^i}}} \leq \frac{1}{2}. \tag{8.14}$$

Note that if $t = 0$—that is, Equation (8.14) already does not hold for $i = 0$, we have

$$\frac{8\ln\left(\epsilon n \cdot \varphi_{\mathcal{F}}\left(c_1 n, \epsilon n\right)\right)}{\epsilon n} > \frac{1}{4},$$

we can simply return P as the ϵ-net N. Then for each $p \in X$,

$$\Pr\left[p \in N\right] = 1 \leq \frac{32\ln\left(\epsilon n \cdot \varphi_{\mathcal{F}}\left(c_1 n, \epsilon n\right)\right)}{\epsilon n} = O\left(\frac{1}{\epsilon n}\ln \varphi_{\mathcal{F}}\left(c_1 n, \epsilon n\right)\right),$$

as required.

We will also assume that $\frac{\epsilon n}{c_2} \geq 8$—otherwise return P as the ϵ-net N, with each element being picked into N with probability $1 < \frac{8c_2}{\epsilon n} = O\left(\frac{c_2}{\epsilon n}\right)$.

As either $t = \log \frac{\epsilon n}{c_2}$ or Equation (8.14) is violated for some $i < \log \frac{\epsilon n}{c_2}$, we get the following.

$$\text{(8.15)} \quad \frac{1}{2^t} \leq \max\left\{\frac{c_2}{\epsilon n}, \frac{32\ln\left(\frac{\epsilon n}{2^t}\cdot \varphi_{\mathcal{F}}\left(\frac{c_1 n}{2^t}, \frac{\epsilon n}{2^t}\right)\right)}{\epsilon n}\right\} = O\left(\frac{c_2}{\epsilon n} + \frac{\ln \varphi_{\mathcal{F}}\left(\frac{c_1 n}{2^t}, \frac{\epsilon n}{2^t}\right)}{\epsilon n}\right).$$

We will need the following claim.

CLAIM 8.16. For each $i < t$,

$$\sqrt{\frac{8\ln\left(\frac{\epsilon n}{2^{i+1}}\varphi_{\mathcal{F}}\left(\frac{c_1 n}{2^{i+1}}, \frac{\epsilon n}{2^{i+1}}\right)\right)}{\frac{\epsilon n}{2^{i+1}}}} \geq \sqrt{\frac{2}{a}} \cdot \sqrt{\frac{8\ln\left(\frac{\epsilon n}{2^i}\varphi_{\mathcal{F}}\left(\frac{c_1 n}{2^i}, \frac{\epsilon n}{2^i}\right)\right)}{\frac{\epsilon n}{2^i}}}.$$

PROOF. By squaring, removing the logarithms and re-arranging the terms, this is equivalent to

$$\frac{\epsilon n}{2^{i+1}}\varphi_{\mathcal{F}}\left(\frac{c_1 n}{2^{i+1}}, \frac{\epsilon n}{2^{i+1}}\right) \geq \left(\frac{\epsilon n}{2^i}\right)^{\frac{1}{a}}\left(\varphi_{\mathcal{F}}\left(\frac{c_1 n}{2^i}, \frac{\epsilon n}{2^i}\right)\right)^{\frac{1}{a}},$$

which follows from the next two inequalities.

1. $$\frac{\left(\frac{\epsilon n}{2^i}\right)^{\frac{1}{a}}}{\frac{\epsilon n}{2^{i+1}}} = \frac{2\cdot 2^{\left(1-\frac{1}{a}\right)i}}{\left(\epsilon n\right)^{1-\frac{1}{a}}} \leq \frac{2\cdot\left(\frac{\epsilon n}{c_2}\right)^{1-\frac{1}{a}}}{\left(\epsilon n\right)^{1-\frac{1}{a}}} = \frac{2}{c_2^{\frac{a-1}{a}}} \leq 1,$$

since $2^i \leq 2^t \leq \frac{\epsilon n}{c_2}$ and $c_2 \geq 2^{\frac{a}{a-1}}$.

2. $$\left(\varphi_{\mathcal{F}}\left(\frac{c_1 n}{2^{i+1}}, \frac{\epsilon n}{2^{i+1}}\right)\right)^a \geq \left(\varphi_{\mathcal{F}}\left(\frac{c_1 n}{2^i}, \frac{\epsilon n}{2^i}\right)\right),$$

by the definition of (a, b)-well-behaved functions and the fact that $\frac{\epsilon n}{2^i} \geq \frac{\epsilon n}{2^{\log\frac{\epsilon n}{c_2}}} \geq c_2 \geq b$. □

PROOF OF THEOREM 8.11. It is easy to see that $N = X_t \cup Q_1 \cup \cdots \cup Q_t$ is an ϵ-net: for each $S \in \mathcal{F}$ of size at least ϵn, either S was hit by Q_i for some $i \leq t$ or it contains at least $\frac{\epsilon n}{2^t} \geq c_2 \geq 1$ elements of X_t.

It remains to upper bound the probability of any fixed element of X belonging to N.

CLAIM 8.17. For each $i \leq t$, $|X_i| \leq c_1 \cdot \frac{n}{2^i}$. Furthermore, $\Pr\left[p \in X_i\right] \leq c_1 \cdot \frac{1}{2^i}$ for each $p \in X_i$.

PROOF. The proof is by induction on i. Clearly it is true for $i = 0$. So assume that $|X_j| \le c_1 \cdot \frac{n}{2^j}$ for all $j < i$. Applying Lemma 8.13, we get

$$\begin{aligned}
|X_i| &= |X_{i-1}| \cdot \left(\frac{1}{2} + \sqrt{\frac{8 \ln \left(k_{i-1} \cdot \varphi_{\mathcal{F}}\left(\frac{c_1 n}{2^{i-1}}, k_{i-1} \right) \right)}{k_{i-1}}} \right) \\
&= \cdots = n \cdot \prod_{j=0}^{i-1} \left(\frac{1}{2} + \left(\sqrt{\frac{8 \ln \left(k_j \cdot \varphi_{\mathcal{F}}(\frac{c_1 n}{2^j}, k_j) \right)}{k_j}} \right) \right) \\
&= \frac{n}{2^i} \prod_{j=0}^{i-1} \left(1 + 2 \left(\sqrt{\frac{8 \ln \left(k_j \cdot \varphi_{\mathcal{F}}(\frac{c_1 n}{2^j}, k_j) \right)}{k_j}} \right) \right) \\
&\le \frac{n}{2^i} \cdot \exp \left(2 \sum_{j=0}^{i-1} \left(\sqrt{\frac{8 \ln \left(k_j \cdot \varphi_{\mathcal{F}}(\frac{c_1 n}{2^j}, k_j) \right)}{k_j}} \right) \right) \qquad \left(\text{using } 1 + x \le e^x \right) \\
&\le \frac{n}{2^i} \cdot \exp \left(2 \sqrt{\frac{8 \ln \left(k_{i-1} \cdot \varphi_{\mathcal{F}}(\frac{c_1 n}{2^{i-1}}, k_{i-1}) \right)}{k_{i-1}}} \cdot \sum_{j=0}^{i-1} \frac{1}{\sqrt{\frac{2}{a}}^j} \right) \qquad \left(\text{Claim 8.16} \right) \\
&\le \frac{n}{2^i} \cdot \exp \left(\sum_{j=0}^{i-1} \left(\sqrt{\frac{a}{2}} \right)^j \right) \qquad \left(\text{Equation (8.14)} \right) \\
&\le \frac{n}{2^i} \cdot \exp \left(\frac{1}{1 - \frac{\sqrt{a}}{\sqrt{2}}} \right) = c_1 \cdot \frac{n}{2^i}.
\end{aligned}$$

Similarly,

$$\Pr\left[p \in X_i\right] \le \prod_{j=0}^{i-1} \left(\frac{1}{2} + \left(\sqrt{\frac{8 \ln \left(k_j \cdot \varphi_{\mathcal{F}}(\frac{c_1 n}{2^j}, k_j) \right)}{k_j}} \right) \right) \le c_1 \cdot \frac{1}{2^i}.$$

□

Thus,

$$\begin{aligned}
\Pr\left[p \in Q_1 \cup \cdots \cup Q_t\right] &= \sum_{j=1}^{t} \Pr\left[p \in X_{j-1}\right] \cdot \Pr\left[p \in Q_j\right] \\
&= \sum_{j=1}^{t} O\left(\frac{c_1}{2^{j-1}} \right) \cdot O\left(\frac{1}{\left(\frac{\epsilon n}{2^j} \right)^2} \right) \\
&= O\left(\frac{c_1}{(\epsilon n)^2} \right) \sum_{j=1}^{t} 2^j = O\left(\frac{c_1}{(\epsilon n)^2} \right) \cdot O(2^t) \\
&= O\left(\frac{c_1}{(\epsilon n)^2} \cdot \frac{\epsilon n}{c_2} \right) = O\left(\frac{c_1}{\epsilon n} \right).
\end{aligned}$$

Finally, using Equation (8.15),

$$\Pr\left[p \in X_t\right] = O\left(\frac{c_1}{2^t} \right) = O\left(\frac{c_1 \, c_2}{\epsilon n} + \frac{c_1 \ln \varphi_{\mathcal{F}}\left(\frac{c_1 n}{2^t}, \frac{\epsilon n}{2^t} \right)}{\epsilon n} \right).$$

This finishes the proof of the first part of Theorem 8.11. The second part follows from the first and linearity of expectation. □

LEMMA 8.13. *Let $(X, \mathcal{F})$ be a set system and k a positive integer such that $|S| \geq k$ for all $S \in \mathcal{F}$, and where $\sqrt{\frac{8\ln(k\cdot\varphi_{\mathcal{F}}(|X|,k))}{k}} \leq \frac{1}{2}$. Let Y be a uniform random sample of X of size*

$$|X| \cdot \left(\frac{1}{2} + \sqrt{\frac{8\ln\left(k \cdot \varphi_{\mathcal{F}}\left(|X|, k\right)\right)}{k}}\right).$$

Then one can construct a set $Q \subseteq X \setminus Y$ such that

(1) *for each $S \in \mathcal{F}$, either $|S \cap Y| \geq \frac{k}{2}$ or $S \cap Q \neq \emptyset$, and*
(2) *each $p \in X$ belongs to Q with probability $O\left(\frac{1}{k^2}\right)$, where the probability is over the choice of Y.*

PROOF. Let $\mathcal{F} = \{S_1, \ldots, S_m\}$, $n = |X|$ and for $i \geq 0$ define

$$\mathcal{F}_i = \left\{S \in \mathcal{F} \,:\, 2^i\, k \leq |S| < 2^{i+1}\, k\right\}.$$

LEMMA 8.18. *There is a way to assign each $S \in \mathcal{F}_i$ to an element $p_S \in S$ such that each $p \in X$ is assigned at most*

$$\varphi_{\mathcal{F}}\left(n, 2^{i+1}k\right) \cdot 2^{i+1}k$$

sets of $\mathcal{F}_i$.

PROOF. By the shallow-cell complexity of $\mathcal{F}$ and the assumption that $\varphi(\cdot, \cdot)$ is non-decreasing in both arguments, $|\mathcal{F}_i| \leq n \cdot \varphi_{\mathcal{F}}(n, 2^{i+1}k)$. This implies that

$$\sum_{S \in \mathcal{F}_i} |S| \leq |\mathcal{F}_i| \cdot 2^{i+1}k \leq n \cdot \varphi_{\mathcal{F}}(n, 2^{i+1}k) \cdot 2^{i+1}k.$$

By the pigeonhole principle there exists an element $p \in X$ belonging to at most $\varphi_{\mathcal{F}}(n, 2^{i+1}k) \cdot 2^{i+1}k$ sets of $\mathcal{F}_i$; all these sets are assigned to p. Remove p from X, remove the sets containing p from $\mathcal{F}_i$ and reiterate for the remaining sets.
At the j-th step there are at most

$$(n-j) \cdot \varphi_{\mathcal{F}}(n-j, 2^{i+1}k) \leq (n-j) \cdot \varphi_{\mathcal{F}}\left(n, 2^{i+1}k\right)$$

remaining sets of $\mathcal{F}_i$ over the remaining $(n-j)$ elements of X. Again one of these $n-j$ elements will be contained in at most $\varphi_{\mathcal{F}}\left(n, 2^{i+1}k\right) \cdot 2^{i+1}k$ remaining sets—these are then assigned to it.
One can continue this process until all sets of $\mathcal{F}_i$ have been assigned to some element of X. □

Our construction of Q is simple: if $S \in \mathcal{F}$ violates the first property—that is, $|S \cap Y| < \frac{k}{2}$—then we add $p_S \in S$ to Q. Next, we upper bound the probability that an element $p_S \in X$ is forced to be picked in Q due to a set $S \in \mathcal{F}$. Set

$$f(k) = \sqrt{\frac{8\ln\left(k \cdot \varphi_{\mathcal{F}}\left(n, k\right)\right)}{k}}.$$

CLAIM 8.19. For any $S \in \mathcal{F}$,

$$\Pr\left[|S \cap Y| < \frac{k}{2}\right] \leq \exp\left(-|S| \cdot f(k)^2\right).$$

PROOF. Chernoff's bound (Corollary 1.24) implies that

$$\begin{aligned}
\Pr\left[|S \cap Y| < \frac{k}{2}\right] &= \Pr\left[|S \cap Y| < \left(\frac{k}{|S|(1+2f(k))}\right) \cdot |S|\left(\frac{1}{2} + f(k)\right)\right] \\
&= \Pr\left[|S \cap Y| < \left(1 - \frac{|S|(1+2f(k)) - k}{|S|(1+2f(k))}\right) \cdot |S|\left(\frac{1}{2} + f(k)\right)\right] \\
&\leq \exp\left(-\frac{1}{2}\left(\frac{|S|(1+2f(k)) - k}{|S|(1+2f(k))}\right)^2 \cdot |S|\left(\frac{1}{2} + f(k)\right)\right) \\
&= \exp\left(-\frac{(|S| + |S|2f(k) - k)^2}{4|S|(1+2f(k))}\right) \\
&\leq \exp\left(-\frac{(|S| + |S|2f(k) - k)^2}{8|S|}\right) \\
&\leq \exp\left(-\frac{(|S|2f(k))^2}{8|S|}\right) = \exp\left(-\frac{1}{2}|S| \cdot f(k)^2\right).
\end{aligned}$$

The third-to-last step used the fact that $f(k) \leq \frac{1}{2}$ and the second-to-last step that $|S| \geq k$. □

Now one can calculate the probability that any $p \in X$ is added to Q:

$$\begin{aligned}
\Pr[p \in Q] &\leq \sum_{i=0}^{\lfloor \log \frac{n}{k} \rfloor} \sum_{\substack{S \in \mathcal{F}_i \\ p_S = p}} \Pr\left[|S \cap Y| < \frac{k}{2}\right] \leq \sum_{i=0}^{\lfloor \log \frac{n}{k} \rfloor} \sum_{\substack{S \in \mathcal{F}_i \\ p_S = p}} \exp\left(-\frac{1}{2}|S| \cdot f(k)^2\right) \\
&\leq \sum_{i=0}^{\lfloor \log \frac{n}{k} \rfloor} \sum_{\substack{S \in \mathcal{F}_i \\ p_S = p}} \exp\left(-2^{i-1} k \cdot f(k)^2\right) \\
&\leq \sum_{i=0}^{\lfloor \log \frac{n}{k} \rfloor} \varphi_{\mathcal{F}}\left(n, 2^{i+1}k\right) 2^{i+1}k \cdot \exp\left(-2^{i-1}k \cdot \frac{8\ln\left(k \cdot \varphi_{\mathcal{F}}(n,k)\right)}{k}\right) \\
&\leq \sum_{i=0}^{\lfloor \log \frac{n}{k} \rfloor} \frac{2^{i+1}k}{k^{2i+2}} \frac{\varphi_{\mathcal{F}}\left(n, 2^{i+1}k\right)}{\left(\varphi_{\mathcal{F}}(n,k)\right)^{2i+2}} \leq \sum_{i=0}^{\lfloor \log \frac{n}{k} \rfloor} \frac{2^{i+1}k}{k^{2i+2}} \frac{\left(\varphi_{\mathcal{F}}(n,k)\right)^{2(i+1)}}{\left(\varphi_{\mathcal{F}}(n,k)\right)^{2i+2}} \\
&\leq \sum_{i=0}^{\infty} \frac{2^{i+1}k}{k^{2i+2}} = O\left(\frac{1}{k^2}\right),
\end{aligned}$$

where the third-to-last step uses the fact that $\varphi_{\mathcal{F}}$ is well-behaved and that $k \geq \frac{\epsilon n}{2^i} \geq b$.

This completes the proof of the lemma. □

Bibliography and discussion. The proof follows the ones in [**Var10, Cha+12**], though we stated and proved it in a slightly different setting.

[Cha+12] T. M. Chan, E. Grant, J. Könemann, and M. Sharpe, *Weighted capacitated, priority, and geometric set cover via improved quasi-uniform sampling*, Proceedings of the Twenty-Third Annual ACM-SIAM Symposium on Discrete Algorithms, ACM, New York, 2012, pp. 1576–1585. MR3205315

[Var10] K. Varadarajan, *Weighted geometric set cover via quasi-uniform sampling*, STOC'10—Proceedings of the 2010 ACM International Symposium on Theory of Computing, ACM, New York, 2010, pp. 641–647. MR2743313

3. Application: Rounding Linear Programs

The aim of theory really is, to a great extent, that of systematically organising past experience in such a way that the next generation, our students and their students and so on, will be able to absorb the essential aspects in as painless a way as possible, and this is the only way in which you can go on cumulatively building up any kind of scientific activity without eventually coming to a dead end.

Michael Atiyah

Consider the weighted minimum hitting set problem for disks.

> Let P be a set of n points in $\mathbb{R}^2$, $w : P \to \mathbb{R}^+$ a weight function on P and $\mathcal{R}$ a collection of m subsets of P induced by disks in the plane. Then the minimum-weight hitting set problem asks for a hitting set $Q \subseteq P$ of $\mathcal{R}$ of minimum weight.

The main theorem of this section is the following.

THEOREM 8.20. *Let P be a set of n points in $\mathbb{R}^2$, $w : P \to \mathbb{R}^+$ a weight function and $\mathcal{R}$ a collection of m subsets of P induced by disks in the plane. Then there is a polynomial-time algorithm to compute a $O(1)$-approximation to the weighted minimum hitting set problem on P and $\mathcal{R}$.*

PROOF. The linear programming relaxation (LP) for this problem assigns a variable x_p to each $p \in P$, with the objective function being to minimize the weighted sum of the x_p's. See the LP below.

$$\text{Minimize } \sum_{p \in P} w(p) \cdot x_p$$

subject to

(C1) for each $R \in \mathcal{R}$: $\sum_{p \in R} x_p \geq 1$,

(C2) for each $p \in P$: $0 \leq x_p \leq 1$.

This LP can be solved in polynomial time using standard algorithms (see discussion). Furthermore, the x_p variables are rational and so there exists a sufficiently large integer Δ such that each scaled weight $\Delta \cdot x_p$ is an integer.
Construct a new set system $(P', \mathcal{R}')$ as follows. For each $p \in P$, let X_p be a set of $\Delta \cdot x_p$ new points, each with the same coordinates and weight as p. Then set

$$P' = \bigcup_{p \in P} X_p,$$

$$\mathcal{R}' = \left\{ \bigcup_{p \in R} X_p \ : \ R \in \mathcal{R} \right\}.$$

Note that a hitting set of $\mathcal{R}'$ gives a hitting set of the same weight of $\mathcal{R}$. Furthermore,

$$|P'| = \sum_{p \in P} \Delta \cdot x_p, \quad \text{and} \quad \forall\, R' \in \mathcal{R}', \quad |R'| = \sum_{p \in R} \Delta \cdot x_p \geq \Delta,$$

where the last inequality follows from the LP constraint (C1). As $|R'| \geq \Delta$ for each $R' \in \mathcal{R}'$, to compute a hitting set of $\mathcal{R}'$, it suffices to compute an ϵ'-net of $\mathcal{R}'$, where

$$\epsilon' = \frac{\Delta}{|P'|} = \frac{\Delta}{\sum_{p \in P} \Delta \cdot x_p} = \frac{1}{\sum_{p \in P} x_p}.$$

Apply Theorem 8.2 to $(P', \mathcal{R}')$ to get an ϵ'-net N such that for all $p' \in P'$,

$$\Pr\left[p' \in N\right] = O\left(\frac{1}{\epsilon' |P'|}\right).$$

By the linearity of expectation, the expected total weight of N can be upper bounded as

$$\begin{aligned}\sum_{p' \in P'} \Pr\left[p' \in N\right] \cdot w\left(p'\right) &= O\left(\frac{1}{\epsilon' |P'|}\right) \sum_{p' \in P'} w\left(p'\right) \\ &= O\left(\frac{\sum_{p \in P} w\left(p\right) \cdot \Delta \cdot x_p}{\frac{1}{\sum_{p \in P} x_p} \cdot \sum_{p \in P} \Delta \cdot x_p}\right) = O\left(\sum_{p \in P} w\left(p\right) \cdot x_p\right).\end{aligned}$$

Recall that $\sum_{p \in P} w(p) \cdot x_p$ is the value of the solution returned by the LP. Therefore the set N with the above weight—after discarding multiple copies of the same point of P—is a hitting set that is, in expectation, a constant-factor approximation to the weighted minimum hitting set of $\mathcal{R}$.

This completes the proof. □

A similar statement follows for more general set systems as a function of their shallow-cell complexity, by using Theorem 8.11.

Bibliography and discussion. The proof of this section is from Varadarajan [**Var10**], and follows along the steps of the proof for the unweighted case given in [**ERS05, Lon01**]. For linear programming algorithms and related literature, we refer the reader to [**Dye+18, MG07**].

[Dye+18] M. Dyer, B. Gärtner, N. Megiddo and E. Welzl. *Linear programming. Handbook of Discrete and Computational Geometry.* CRC Press, 2018. pp. 1291–1310.

[ERS05] G. Even, D. Rawitz, and S. Shahar, *Hitting sets when the VC-dimension is small*, Inform. Process. Lett. **95** (2005), no. 2, 358–362, DOI 10.1016/j.ipl.2005.03.010. MR2149262

[Lon01] P. M. Long, *Using the pseudo-dimension to analyze approximation algorithms for integer programming*, Algorithms and data structures (Providence, RI, 2001), Lecture Notes in Comput. Sci., vol. 2125, Springer, Berlin, 2001, pp. 26–37, DOI 10.1007/3-540-44634-6_4. MR1936398

[MG07] J. Matoušek and B. Gärtner. *Understanding and Using Linear Programming.* Springer, 2007.

[Var10] K. Varadarajan, *Weighted geometric set cover via quasi-uniform sampling*, STOC'10—Proceedings of the 2010 ACM International Symposium on Theory of Computing, ACM, New York, 2010, pp. 641–647. MR2743313

CHAPTER 9

Epsilon-Nets: Convex Sets

Let $d \geq 2$ be an integer, and $\mathcal{C}$ the family of all convex sets in $\mathbb{R}^d$. Let $\mathcal{R}$ be the primal set system induced by $\mathcal{C}$ on a set P of n points in $\mathbb{R}^d$; that is,

$$\mathcal{R} = \{C \cap P \, : \, C \in \mathcal{C}\}.$$

Observe that if P is in convex position, then $\mathcal{R} = 2^P$. This implies that VC-dim $(\mathcal{R})$ is unbounded: for any positive integer n, a set of n points in $\mathbb{R}^d$ in convex position is shattered by $\mathcal{C}$ (see Definition 4.2). In particular, any ϵ-net for $\mathcal{R}$ must have size at least $n - \epsilon n$, that is, almost all points of P!

This leads naturally to the notion of a *weak ϵ-net*, where one is allowed to construct a net by picking points in $\mathbb{R}^d$ and not just from P.

DEFINITION 9.1. Let P be a set of n points in $\mathbb{R}^d$. For a parameter $\epsilon > 0$, a set $Q \subseteq \mathbb{R}^d$ is a weak ϵ-net of P with respect to convex sets if $C \cap Q \neq \emptyset$ for any convex set $C \subset \mathbb{R}^d$ containing at least ϵn points of P.

It turns out that it is possible to construct weak ϵ-nets with respect to convex sets of size *independent* of $|P|$ and depending only on ϵ and d. The first part of this chapter will present the proof for the current-best bound for any $\mathbb{R}^d$. Unlike the case of ϵ-nets of set systems with bounded VC-dimension, there is a large gap between the best-known upper and lower bounds for weak ϵ-nets.

The second main result of this chapter is a nice application of ϵ-nets to show that any convex set can be approximated by a convex polytope with few vertices.

1. Weak ϵ-nets

One almost wouldn't believe that after thousands of years of geometry, it is still possible to discover such pretty theorems [E. Welzl's theorem on matchings with low crossing number] about points in the plane.

Jiří Matoušek

The main theorem that we will prove is the following.

THEOREM 9.2. *Let P be a set of n points in $\mathbb{R}^d$, $d \geq 2$, and $\epsilon \in \left(0, \frac{1}{2}\right]$ a given parameter. Then there exists a weak ϵ-net N with respect to convex sets in $\mathbb{R}^d$ of P, with*

$$|N| = O\left(\frac{1}{\epsilon^d} \log^a \left(\frac{1}{\epsilon}\right)\right),$$

where $a = \Theta\left(d^3 \ln d\right)$ and the constant in the asymptotic notation depends on d.

To simplify the presentation, we will assume that P is in general position.

Overview of ideas. As a warm-up, consider the following proof demonstrating the existence of a weak ϵ-net N of size $O\left(\frac{1}{\epsilon^2}\right)$ for the two-dimensional case.

Let l be a vertical line with $\frac{n}{2}$ points of P on each side (assume n is even), and denote these two sets by P_1 and P_2. Order the $\left(\frac{n}{2}\right)^2$ intersection points of l with the line segments spanned by a point of P_1 and a point of P_2, by their y-coordinate and pick every $\left\lfloor \left(\frac{\epsilon n}{4}\right)^2 \right\rfloor$-th point to form the set N_l, with

$$|N_l| = \frac{\left(\frac{n}{2}\right)^2}{\left\lfloor \frac{\epsilon^2 n^2}{16} \right\rfloor} = O\left(\frac{1}{\epsilon^2}\right).$$

Next, recursively compute a weak $\frac{3}{2}\epsilon$-net N_1 of P_1 and separately a weak $\frac{3}{2}\epsilon$-net N_2 of P_2, and set $N = N_l \cup N_1 \cup N_2$. Letting $f\left(\frac{1}{\epsilon}\right)$ denote the size of N, we have

$$f\left(\frac{1}{\epsilon}\right) = 2f\left(\frac{2}{3\epsilon}\right) + O\left(\frac{1}{\epsilon^2}\right), \quad \text{which solves to} \quad f\left(\frac{1}{\epsilon}\right) = O\left(\frac{1}{\epsilon^2}\right).$$

To see that N is a weak ϵ-net of P, let C be any convex set containing at least ϵn points of P and consider these two cases:

C contains at least $\frac{\epsilon n}{4}$ points from both P_1 and P_2: Then the interval $C \cap l$ contains at least $\left(\frac{\epsilon n}{4}\right)^2$ intersections between l and the segments spanned by $C \cap P$, and so C must contain a point of N_l.

C contains at least $\frac{3\epsilon n}{4}$ points from one of P_1 or P_2: Say from P_1; then a weak ϵ'-net of P_1 would hit C if

$$\frac{3}{4}\epsilon n \geq \epsilon' \cdot \frac{n}{2} \qquad \Longrightarrow \qquad \epsilon' \leq \frac{3\epsilon n/4}{(n/2)} = \frac{3}{2}\epsilon.$$

Thus C is hit by N_1.

We remark here that the natural way to extend the above construction to $\mathbb{R}^d$ gives a weak ϵ-net of size $O\left(\frac{1}{\epsilon^{2d}}\right)$—adding many points lying on the same hyperplane is just too wasteful.

Almost all known weak ϵ-net constructions, including the proof of Theorem 9.2, are recursive. That is, for carefully chosen integers r and t, we show the existence of a partition of P into t equal-sized sets and a set $Q \subset \mathbb{R}^d$ such that any convex set containing points from greater than r sets of the partition must contain a point of Q. Any remaining unhit convex set must then contain at least $\frac{\epsilon n}{r}$ points from one of the sets of the partition. Thus Q together with the weak $\frac{\epsilon t}{r}$-nets constructed recursively for each set of the partition is a weak ϵ-net of P. The method of the partitioning, construction of Q and the parameters t, r involve trade-offs, which then dictate the size of the final weak ϵ-net.

The proof of Theorem 9.2 relies on two key structural results, which we first present.

THEOREM 9.3. *Let P be a set of n points in $\mathbb{R}^d$ in general position. Define the convex polytope*

$$\text{centerpolytope}(P) = \bigcap_{\substack{P' \subseteq P \\ |P'| > \frac{d}{(d+1)} \cdot |P|}} \text{conv}(P').$$

Then centerpolytope(P) *is non-empty.*
Each $q \in$ centerpolytope(P) *is called a centerpoint of P, and has the useful property that any closed half-space containing q contains at least $\frac{|P|}{(d+1)}$ points of P.*

We sketch a proof of the fact that centerpolytope(P) is non-empty for the two-dimensional case; the general case follows similarly. Let

$$\mathcal{C}_P = \left\{ \text{conv}(P') \ : \ P' \subseteq P, \ |P'| > \frac{2}{3}|P| \right\}.$$

For any $C, C' \in \mathcal{C}_P$, let

$$f(C, C') = \min_{q \in C \cap C'} y\text{-coordinate}(q).$$

Let $C_1, C_2 \in \mathcal{C}_P$ be the pair *maximizing* $f(\cdot, \cdot)$ over all pairs of sets in $\mathcal{C}_P$ and let $q \in C_1 \cap C_2$ be the point realizing $f(C_1, C_2)$ (see figure). We claim that q is contained in all sets of $\mathcal{C}_P$ and thus centerpolytope(P) is non-empty.

Otherwise let $C_3 \in \mathcal{C}_P$ be a set not containing q. Now C_3 must intersect $C_1 \cap C_2$—as every 3-tuple of sets of $\mathcal{C}_P$ must have a point of P in common as their union contains greater than $2n$ points of P—and so either $f(C_1, C_3) > f(C_1, C_2)$ or $f(C_2, C_3) > f(C_1, C_2)$, contradicting the extremal choice of $\{C_1, C_2\}$.

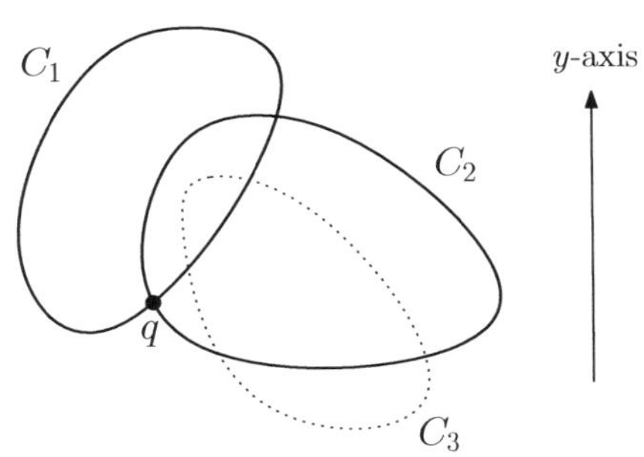

COROLLARY 9.4. *Let P' be a set of points in $\mathbb{R}^d$ in general position. Then there exists a set $Q \subseteq \mathbb{R}^d$ of size at most $|P'|^{d^2}$ such that for every $S \subseteq P'$, a centerpoint of S belongs to Q.*

PROOF. For any $S \subseteq P'$, apply Theorem 9.3 to get centerpolytope(S). Each vertex of centerpolytope(S) is a centerpoint of S and, by our general position assumption, is the intersection of d hyperplanes, each spanned by d points of S.
Thus the set of all distinct vertices of centerpolytope(S), over all $S \subseteq P'$, forms the required set Q. As Q consists of the intersections of all d-tuples of hyperplanes, each of which is spanned by d points of P', $|Q| \le \binom{|P'|}{d}^d \le |P'|^{d^2}$. □

The second structural result needed is Matoušek's simplicial partition theorem[1], stated without proof.

THEOREM 9.5. *Let P be a set of n points in $\mathbb{R}^d$ and $t \in \left[2, \frac{n}{2}\right]$ a given integer parameter. Then there exists a partition $\{P_1, \ldots, P_t\}$ of P with $|P_i| = \left\lfloor \frac{n}{t} \right\rfloor$ for $i = 1, \ldots, t-1$, such that any hyperplane in $\mathbb{R}^d$ intersects the convex-hull of at most $C_5 \cdot t^{1-\frac{1}{d}}$ sets of this partition. Here C_5 is a constant depending only on d.*

PROOF OF THEOREM 9.2. Let $t \in \left[2, \frac{n}{2}\right]$ be an integer parameter whose value is independent of n and will be set later. At the cost of worse constants, we will assume for simplicity that n is divisible by t. If $\epsilon \le \frac{1}{n}$, we can return P as a weak ϵ-net, of size $n \le \frac{1}{\epsilon}$; otherwise our construction is the following.

Weak Net-Finder Algorithm $\left(P \subset \mathbb{R}^d, \epsilon > 0\right)$.

Partition P into t sets $\Xi = \{P_1, \ldots, P_t\}$ using Theorem 9.5.
Pick an arbitrary point from each $P_i \in \Xi$ and let P' be this set of t points.
Apply Corollary 9.4 to P' to get a set $Q \subseteq \mathbb{R}^d$ of size at most t^{d^2}.
for $i = 1, \ldots, t$ **do**
  recursively construct a weak ϵ'-net N_i of P_i, where $\epsilon' = \frac{\epsilon\, t^{\frac{1}{d}}}{C_5(d+1)}$.
return $N = N_1 \cup \cdots \cup N_t \cup Q$.

CLAIM 9.6. The set N is a weak ϵ-net for P.

PROOF. Let C be a convex set containing at least ϵn points of P. If C contains at least $\epsilon' \cdot \frac{n}{t}$ points from some $P_i \in \Xi$, then C contains a point of N_i and we are done.
So assume that C contains less than $\epsilon' \cdot \frac{n}{t}$ points from each set of Ξ. Let Ξ_C be the sets of Ξ with non-empty intersection with C and let $P_C \subseteq P'$ be the points chosen from the sets of Ξ_C. Note that

$$|P_C| = |\Xi_C| > \frac{\epsilon n}{(\epsilon' n/t)} = \frac{\epsilon t}{\epsilon'}.$$

Let $q \in Q$ be a centerpoint of P_C. We claim that C must contain q. Otherwise, there exists a hyperplane h separating C from q. Let P_C^h be the points of P_C lying

[1]Matoušek's work is built on a stunning insight due to E. Welzl on matchings with low crossing numbers.

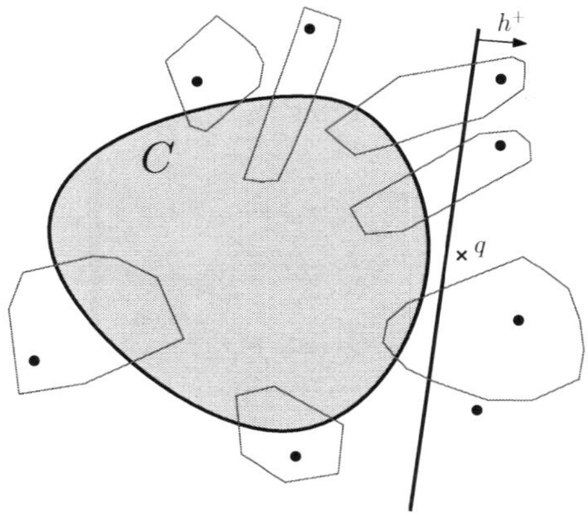

in the closed half-space h^+ containing q and not containing C and let Ξ_C^h denote their corresponding sets. See figure. The fact that q is a centerpoint of P_C implies that

$$|\Xi_C^h| = |P_C^h| \geq \frac{|P_C|}{(d+1)} > \frac{\epsilon t}{\epsilon' (d+1)}. \tag{9.7}$$

However, observe that the convex-hull of each set of Ξ_C^h must intersect h—each such set intersects C and also contains a point of P_C^h, and these two lie on different sides of h. Thus Theorem 9.5 together with Equation (9.7) implies that

$$C_5 \cdot t^{1-\frac{1}{d}} \geq \left|\Xi_C^h\right| > \frac{\epsilon t}{\epsilon' (d+1)}, \quad \text{and hence} \quad \epsilon' > \frac{\epsilon\, t^{\frac{1}{d}}}{C_5 (d+1)},$$

contradicting the value of ϵ' set in the algorithm. □

It remains to upper bound the size of the weak ϵ-net.

CLAIM 9.8.

$$|N| \leq \frac{b}{\epsilon^d} \log^a \frac{1}{\epsilon},$$

where $a = 2d^3 \ln (C_5 (d+1))$ and $b = (C_5 (d+1))^{9d^4}$.

PROOF. Let $f\left(\frac{1}{\epsilon}\right)$ be the size of the weak ϵ-net constructed by the algorithm. Then for any integer t,

$$|N| = f\left(\frac{1}{\epsilon}\right) \leq t \cdot f\left(\frac{1}{\epsilon'}\right) + |Q| \leq t \cdot f\left(\frac{C_5 (d+1)}{\epsilon\, t^{\frac{1}{d}}}\right) + t^{d^2}. \tag{9.9}$$

We show by induction that $f\left(\frac{1}{\epsilon}\right) \leq \frac{b}{\epsilon^d} \log^a \frac{1}{\epsilon}$ for an appropriate value of t.
The construction for the base case, when $\epsilon > \frac{1}{(C_5(d+1))^{8d^2}}$, is as follows:

> Use Theorem 9.5 with $t' = \left\lceil \left(\frac{C_5(d+1)}{\epsilon}\right)^d + 1 \right\rceil$ to get a partition Ξ' of t' sets. Let R be a set of t' points, one from each set of Ξ', and let Q' be a set containing a centerpoint of each subset of R. Then we claim that Q' is a weak ϵ-net of P. To see this, let C be a convex set containing at least ϵn points of P and intersecting the sets $\Xi'_C \subseteq \Xi'$. Note that $|\Xi'_C| \geq \frac{\epsilon n}{(n/t')} = \epsilon t'$. We show that a centerpoint $c' \in Q'$ of Ξ'_C must hit C. Otherwise, the hyperplane separating

c' from C intersects the convex-hull of at least $\frac{\epsilon t'}{(d+1)}$ sets of Ξ'_C and at most $C_5 \cdot t'^{1-\frac{1}{d}}$ by Theorem 9.5. The value of t' was chosen to get a contradiction here. Finally, by Corollary 9.4 and the fact that $\epsilon > \frac{1}{(C_5(d+1))^{8d^2}}$,

$$|Q'| \leq \left\lceil \left(\frac{C_5(d+1)}{\epsilon} \right)^d + 1 \right\rceil^{d^2} \leq \frac{1}{\epsilon^d} \left(\frac{(2C_5(d+1))^{d^3}}{\epsilon^{d^2}} \right)$$
$$\leq \frac{1}{\epsilon^d} \left((C_5(d+1))^{9d^4} \right) = \frac{b}{\epsilon^d}.$$

Returning to the inductive step, note that to get an upper bound of $\tilde{O}\left(\frac{1}{\epsilon^d}\right)$, the t^{d^2} term in Equation (9.9) must be $\tilde{O}\left(\frac{1}{\epsilon^d}\right)$, which forces us to set $t = \frac{1}{\epsilon^{1/d}}$[2]. Applying the inductive upper bound on $f\left(\frac{1}{\epsilon'}\right)$,

$$f\left(\frac{1}{\epsilon}\right) \leq t \cdot \frac{b\,(C_5(d+1))^d}{\epsilon^d t} \log^a \frac{C_5(d+1)}{\epsilon\, t^{\frac{1}{d}}} + \frac{1}{\epsilon^d}$$
$$= \frac{b}{\epsilon^d} \underbrace{\left((C_5(d+1))^d \log^a \frac{C_5(d+1)}{\epsilon^{1-\frac{1}{d^2}}} \right)} + \frac{1}{\epsilon^d}.$$

The underlined expression can be upper bounded as

$$\leq (C_5(d+1))^d \left(1 - \frac{1}{d^2}\right)^a \log^a \frac{(C_5(d+1))^2}{\epsilon} \quad \left(\text{since } d \geq 2\right)$$
$$\leq (C_5(d+1))^d\, e^{-a/d^2} \log^a \frac{(C_5(d+1))^2}{\epsilon}$$
$$= \frac{1}{(C_5(d+1))^d} \left(\log \frac{1}{\epsilon} + \log (C_5(d+1))^2 \right)^a \quad \left(\text{substituting the value of } a\right)$$
$$= \frac{1}{(C_5(d+1))^d} \cdot \log^a \frac{1}{\epsilon} \cdot \exp\left(\frac{\log (C_5(d+1))^2}{\log \frac{1}{\epsilon}} \cdot 2d^3 \ln (C_5(d+1)) \right)$$
$$= \frac{1}{(C_5(d+1))^d} \cdot \log^a \frac{1}{\epsilon} \cdot \exp\left(\frac{\log (C_5(d+1))^2}{8d^2 \log (C_5(d+1))} \cdot 2d^3 \ln (C_5(d+1)) \right)$$
$$= \frac{1}{(C_5(d+1))^d} \cdot \log^a \frac{1}{\epsilon} \cdot \exp\left(\frac{1}{2} \cdot d \ln (C_5(d+1)) \right) \leq \frac{1}{2} \log^a \frac{1}{\epsilon}.$$

Thus we can conclude that

$$f\left(\frac{1}{\epsilon}\right) \leq \frac{b}{\epsilon^d} \left(\frac{1}{2} \log^a \frac{1}{\epsilon} + \frac{1}{b} \right) \leq \frac{b}{\epsilon^d} \log^a \frac{1}{\epsilon}.$$

□

This completes the proof of Theorem 9.2. □

[2]Theorem 9.5 can be applied for this value of t since $\epsilon \in \left[\frac{1}{n}, \frac{1}{2^d}\right]$ implies that $t \in \left[2, \frac{n}{2}\right]$. Furthermore $\epsilon' \in \left(0, \frac{1}{2}\right]$ for this value of t.

Bibliography and discussion. The proof in $\mathbb{R}^2$ is from [**Alo+92**]. The beautiful proof of the main result of this chapter is from [**MW04**], simplifying and improving an earlier bound in [**Cha+95**]. By setting the parameters more carefully and using a better bound for the base case, one can improve the constant hidden in the asymptotic notation as well as the exponent of the logarithmic term by a factor of d^2 (see [**MW04**]). The detailed proof of Theorem 9.3 via induction on the dimension d can be found in [**MR09**]. For further properties of centerpolytope (P) we refer the reader to the survey [**Loe+19**]. Matoušek's simplicial partition theorem, a fundamental tool in combinatorial and computational geometry, is from [**Mat92**]. In a recent breakthrough, the bound in $\mathbb{R}^2$ was improved to roughly $O\left(\epsilon^{-3/2}\right)$ [**Rub18**]. When the input set is in convex position in $\mathbb{R}^2$, the bound can be further improved to $O\left(\frac{1}{\epsilon}\alpha\left(\frac{1}{\epsilon}\right)\right)$, where $\alpha(\cdot)$ is the inverse Ackermann function [**Alo+08**]. For general d, the best-known lower bound is $\Omega\left(\frac{1}{\epsilon}\log^{d-1}\frac{1}{\epsilon}\right)$ [**BMN11**].

[Alo+08] N. Alon, H. Kaplan, G. Nivasch, M. Sharir, and S. Smorodinsky, *Weak ϵ-nets and interval chains*, J. ACM **55** (2008), no. 6, Art. 28, 32, DOI 10.1145/1455248.1455252. MR2477488

[Alo+92] N. Alon, I. Bárány, Z. Füredi, and D. J. Kleitman, *Point selections and weak ϵ-nets for convex hulls*, Combin. Probab. Comput. **1** (1992), no. 3, 189–200, DOI 10.1017/S0963548300000225. MR1208800

[BMN11] B. Bukh, J. Matoušek, and G. Nivasch, *Lower bounds for weak epsilon-nets and stair-convexity*, Israel J. Math. **182** (2011), 199–208, DOI 10.1007/s11856-011-0029-1. MR2783971

[Cha+95] B. Chazelle, H. Edelsbrunner, M. Grigni, L. Guibas, M. Sharir, and E. Welzl, *Improved bounds on weak ϵ-nets for convex sets*, Discrete Comput. Geom. **13** (1995), no. 1, 1–15, DOI 10.1007/BF02574025. MR1300506

[Loe+19] J. A. De Loera, X. Goaoc, F. Meunier, and N. H. Mustafa, *The discrete yet ubiquitous theorems of Carathéodory, Helly, Sperner, Tucker, and Tverberg*, Bull. Amer. Math. Soc. (N.S.) **56** (2019), no. 3, 415–511, DOI 10.1090/bull/1653. MR3974609

[MR09] N. H. Mustafa and S. Ray, *An optimal extension of the centerpoint theorem*, Comput. Geom. **42** (2009), no. 6-7, 505–510, DOI 10.1016/j.comgeo.2007.10.004. MR2519371

[MW04] J. Matoušek and U. Wagner, *New constructions of weak ϵ-nets*, Discrete Comput. Geom. **32** (2004), no. 2, 195–206, DOI 10.1007/s00454-004-1116-4. MR2074337

[Mat92] J. Matoušek, *Efficient partition trees*, Discrete Comput. Geom. **8** (1992), no. 3, 315–334, DOI 10.1007/BF02293051. ACM Symposium on Computational Geometry (North Conway, NH, 1991). MR1174360

[Rub18] N. Rubin, *An improved bound for weak epsilon-nets in the plane*, 59th Annual IEEE Symposium on Foundations of Computer Science—FOCS 2018, IEEE Computer Soc., Los Alamitos, CA, 2018, pp. 224–235, DOI 10.1109/FOCS.2018.00030. MR3899592

2. Application: Polytope Approximation

I want to express a radical alternative that I learned from Sir Michael Atiyah. His view was that the most significant aspects of a new idea are often not contained in the deepest or most general theorem which they lead to. Instead, they are often embodied in the simplest examples, the simplest definition and their first consequences.

David Mumford

Let K be a convex set in $\mathbb{R}^d$. For a parameter $\vartheta \in (0,1)$ and a point $q \in K$, define ϑK_q to be the convex set obtained by 'shrinking' K by a factor of ϑ radially around q. Formally, for a point $q \in \mathbb{R}^d$ and a parameter $\vartheta \in (0,1)$, define the bijective function

$$f_{q,\vartheta}(p) = \vartheta \cdot p + (1-\vartheta) \cdot q.$$

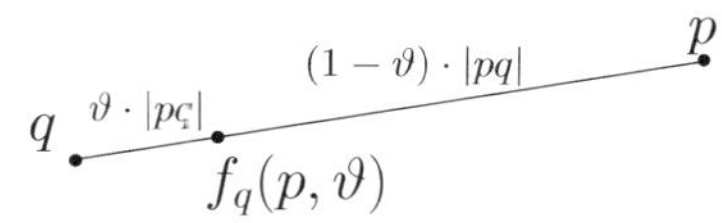

Then

$$\vartheta K_q = \left\{ f_{q,\vartheta}(p) \;:\; p \in K \right\}.$$

Note that $\mathrm{vol}(\vartheta K_q) = \vartheta^d \cdot \mathrm{vol}(K)$.
The main theorem of this section shows that any convex set K can be approximated by a convex polytope with few vertices (see figure).

THEOREM 9.10. *Let K be a convex set in $\mathbb{R}^d$, $\alpha \in \left(0, \frac{1}{2}\right)$ and $q \in K$ a point such that any half-space containing q contains at least $\alpha \cdot \mathrm{vol}(K)$ volume of K (we say that q has half-space depth α with respect to K). Then for any parameter $\vartheta \in (0,1)$ one can find a convex polytope C with $O\left(\frac{d}{\alpha(1-\vartheta)^d} \ln \frac{1}{\alpha(1-\vartheta)^d}\right)$ vertices such that*

$$\vartheta K_q \subseteq C \subseteq K.$$

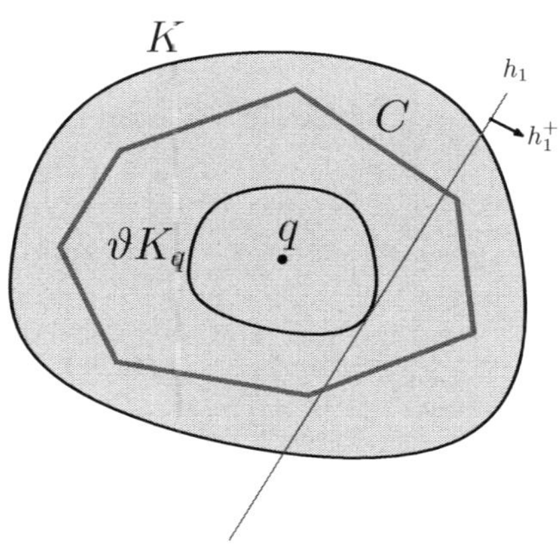

Overview of ideas. The fact that any half-space containing q contains at least $\alpha \cdot \mathrm{vol}(K)$ volume of K will be used to show that ϑK_q must be 'in the middle' of K. More precisely, consider any hyperplane h_1 tangent to ϑK_q (see the figure). Then we will show that the half-space h_1^+ (the side *not* containing ϑK_q) contains a large fraction of the volume of K, namely

$$\mathrm{vol}\left(K \cap h_1^+\right) \geq \epsilon' \cdot \mathrm{vol}(K),$$

where ϵ' depends only on α, ϑ and d. Then the convex-hull of an ϵ'-net of the primal set system induced on K by half-spaces must contain ϑK_q and thus is the required convex set C.

PROOF OF THEOREM 9.10. The main claim is the following.

CLAIM 9.11. Let h_1^+ be a half-space with a non-empty intersection with ϑK_q. Then

$$\mathrm{vol}\left(K \cap h_1^+\right) \geq (1-\vartheta)^d \cdot \alpha \cdot \mathrm{vol}(K).$$

PROOF. By translating h_1^+ 'outwards', we can assume that h_1 is tangent to ϑK_q, say at the point q_1, and where h_1^+ does not contain q. Further let h_0, h_2 be two hyperplanes—both parallel to h_1—such that h_0 contains q and h_2 is tangent to K at a point q_2. See the figure.
To lower bound the volume of $K \cap h_1^+$ (the patterned region in the figure), we are going to argue that a copy K' of $K \cap h_0^+$—scaled by a factor of $(1-\vartheta)$—lies inside $K \cap h_1^+$. Then we have

$$\mathrm{vol}\left(K \cap h_1^+\right) \geq \mathrm{vol}(K') = (1-\vartheta)^d \cdot \mathrm{vol}\left(K \cap h_0^+\right) \geq (1-\vartheta)^d \cdot \alpha \cdot \mathrm{vol}(K),$$

and we are done.
It remains to construct such a K'. The trick is to scale $K \cap h_0^+$ with respect to the point q_2. Namely set

$$K' = \left\{f_{q_2,(1-\vartheta)}(p) \ : \ p \in K \cap h_0^+\right\}.$$

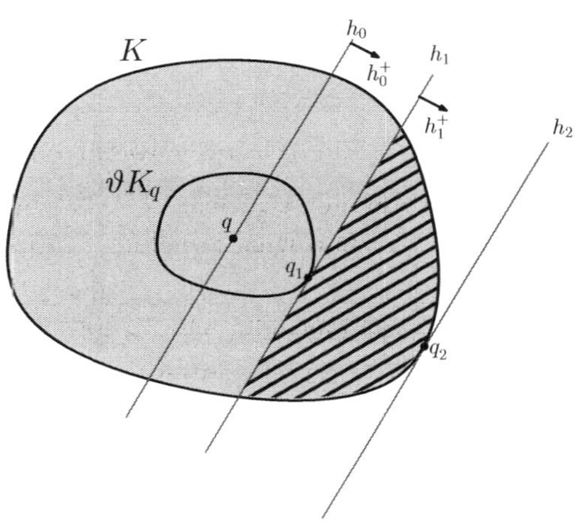

As $f(\cdot)$ maps a hyperplane to a parallel hyperplane (for completeness we give a proof of this fact at the end), we show that in the mapping $K \cap h_0^+ \to K'$, h_0 gets mapped to h_1.

(1) In the mapping $K \to \vartheta K_q$, the hyperplane h_2 must map to h_1 under $f_{q,\vartheta}(\cdot)$ and the point q_2 is mapped to q_1; that is, $f_{q,\vartheta}(q_2) = q_1$.
(2) Then $f_{q_2,(1-\vartheta)}(q) = q_1$; that is, the mapping $K \cap h_0^+ \to K'$ maps q to q_1, since

$$f_{q_2,(1-\vartheta)}(q) = \vartheta \cdot q_2 + (1-\vartheta) \cdot q = f_{q,\vartheta}(q_2) = q_1.$$

This implies that h_0 is mapped to h_1 under $f_{q_2,(1-\vartheta)}(\cdot)$.

Thus each point in $K \cap h_0^+$ gets mapped to a point of $K \cap h_1^+$ and so $K' \subseteq K \cap h_1^+$. □

We can now conclude the proof of Theorem 9.10. Set $\epsilon' = (1-\vartheta)^d \cdot \alpha$. By Claim 9.11, each half-space intersecting ϑK_q contains at least ϵ' fraction of the volume of K. Let $N \subset K$ be an ϵ'-net of the primal set system induced on K by half-spaces in $\mathbb{R}^d$. The fact that N is an ϵ'-net implies that the convex-hull of N must contain ϑK_q, as required.

The number of vertices of C can be upper bounded using Theorem 6.10:

$$|C| \le |N| = O\left(\frac{d}{\epsilon'} \ln \frac{1}{\epsilon'}\right) = O\left(\frac{d}{\alpha(1-\vartheta)^d} \ln \frac{1}{\alpha(1-\vartheta)^d}\right).$$

□

Theorem 9.3 together with Theorem 9.10 implies the following.

COROLLARY 9.12. *Let K be a convex set in $\mathbb{R}^d$ and $\vartheta \in [0,1]$ a given parameter. Then one can find a convex polytope C with $O\left(\frac{d^2}{(1-\vartheta)^d} \log \frac{d}{(1-\vartheta)^d}\right)$ vertices such that*

$$\vartheta K_q \subseteq C \subseteq K.$$

The above corollary can be improved further using a beautiful fact—called Grünbaum's inequality—that for any convex set K in $\mathbb{R}^d$ there exists a point $q \in K$ (in fact, the centroid of K) with half-space depth $\frac{1}{e}$ with respect to K. The fact that this is independent of d is quite amazing.

For completeness, we show that the function $f_{q,\vartheta}(\cdot)$ maps a hyperplane to a parallel hyperplane.

CLAIM 9.13. Let $q \in \mathbb{R}^d$ and $\vartheta \in (0,1)$ be a given parameter. Let h be a hyperplane passing through a point $p \in \mathbb{R}^d$ and with normal N. Then the function $f_{q,\vartheta}(\cdot)$ maps h to the hyperplane that is parallel to h and contains the point $f_{q,\vartheta}(p)$.

PROOF. Let $r \neq p$ be a point lying on h, that is, $(r-p) \cdot N = 0$, where N is a normal unit vector of h. Then $f_{q,\vartheta}(r)$ lies on the hyperplane h' passing through $f_{q,\vartheta}(p)$ and with normal N, since

$$\big(f_{q,\vartheta}(r) - f_{q,\vartheta}(p)\big) \cdot N = (\vartheta r - \vartheta p) \cdot N = \vartheta(r-p) \cdot N = 0.$$

□

Bibliography and discussion. The theorem and its proof is from Naszódi [**Nas19**]. There has been a long sequence of beautiful work on approximations of polytopes (e.g., see [**Ary+20**]).

[Ary+20] R. Arya, S. Arya, G. D. da Fonseca, and D. M. Mount, *Optimal bound on the combinatorial complexity of approximating polytopes*, Proceedings of the 2020 ACM-SIAM Symposium on Discrete Algorithms, SIAM, Philadelphia, PA, 2020, pp. 786–805. MR4141230

[Nas19] M. Naszódi, *Approximating a convex body by a polytope using the epsilon-net theorem*, Discrete Comput. Geom. **61** (2019), no. 3, 686–693, DOI 10.1007/s00454-018-9977-0. MR3918553

CHAPTER 10

VC-dimension of k-Fold Unions: Basic Case

We now turn towards lower bounds on sizes of ϵ-nets for both abstract and geometric set systems. The starting point is an observation due to N. Alon, which we illustrate with the following theorem from an area of combinatorics called Ramsey theory (there are quantitatively precise versions of this statement but this suffices to illustrate the main idea).

THEOREM. *There exists a function $f\colon \mathbb{Z}^+ \to \mathbb{Z}^+$ such that the following is true. Given any positive integer k, any subset of $\{1, \ldots, f(k)\}$ of size at least $\frac{f(k)}{2}$ contains an arithmetic progression of size k.*

Given a positive integer k, let $X = \{1, \ldots, f(k)\}$. Now suppose we want to hit all arithmetic progressions of size at least k in X. That is, we seek an ϵ'-net, $\epsilon' = \frac{k}{f(k)}$, of the set system $(X, \mathcal{F})$, where

$$\mathcal{F} = \{R \subseteq X \colon R \text{ is an arithmetic progression}\}.$$

Then any ϵ'-net N of $\mathcal{F}$ must have size at least $\frac{f(k)}{2}$, since otherwise

> the set $X \setminus N$ has size at least $\frac{f(k)}{2}$ and by the above Theorem, $X \setminus N$ contains an arithmetic progression R of size k. But R is not hit by N, a contradiction!

Now observe that the lower bound of $\frac{f(k)}{2}$ on the size of any ϵ'-net N is a *super-linear* function of $\frac{1}{\epsilon'}$. That is, for any ϵ'-net N,

$$|N| \geq \frac{f(k)}{2} = \frac{k}{2} \cdot \frac{f(k)}{k} = \frac{k}{2} \cdot \frac{1}{\epsilon'}.$$

One can re-write the lower bound on $|N|$ purely in terms of $\frac{1}{\epsilon'}$ by writing k in terms of $\frac{1}{\epsilon'}$, but for that one needs to know the function $f(\cdot)$.

> Further observe that $(X, \mathcal{F})$ can be realized as a geometric set system induced on a set of points in the plane—e.g., by mapping X to a set P of points in $\mathbb{R}^2$ with the map $x \to (x, 0)$. Then the set system $\mathcal{F}$ induced on X by arithmetic progressions is contained in the primal set system induced on P by sine curves in the plane.

A few years after N. Alon's work, in a major breakthrough, J. Pach and G. Tardos gave a tight lower bound on the size of ϵ-nets of primal set systems induced by half-spaces in $\mathbb{R}^4$. Their technique was later used to derive tight lower bounds for several other basic geometric set systems, including those induced by half-spaces in $\mathbb{R}^d$.

Rather than proving these lower bounds directly, we will show that a more powerful structural property is true, which then immediately implies lower bounds on sizes of ϵ-nets. For that we define the notion of the k-fold union of a set system.

DEFINITION 10.1. Given a set system $(X, \mathcal{F})$ and an integer $k \geq 1$, the *k-fold union* of $\mathcal{F}$, denoted by $\mathcal{F}^k$, is the set system obtained by adding the union of at most k sets of $\mathcal{F}$ to $\mathcal{F}^k$. That is,

$$\mathcal{F}^k = \Big\{ R_{i_1} \cup \cdots \cup R_{i_k} \ : \ R_{i_j} \in \mathcal{F} \ \text{ for all } \ 1 \leq j \leq k \Big\}.$$

As the k sets need not be distinct, we have $\mathcal{F} \subseteq \mathcal{F}^k$. Similarly one can define k-fold intersections of a given set system.

The topic of Chapters 10 and 11 is to study lower bounds on the VC-dimension of k-fold unions of both abstract and geometric set systems. These will then be used to derive lower bounds on sizes of ϵ-nets.

1. Abstract Set Systems

If one is trying to maximize the size of some structure under certain constraints, and if the constraints seem to force the extremal examples to be spread about in a uniform sort of way, then choosing an example randomly is likely to give a good answer.

Timothy Gowers

We will prove the following theorem in this section.

THEOREM 10.2. *There exists a constant $C_9 \geq 1$ such that for any integer $k \geq C_9$, there exists a set system $(X, \mathcal{F})$ with $|X| = \lfloor k \log k \rfloor$ such that*

(1) $\text{VC-dim}(\mathcal{F}) \leq 5$, *and*
(2) $\mathcal{F}^{3k}$ *shatters X. That is,* $\text{VC-dim}\left(\mathcal{F}^{3k}\right) \geq \lfloor k \log k \rfloor$.

Note that an *upper bound* on the VC-dimension of $\mathcal{F}^k$ in terms of k and the VC-dimension of $\mathcal{F}$ follows by a direct counting argument.

LEMMA 10.3. *Let $(X, \mathcal{F})$ be a set system with* $\text{VC-dim}(\mathcal{F}) \leq d$ *and let $k \geq 2$ be an integer. Then* $\text{VC-dim}\left(\mathcal{F}^k\right) = O(d\, k \log k)$.

PROOF. For any $Y \subseteq X$ with $|Y| \geq d$, we will show that $\left|\mathcal{F}^k|_Y\right| \leq \left(\frac{e|Y|}{d}\right)^{dk}$. Lemma 4.6 then implies that $\text{VC-dim}\left(\mathcal{F}^k\right) = O(d\, k \log k)$, as required.
Fix any $Y \subseteq X$. For each $R_Y \in \mathcal{F}^k|_Y$ there exists a $R \in \mathcal{F}^k$ such that $R_Y = R \cap Y$, with R being the union of k sets of $\mathcal{F}$. Say $R = R_1 \cup \cdots \cup R_k$, where $R_i \in \mathcal{F}$. As

$$R_Y = R \cap Y = (R_1 \cap Y) \cup \cdots \cup (R_k \cap Y),$$

each set in $\mathcal{F}^k|_Y$ is the union of at most k sets of $\mathcal{F}|_Y$. As $|\mathcal{F}|_Y| \leq \left(\frac{e|Y|}{d}\right)^d$ by Lemma 4.3,

$$\left|\mathcal{F}^k|_Y\right| \leq (|\mathcal{F}|_Y|)^k \leq \left(\frac{e|Y|}{d}\right)^{dk},$$

as required. □

The key to proving Theorem 10.2 is the following set system.

LEMMA 10.4. *There exists a constant $C_9 \geq 1$ such that for any integer $k \geq C_9$, there exists a set system $(X, \mathcal{F})$ with $|X| = \lfloor k \log k \rfloor$ such that*

(P1) $|S \cap S'| \leq 4$ *for all $S, S' \in \mathcal{F}$, and*
(P2) *for every integer $b = 1, \ldots, \lceil \log \log k \rceil$, the following holds:*
any $Y \subseteq X$ of size at least $\frac{k \log k}{2^b}$ contains a set $S \in \mathcal{F}$ of size at least $\frac{\log k}{b}$.

We use this to first complete the proof of Theorem 10.2, and then return to the proof of Lemma 10.4.

PROOF OF THEOREM 10.2. Let $(X, \mathcal{F})$ be the set system given by Lemma 10.4. Add to $\mathcal{F}$ all the $|X|$ singleton sets. We show that this is the desired set system. Clearly we still have $|S \cap S'| \leq 4$ for all $S, S' \in \mathcal{F}$.

CLAIM 10.5. VC-dim $(\mathcal{F}) \leq 5$.

PROOF. If a set $Y \subseteq X$ of size 6 is shattered by $\mathcal{F}$, then $\mathcal{F}$ must contain two sets having 5 elements of Y in common, a contradiction to property (P1). □

CLAIM 10.6. $\mathcal{F}^{3k}$ shatters X. In other words, any set $Y \subseteq X$ can be expressed as a union of at most $3k$ sets of $\mathcal{F}$.

PROOF. Let $Y \subseteq X$ be any subset of X. If $|Y| \leq k$, then Y is the union of at most k singleton sets of $\mathcal{F}$ and we're done. So assume that $|Y| > k$ and we now show that there exist at most $3k$ sets in $\mathcal{F}$ whose union is Y.

Let $b \in [1, \lceil \log \log k \rceil]$ be an integer such that

$$\frac{k \log k}{2^b} < |Y| \leq \frac{k \log k}{2^{b-1}}. \tag{10.7}$$

By property (P2), there exists a set $S \in \mathcal{F}$ of size at least $\frac{\log k}{b}$ such that $S \subseteq Y$. Set $Y = Y \setminus S$. If Y still satisfies Equation (10.7), we re-iterate the previous step (for the same value of b). This can continue for at most

$$\frac{\frac{k \log k}{2^{b-1}} - \frac{k \log k}{2^b}}{\frac{\log k}{b}} = \frac{\frac{k \log k}{2^b}}{\frac{\log k}{b}} = k \cdot \frac{b}{2^b} \quad \text{steps,} \tag{10.8}$$

each of which consists of choosing a set of $\mathcal{F}$ covering at least $\frac{\log k}{b}$ remaining elements of Y.

After these steps, we have $Y| \leq \frac{k \log k}{2^b}$. Again let $b' > b$ be an integer such that

$$\frac{k \log k}{2^{b'}} < |Y| \leq \frac{k \log k}{2^{b'-1}}.$$

As before, we continue to find sets of $\mathcal{F}$ of size at least $\frac{\log k}{b'}$ contained in Y.

This process continues as long as $|Y| > \frac{k \log k}{2^{\log \log k}} = k$. When we finally get $|Y| \leq k$, we can add k singletons to cover the remaining elements.

The total number of sets added can be upper bounded, using Equation (10.8), by

$$\sum_{i=1}^{\lceil \log \log k \rceil} k \cdot \frac{i}{2^i} \leq k \sum_{i=0}^{\infty} \frac{i}{2^i} = 2k.^1$$

Thus in total we were able to cover Y with at most $3k$ sets of $\mathcal{F}$, as required. □

This completes the proof of Theorem 10.2. □

It remains to prove the following.

[1] $\sum_{i=0}^{\infty} \frac{i}{2^i} = \left(\frac{1}{2}\right) + \left(\frac{1}{4} + \frac{1}{4}\right) + \left(\frac{1}{8} + \frac{1}{8} + \frac{1}{8}\right) + \cdots$. The sum of the first terms from each parenthesis gives 1, the sum of all the second terms give $\frac{1}{2}$ and so on, to get the total of 2.

LEMMA 10.4. *There exists a constant $C_9 \geq 1$ such that for any integer $k \geq C_9$, there exists a set system $(X, \mathcal{F})$ with $|X| = \lfloor k \log k \rfloor$ such that*

(P1) *$|S \cap S'| \leq 4$ for all $S, S' \in \mathcal{F}$, and*
(P2) *for every integer $b = 1, \ldots, \lceil \log \log k \rceil$, the following holds:*

any $Y \subseteq X$ of size at least $\frac{k \log k}{2^b}$ contains a set $S \in \mathcal{F}$ of size at least $\frac{\log k}{b}$.

PROOF. Let X be a set of $\lfloor k \log k \rfloor$ elements. We will construct the set system $\mathcal{F}$ on X probabilistically and show that, with non-zero probability, $\mathcal{F}$ satisfies both the properties (P1) and (P2) simultaneously.
Fix an integer $t = \lceil 56\, k^2 \log^2 k \rceil$. For each $b = 1, \ldots, \lceil \log \log k \rceil$, let

$$\mathcal{F}_b\colon t \text{ sets of size } \left\lceil \frac{\log k}{b} \right\rceil \text{ chosen at random (with replacement) from } \binom{X}{\left\lceil \frac{\log k}{b} \right\rceil}.$$

Our set system is then

$$\mathcal{F} = \bigcup_{b=1}^{\lceil \log \log k \rceil} \mathcal{F}_b.$$

Discarding the duplicate sets in $\mathcal{F}$, we will show that $\mathcal{F}$ satisfies both the properties (P1) and (P2) simultaneously with positive probability.

As we are only interested in demonstrating the existence of the required set system, the proof does not attempt to optimize the various probability calculations. A more careful calculation gives a lower value of C_9.

LEMMA 10.9. *$\mathcal{F}$ fails to satisfy property (P2) with probability at most $\frac{1}{k}$.*

PROOF. Let $b \in \left[1, \lceil \log \log k \rceil\right]$ be a fixed integer and $Y \subseteq X$ with $|Y| \geq \frac{k \log k}{2^b}$. Then

$$\Pr\left[\text{ a uniform random subset of size } \left\lceil \frac{\log k}{b} \right\rceil \text{ is contained in } Y \right] = \frac{\binom{|Y|}{\left\lceil \frac{\log k}{b} \right\rceil}}{\binom{|X|}{\left\lceil \frac{\log k}{b} \right\rceil}}$$

$$\geq \frac{\left\lceil \frac{k \log k}{2^b} \right\rceil \cdots \left(\left\lceil \frac{k \log k}{2^b} \right\rceil - \left\lceil \frac{\log k}{b} \right\rceil + 1 \right)}{\lfloor k \log k \rfloor \cdots \left(\lfloor k \log k \rfloor - \left\lceil \frac{\log k}{b} \right\rceil + 1 \right)} \geq \left(\frac{\left\lceil \frac{k \log k}{2^b} \right\rceil - \frac{\log k}{b}}{\lfloor k \log k \rfloor} \right)^{\left\lceil \frac{\log k}{b} \right\rceil}$$

$$\geq \frac{1}{2^{\log k + b}} \left(1 - \frac{2^b}{b\, k} \right)^{\left\lceil \frac{\log k}{b} \right\rceil} \geq \frac{1}{2\, k \log k} \left(1 - \frac{2 \log k}{k} \right)^{\lceil \log k \rceil} \geq \frac{1}{28\, k \log k},$$

where the last step used the fact that $\left(1 - \frac{2 \log k}{k}\right)^{\lceil \log k \rceil} \geq \frac{1}{14}$ for $k \geq 40$.

Thus the probability that there exists a set Y that violates property (P2) can be upper bounded as

$$2^{|X|}\cdot\left(1-\frac{1}{28\,k\log k}\right)^t \le 2^{k\log k}\cdot\exp\left(-\frac{t}{28\,k\log k}\right)\le\exp\left(k\log k-\frac{t}{28\,k\log k}\right)$$
$$\le\frac{1}{k},$$

recalling that $t=\lceil 56\,k^2\log^2 k\rceil$. □

LEMMA 10.10. *$\mathcal{F}$ fails to satisfy property (P1) with probability $O\left(\frac{\log^{12}k}{k}\right)$.*

PROOF. Recall that $\mathcal{F}=\bigcup_b\mathcal{F}_b$, where each set in $\mathcal{F}_b$ has size $\left\lceil\frac{\log k}{b}\right\rceil$ and $|\mathcal{F}_b|\le t$ (ignoring duplicate sets).

The probability that two random sets of $\mathcal{F}$ share at least 5 elements of X increases with the sizes of the sets, so we will do the calculations setting the size of each set in $\mathcal{F}$ to the maximum value of $\lceil\log k\rceil$. This will then imply the same upper bound for sets of smaller size.

Fix a set S of size $\lceil\log k\rceil$. We first upper bound the probability that a uniform random set S' of size $\lceil\log k\rceil$ intersects S in at least 5 elements of X.

(10.11)
$$\Pr\Big[|S\cap S'|\ge 5\Big]=\sum_{i=5}^{\lceil\log k\rceil}\frac{\binom{|S|}{i}\cdot\binom{|X|-|S|}{\lceil\log k\rceil-i}}{\binom{|X|}{\lceil\log k\rceil}}=\sum_{i=5}^{\lceil\log k\rceil}\frac{\binom{\lceil\log k\rceil}{i}\cdot\binom{\lfloor k\log k\rfloor-\lceil\log k\rceil}{\lceil\log k\rceil-i}}{\binom{\lfloor k\log k\rfloor}{\lceil\log k\rceil}}.$$

We will need the following fact:

for any two integers a,b with $a>2b\ge 10$,
$$\frac{\binom{b}{5}\binom{a-b}{b-5}}{\binom{a}{b}}\le b^5\cdot\frac{(a-b)!}{(b-5)!\,(a-2b+5)!}\cdot\frac{b!\,(a-b)!}{a!}\le b^{10}\cdot\frac{(a-b)^{b-5}}{(a-b)^b}\le\frac{b^{10}}{\left(\frac{a}{2}\right)^5}.$$

It is not hard to see that the numerator in Equation (10.11) is a decreasing function of i: as $\mathrm{E}\left[|S\cap S'|\right]=|S|\cdot\frac{\lceil\log k\rceil}{\lfloor k\log k\rfloor}<1$, the probability that S and S' share i elements decreases with i. As $\lfloor k\log k\rfloor>2\lceil\log k\rceil\ge 10$ for $k\ge 32$,

$$\Pr\Big[|S\cap S'|\ge 5\Big]\le\lceil\log k\rceil\cdot\frac{\binom{\lceil\log k\rceil}{5}\cdot\binom{\lfloor k\log k\rfloor-\lceil\log k\rceil}{\lceil\log k\rceil-5}}{\binom{\lfloor k\log k\rfloor}{\lceil\log k\rceil}}\le\lceil\log k\rceil\cdot\frac{2^5\lceil\log k\rceil^{10}}{\lfloor k\log k\rfloor^5}.$$

Thus the probability that there exist two sets of $\mathcal{F}$ sharing at least 5 elements of X can be upper bounded, using the union bound, by

$$\binom{|\mathcal{F}|}{2}\cdot\lceil\log k\rceil\cdot\frac{2^5\lceil\log k\rceil^{10}}{\lfloor k\log k\rfloor^5}\le\Big(t\cdot\lceil\log\log k\rceil\Big)^2\cdot\frac{32\lceil\log k\rceil^{11}}{\lfloor k\log k\rfloor^5}=O\left(\frac{\log^{12}k}{k}\right),$$

recalling that $t=\lceil 56\,k^2\log^2 k\rceil$. □

Lemma 10.9 and Lemma 10.10 imply that both properties (P1) and (P2) hold for $\mathcal{F}$ with probability at least $1-O\left(\frac{\log^{12}k}{k}\right)$. There exists a sufficiently large constant $C_9\ge 40$ such that this probability is positive for any $k\ge C_9$, implying the existence of such a set system. □

Bibliography and discussion. The proof in this section is essentially that of Eisenstat and Angluin [**EA0710a**]. By a more careful construction one can improve the bound on the VC-dimension from 5 to 2 [**Eis09**]. While we will see a more powerful lower bound construction with better bounds later, we presented this proof as it is a nice illustration of a probabilistic construction for proving a lower bound.

[EA0710a] D. Eisenstat and D. Angluin, *The VC dimension of k-fold union*, Inform. Process. Lett. **101** (2007), no. 5, 181–184, DOI 10.1016/j.ipl.2006.10.004. MR2291190

[Eis09] D. Eisenstat, *k-fold unions of low-dimensional concept classes*, Inform. Process. Lett. **109** (2009), no. 23-24, 1232–1234, DOI 10.1016/j.ipl.2009.09.005. MR2571754

2. Axis-Aligned Rectangles in $\mathbb{R}^2$

Theories come and go, but examples stay forever.

Israel Gelfand

Let $\mathcal{R}$ be a set of axis-aligned rectangles in the plane. For any point $p \in \mathbb{R}^2$, the set of rectangles of $\mathcal{R}$ containing p is denoted by $\mathcal{R}_p$. That is,

$$\mathcal{R}_p = \{R \in \mathcal{R} : R \ni p\}.$$

The main theorem of this section shows that, surprisingly, the upper bound on the VC-dimension of k-fold unions—Lemma 10.3—is already tight for the dual set system induced by axis-aligned rectangles in the plane.

THEOREM 10.12. *For any integer $r \geq 3$, there exists a set $\mathcal{R}$ of $(r-1)\,2^{r-2}$ axis-aligned rectangles in the plane such that*

> *for any $\mathcal{S} \subseteq \mathcal{R}$, there exists a set $Q \subseteq \mathbb{R}^2$ of size 2^{r-1} such that $\mathcal{S} = \bigcup_{q \in Q} \mathcal{R}_q$.*

In particular, since for any integer $k \geq 4$ there exists an integer $r \geq 3$ such that $2^{r-1} \leq k < 2^r$, the above gives a set $\mathcal{R}$ of size $(r-1)\,2^{r-2} \geq \frac{k}{4} \log \frac{k}{2}$ and Q of size $2^{r-1} \leq k$.

In other words, if $\mathcal{F}$ is the dual set system induced by axis-aligned rectangles in $\mathbb{R}^2$, then

$$\text{VC-dim}\left(\mathcal{F}^k\right) \geq \frac{k}{4} \log \frac{k}{2}.$$

The proof of Theorem 10.12 is based on a parity argument. While technically simple and quite 'thematic', it is ingenious and requires some time to appreciate and digest (and is worth remembering!). Most of the proof will be spent on understanding the construction of $\mathcal{R}$ and its properties. The final proof will then follow naturally.

We first construct a larger set $\mathcal{T}$ of *canonical* rectangles contained in the square $\mathcal{U} = [0, 2^r] \times [0, 2^r]$. $\mathcal{T}$ will be partitioned into $(r-1)$ *levels*, where $\mathcal{T}^i \subseteq \mathcal{T}$ denotes the set of rectangles at level i, for $i = 1, \dots, (r-1)$.

Rectangles of $\mathcal{T}^i$: To construct the rectangles of level i, consider the grid formed by partitioning $\mathcal{U}$ with 2^i equally-spaced vertical strips of width 2^{r-i} and 2^{r-i} equally-spaced horizontal strips of height 2^i. The set $\mathcal{T}^i$ consists of the rectangles induced by the pairwise intersection of these vertical and horizontal strips inside $\mathcal{U}$. These rectangles are interior-disjoint and their union covers $\mathcal{U}$.
There are 2^r rectangles in $\mathcal{T}^i$, each of width 2^{r-i}, height 2^i and area 2^r.
Each rectangle of $\mathcal{T}$ will be an *open* rectangle. For the final construction, one can slightly shrink these rectangles to get a set of closed rectangles.

Binary representation of $\mathcal{T}$: Each rectangle in $\mathcal{T}^i$ is the product of two open intervals, a 'horizontal' interval of length 2^{r-i} and a 'vertical' interval of length 2^i. For notational convenience, we will represent each such interval by a sequence of bits, as follows.
At level i, there are 2^i horizontal intervals, each of length 2^{r-i}. Thus each such interval can be represented by a sequence of i bits. We will use the 'binary-tree labeling' of the intervals: if the sequence $a = a_1 \cdots a_i$ represents the horizontal

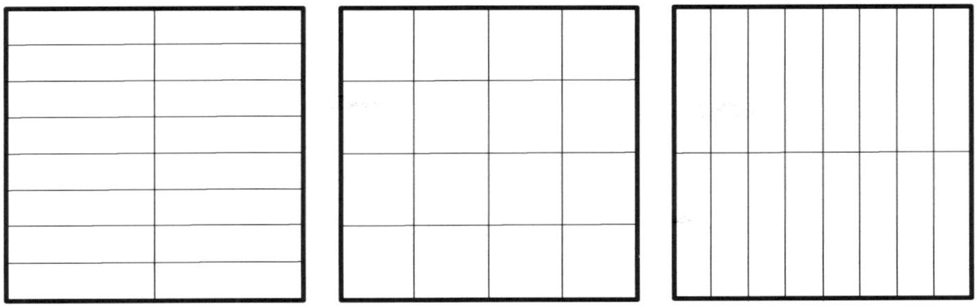

$r = 4, \quad i = 1$ $\qquad$ $r = 4, \quad i = 2$ $\qquad$ $r = 4, \quad i = 3$

interval I_a, then the sequence $a \cdot 0$ is the representation of the first half of I_a, and $a \cdot 1$ the second half of I_a. When it is clear, we will use the sequence a to denote the interval I_a. See figure below.

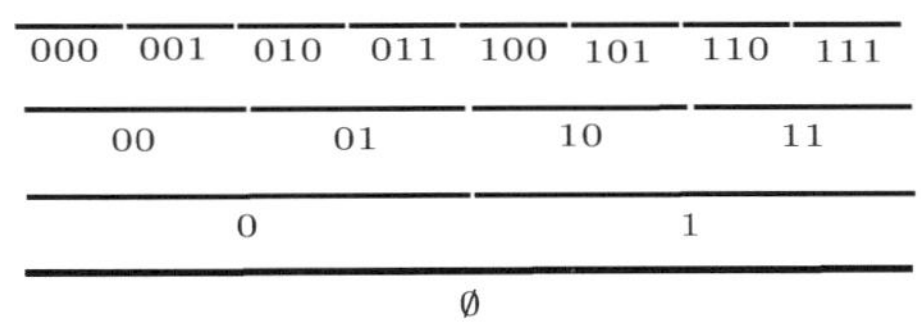

Formally, the bits $a = a_1 \cdots a_i$ denote the interval

$$I_a = \big(s, s + 2^{r-i}\big), \qquad \text{where} \qquad s = \sum_{j=1}^{i} 2^{r-j} \cdot a_j.$$

Similarly, each of the 2^{r-i} vertical intervals, each of length 2^i, can be represented by a sequence of $(r-i)$ bits.
Each rectangle of $\mathcal{T}^i$ can then be seen as the product of two intervals:

$$\mathcal{T}^i = \Big\{a \times b \;:\; a \in \{0,1\}^i,\; b \in \{0,1\}^{r-i}\Big\}.$$

Set $\mathcal{T} = \bigcup_{i=1}^{r-1} \mathcal{T}^i$.

The key properties of the rectangles of $\mathcal{T}$ are:

(1) Any two intervals I_a and $I_{a'}$, represented by $a = a_1 \cdots a_i$ and $a' = a'_1 \cdots a'_j$, are either disjoint or one contains the other; in particular, I_a and $I_{a'}$ intersect if and only if a is a prefix of a' or a' is a prefix of a.

(2) Let $R = a \times b$ and $R' = a' \times b'$ be two rectangles of $\mathcal{T}$; that is, a, b together have r bits and a', b' together have r bits. Then R and R' intersect if and only if I_a intersects $I_{a'}$ *and* I_b intersects $I_{b'}$. Equivalently, R and R' intersect if and only if a is a prefix of a' and b' is a prefix of b, or the other way around.

(3) For $i < j$, if $R = a \times b \in \mathcal{T}^i$ and $R' = a' \times b' \in \mathcal{T}^j$ are two intersecting rectangles, then a is a prefix of a' and b' is a prefix of b. This implies that if $R_i \in \mathcal{T}^i$ intersects $R_j \in \mathcal{T}^j$, which intersects $R_k \in \mathcal{T}^k$, for $i < j < k$, then R_i intersects R_k and there is a point in $\mathbb{R}^2$ common to the interior of R_i, R_j and R_k.

This is true more generally for axis-aligned rectangles in the plane: if every pair of a given set of rectangles has a non-empty intersection, then all of them together have a non-empty intersection.

However, unlike the general case, $R_i \cap R_k$ is in fact contained in R_j. This implies that the VC-dimension of the dual set system induced by $\mathcal{T}$ is 2, as it is not possible to find a point $q \in \mathbb{R}^2$ contained in R_i and R_k but not R_j.

The proof of the next two claims follow immediately from above considerations.

Claim 10.13. Let $1 \leq i_1 < i_2 < \cdots < i_k \leq (r-1)$ be k integers and let $R_{i_1} \in \mathcal{T}^{i_1}, R_{i_2} \in \mathcal{T}^{i_2}, \ldots, R_{i_k} \in \mathcal{T}^{i_k}$ be k rectangles of $\mathcal{T}$. If R_{i_j} intersects $R_{i_{j+1}}$ for all $j = 1, \ldots, k-1$, then the rectangles of $\{R_{i_1}, \ldots, R_{i_k}\}$ are pairwise intersecting and there exists a point in $\mathbb{R}^2$ lying in the interior of all of them.

Claim 10.14. A rectangle $a_1 \cdots a_i \times b_1 \cdots b_{r-i} \in \mathcal{T}^i$ intersects precisely these two rectangles of $\mathcal{T}^{i+1}$:

$$a_1 \cdots a_i \cdot 0 \times b_1 \cdots b_{r-i-1} \qquad \text{and} \qquad a_1 \cdots a_i \cdot 1 \times b_1 \cdots b_{r-i-1}.$$

Similarly, a rectangle $a_1 \cdots a_{i+1} \times b_1 \cdots b_{r-i-1} \in \mathcal{T}^{i+1}$ intersects precisely these two rectangles of $\mathcal{T}^i$:

$$a_1 \cdots a_i \times b_1 \cdots b_{r-i-1} \cdot 0 \qquad \text{and} \qquad a_1 \cdots a_i \times b_1 \cdots b_{r-i-1} \cdot 1.$$

Orthogonal rectangles

Let $a \in \{0,1\}^{i-1}$ and $b \in \{0,1\}^{r-i-1}$. Note that when $i = 1$, I_a is the interval $(0, 2^r)$ and when $i = r-1$, I_b is the interval $(0, 2^r)$. Then the rectangle $a \times b$ contains these four rectangles of $\mathcal{T}^i$:

$$a \cdot 0 \times b \cdot 0, \qquad a \cdot 0 \times b \cdot 1, \qquad a \cdot 1 \times b \cdot 0, \qquad a \cdot 1 \times b \cdot 1.$$

See the figure. This set of four rectangles of $\mathcal{T}^i$ is called the *cluster* of $a \times b$.

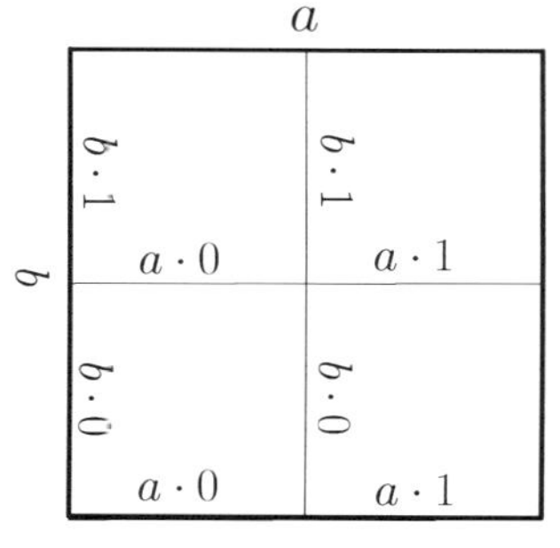

Call a set $\mathcal{S}^i \subset \mathcal{T}^i$ *orthogonal* if and only if for each $a \in \{0,1\}^{i-1}$ and $b \in \{0,1\}^{r-i-1}$, $\mathcal{S}^i$ contains either the two rectangles $\{a \cdot 0 \times b \cdot 0, a \cdot 1 \times b \cdot 1\}$ or the two rectangles $\{a \cdot 0 \times b \cdot 1, a \cdot 1 \times b \cdot 0\}$ from the four rectangles present in the cluster of $a \times b$.

Note that if $\mathcal{S}^i$ is orthogonal, then $|\mathcal{S}^i| = \frac{|\mathcal{T}^i|}{2} = 2^{r-1}$.

The main lemma containing the parity argument is the following.

LEMMA 10.15. *Let $\mathcal{S}^1 \subset \mathcal{T}^1, \ldots, \mathcal{S}^{r-1} \subset \mathcal{T}^{r-1}$ be orthogonal sets. Then there exists a set Q of 2^{r-1} points such that*

$$\mathcal{S}^1 \cup \cdots \cup \mathcal{S}^{r-1} = \bigcup_{q \in Q} \mathcal{T}_q,$$

where $\mathcal{T}_q$ is the set of rectangles of $\mathcal{T}$ containing the point q. In other words, the points in Q hit precisely the rectangles in $\mathcal{S}^1 \cup \cdots \cup \mathcal{S}^{r-1}$.

PROOF. Observe that as $\mathcal{S}^i$ consists of 2^{r-1} disjoint rectangles, we need 2^{r-1} points just to hit all the rectangles in $\mathcal{S}^i$. This immediately implies two things:

(1) If a point set Q, $|Q| = 2^{r-1}$, hits all rectangles in $\mathcal{S}^1 \cup \cdots \cup \mathcal{S}^{r-1}$, it cannot intersect any rectangle in $\mathcal{T} \setminus \{\mathcal{S}^1 \cup \cdots \cup \mathcal{S}^{r-1}\}$.
(2) If the lemma is true, it must be that $\mathcal{S}^1 \cup \cdots \cup \mathcal{S}^{r-1}$ can be partitioned into 2^{r-1} groups where each group contains *precisely* one rectangle from each level and the rectangles in each group have precisely one point of Q in common.

Thus it suffices to show that for each $i = 1, \ldots, r-2$, each rectangle of $\mathcal{S}^i$ intersects *precisely* one rectangle of $\mathcal{S}^{i+1}$ and each rectangle of $\mathcal{S}^{i+1}$ intersects *precisely* one rectangle of $\mathcal{S}^i$. The proof of Lemma 10.15 then follows by Claim 10.13.

Each rectangle of $\mathcal{S}^i$ intersects precisely one rectangle of $\mathcal{S}^{i+1}$: Let R be the rectangle $R = a_1 \cdots a_i \times b_1 \cdots b_{r-i} \in \mathcal{S}^i$. Then by Claim 10.14, R intersects these two rectangles of $\mathcal{T}^{i+1}$:

$$\Xi_R = \Big\{a_1 \cdots a_i \cdot 0 \times b_1 \cdots b_{r-i-1}, \quad a_1 \cdots a_i \cdot 1 \times b_1 \cdots b_{r-i-1}\Big\}.$$

On the other hand, from the cluster of $a_1 \cdots a_i \times b_1 \cdots b_{r-i-2}$ in $\mathcal{T}^{i+1}$[2], $\mathcal{S}^{i+1}$ contains precisely one of these two sets:

$$\begin{gathered}\{a_1 \cdots a_i \cdot 0 \times b_1 \cdots b_{r-i-2} \cdot 0, \quad a_1 \cdots a_i \cdot 1 \times b_1 \cdots b_{r-i-2} \cdot 1\} \\ \{a_1 \cdots a_i \cdot 0 \times b_1 \cdots b_{r-i-2} \cdot 1, \quad a_1 \cdots a_i \cdot 1 \times b_1 \cdots b_{r-i-2} \cdot 0\}.\end{gathered}$$

In both the above two scenarios, precisely one rectangle of $\mathcal{S}^{i+1}$ is present in Ξ_R and thus intersected by R. See figure where the two possibilities for the orthogonal rectangles are patterned.

Each rectangle of $\mathcal{S}^{i+1}$ intersects precisely one rectangle of $\mathcal{S}^i$: The other direction is similar. Let $R' = a'_1 \cdots a'_{i+1} \times b'_1 \cdots b'_{r-i-1} \in \mathcal{S}^{i+1}$. By Claim 10.14, R' intersects these two rectangles of $\mathcal{T}^i$:

$$\Xi'_{R'} = \Big\{a'_1 \cdots a'_i \times b'_1 \cdots b'_{r-i-1} \cdot 0, \quad a'_1 \cdots a'_i \times b'_1 \cdots b'_{r-i-1} \cdot 1\Big\}.$$

On the other hand, from the cluster of $a'_1 \cdots a'_{i-1} \times b'_1 \cdots b'_{r-i-1}$ in $\mathcal{T}^i$, $\mathcal{S}^i$ contains precisely one of these two sets:

$$\begin{gathered}\{a'_1 \cdots a'_{i-1} \cdot 0 \times b'_1 \cdots b'_{r-i-1} \cdot 0, \quad a'_1 \cdots a'_{i-1} \cdot 1 \times b'_1 \cdots b'_{r-i-1} \cdot 1\} \\ \{a'_1 \cdots a'_{i-1} \cdot 0 \times b'_1 \cdots b'_{r-i-1} \cdot 1, \quad a'_1 \cdots a'_{i-1} \cdot 1 \times b'_1 \cdots b'_{r-i-1} \cdot 0\}.\end{gathered}$$

[2]When $i = r-2$, the interval $b_1 \cdots b_{r-i-2}$ is the interval $(0, 2^r)$.

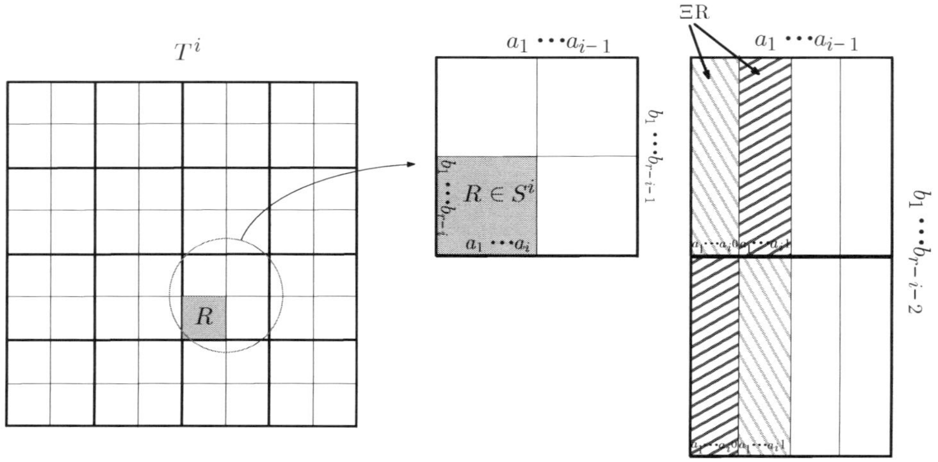

In both the above two scenarios, precisely one rectangle of $\mathcal{S}^i$ is present in $\Xi'_{R'}$ and thus intersected by R'.

□

We have a parity argument at the core of the above proof: in each cluster of four rectangles at every level i, exactly one rectangle from each row and each column of the cluster is present in $\mathcal{S}^i$. For a rectangle $R \in \mathcal{S}^i$, when we consider the level $i+1$, R will be 'split' into two rectangles of half its width and both belonging to the same row in level $i+1$. Exactly one of them will be picked into $\mathcal{S}^{i+1}$ and will be the only rectangle of $\mathcal{S}^{i+1}$ intersecting R.

Finally we return to the proof of the main theorem.

Proof of Theorem 10.12. Our goal is to show that given an integer r, there exists a set $\mathcal{R}$ of $(r-1)\cdot 2^{r-2}$ axis-aligned rectangles in the plane such that for any set $\mathcal{S} \subseteq \mathcal{R}$, there exist a set $Q \subseteq \mathbb{R}^2$ of 2^{r-1} points with $\mathcal{S} = \bigcup_{q\in Q} \mathcal{R}_q$.
$\mathcal{R}$ will be constructed by choosing, from each set $\mathcal{T}^i$, $i = 1 \ldots (r-1)$, the following set $\mathcal{R}^i$ of 2^{r-2} rectangles:

$$\mathcal{R}^i = \Big\{a_1 \cdots a_i \times b_1 \cdots b_{r-i} \in \mathcal{T}^i \ : \ a_i = b_{r-i} = 0\Big\}.$$

See figure. Set $\mathcal{R} = \bigcup_{i=1}^{r-1} \mathcal{R}^i$. Note that $|\mathcal{R}| = (r-1)\cdot 2^{r-2}$.
Let $\mathcal{S} \subseteq \mathcal{R}$. We now show the existence of a set Q of 2^{r-1} points hitting precisely the rectangles in $\mathcal{S}$.

Construct the orthogonal sets $\mathcal{S}^1, \ldots, \mathcal{S}^{r-1}$ as follows:

If $a_1 \cdots a_{i-1} \cdot 0 \times b_1 \cdots b_{r-i-1} \cdot 0 \in \mathcal{S}$, add

$$\left\{\begin{matrix} a_1 \cdots a_{i-1} \cdot 0 \times b_1 \cdots b_{r-i-1} \cdot 0 \\ a_1 \cdots a_{i-1} \cdot 1 \times b_1 \cdots b_{r-i-1} \cdot 1 \end{matrix}\right\} \text{ to } \mathcal{S}^i.$$

Otherwise add

$$\left\{\begin{matrix} a_1 \cdots a_{i-1} \cdot 0 \times b_1 \cdots b_{r-i-1} \cdot 1 \\ a_1 \cdots a_{i-1} \cdot 1 \times b_1 \cdots b_{r-i-1} \cdot 0 \end{matrix}\right\} \text{ to } \mathcal{S}^i.$$

See figure.

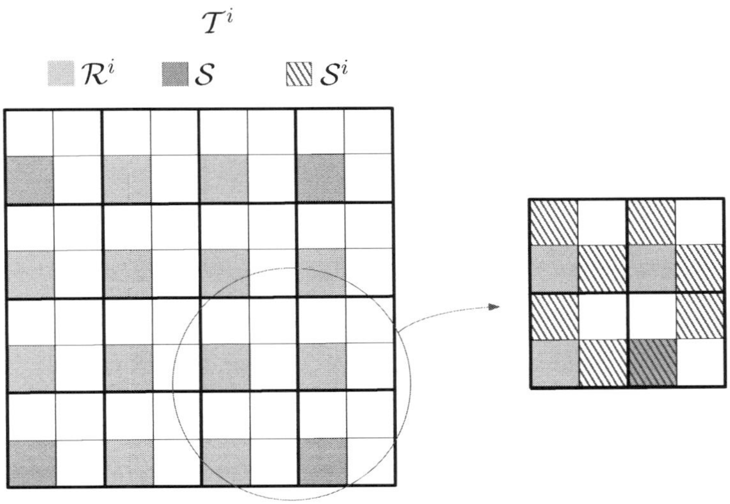

By construction, $\mathcal{S}^1 \cup \cdots \cup \mathcal{S}^{r-1}$ contains every rectangle of $\mathcal{S}$ and no rectangle of $\mathcal{R} \setminus \mathcal{S}$. Furthermore each $\mathcal{S}^i$ is an orthogonal set of $\mathcal{T}^i$.
Applying Lemma 10.15 to $\mathcal{S}^1, \ldots, \mathcal{S}^{r-1}$, we get the required set Q of 2^{r-1} points hitting precisely the rectangles in $\mathcal{S}^1 \cup \cdots \cup \mathcal{S}^{r-1}$ and thus $\mathcal{S}$. $\square$

As noted earlier, while the dual set system induced by rectangles has VC-dimension 4, the ones in the above construction have VC-dimension 2.

Bibliography and discussion. The beautiful proof is from the breakthrough paper [**PT13**] (a more accurate calculation in the paper gives better constant factors), where it was used to derive lower bounds on the size of ϵ-nets of dual set systems induced by axis-parallel rectangles in the plane. The construction is related to the notion of 'orthogonal functions' that were used by Roth to prove lower bounds for problems in geometric discrepancy; we refer the reader to the texts [**BC87, Cha00, Mat99**] for details.

[BC87] J. Beck and W. W. L. Chen, *Irregularities of distribution*, Cambridge Tracts in Mathematics, vol. 89, Cambridge University Press, Cambridge, 1987, DOI 10.1017/CBO9780511565984. MR903025

[Cha00] B. Chazelle, *The discrepancy method*, Cambridge University Press, Cambridge, 2000. Randomness and complexity, DOI 10.1017/CBO9780511626371. MR1779341

[Mat99] J. Matoušek. *Geometric Discrepancy: An Illustrated Guide*. Springer, 1999.

[PT13] J. Pach and G. Tardos, *Tight lower bounds for the size of epsilon-nets*, J. Amer. Math. Soc. **26** (2013), no. 3, 645–658, DOI 10.1090/S0894-0347-2012-00759-0. MR3037784

CHAPTER 11

VC-dimension of k-Fold Unions: General Case

Given a set system $(X, \mathcal{F})$ with VC-dimension d, Theorem 6.1 states the existence of an ϵ-net of $\mathcal{F}$ of size

$$O\left(\frac{d}{\epsilon}\log\frac{1}{\epsilon}\right).$$

This is not an 'if and only if' statement—that is, it is not true that if $\mathcal{F}$ has VC-dimension at least d, then any ϵ-net of $\mathcal{F}$ must have size $\Omega\left(\frac{d}{\epsilon}\log\frac{1}{\epsilon}\right)$. For example, the primal set system induced by half-spaces in $\mathbb{R}^3$ has VC-dimension 4 and yet there exist ϵ-nets of size $O\left(\frac{1}{\epsilon}\right)$.

> It is tempting to conjecture an improved bound is true for primal set systems induced by half-spaces in $\mathbb{R}^d$—that is, given any set P of points in $\mathbb{R}^d$ and $\epsilon > 0$, there exists an ϵ-net $N \subseteq P$ of size $O\left(\frac{1}{\epsilon}\right)$, where the constant hidden in the asymptotic notation may depend on d. The best upper bound we have seen so far is $O\left(\frac{d}{\epsilon}\log\frac{1}{\epsilon}\right)$, implied by Theorem 6.1 and Lemma 4.13.

It was conjectured for many years, when no super-linear lower bound was known, that an improved upper bound for ϵ-nets of half-spaces in $\mathbb{R}^d$, $d \geq 4$, should be possible. Surprisingly, it turned out not to be the case: for any n, $d \geq 4$ and $\epsilon > 0$, there exist a set P of n points in $\mathbb{R}^d$ such that any ϵ-net for the primal set system induced on P by half-spaces in $\mathbb{R}^d$ must have size $\Omega\left(\frac{d}{\epsilon}\log\frac{1}{\epsilon}\right)$. The proof of this fact proceeds in two steps:

(1) we first *lower bound* the VC-dimension of the k-fold union of the primal set system induced by half-spaces in $\mathbb{R}^d$; in particular we will show that this is $\Omega\left(dk\log k\right)$ for $d \geq 4$.

(2) we next observe that a lower bound on the VC-dimension of the k-fold union of a set system is a lower bound on the size of any $\frac{1}{2k}$-net of the set system. Plugging in $k = \frac{2}{\epsilon}$ gives the desired lower bound of $\Omega\left(\frac{d}{\epsilon}\log\frac{1}{\epsilon}\right)$.

This chapter uses the above recipe to derive tight lower bounds on ϵ-net sizes for several basic geometric set systems.

1. Products of Set Systems

Wittgenstein once greeted me with the question: 'Why do people say that it was natural to think that the sun went round the earth rather than that the earth turned on its axis?'
I replied: 'I suppose, because it looked as if the sun went round the earth.'
'Well,' he asked, 'what would it have looked like if it had looked as if the earth turned on its axis?'

Elizabeth Anscombe

Given a set system $(X', \mathcal{F}')$ and an integer $d \geq 1$, the goal of this section is to show that one can construct a set system $(X, \mathcal{F})$ such that

$$\text{VC-dim}(\mathcal{F}) \leq d\cdot\text{VC-dim}(\mathcal{F}') \quad \textbf{and} \quad \text{VC-dim}\left(\mathcal{F}^k\right) \geq d\cdot\text{VC-dim}\left(\mathcal{F}'^k\right).$$

Here $\mathcal{F}'^k$ is the k-fold union of $\mathcal{F}'$, which we recall.

Definition 10.1. Given a set system $(X, \mathcal{F})$ and an integer $k \geq 1$, the *k-fold union* of $\mathcal{F}$, denoted by $\mathcal{F}^k$, is the set system obtained by adding the union of at most k sets of $\mathcal{F}$ to $\mathcal{F}^k$. That is,

$$\mathcal{F}^k = \left\{R_{i_1} \cup \cdots \cup R_{i_k} \ : \ R_{i_j} \in \mathcal{F} \ \text{ for all } \ 1 \leq j \leq k\right\}.$$

As the k sets need not be distinct, we have $\mathcal{F} \subseteq \mathcal{F}^k$. Similarly one can define k-fold intersections of a given set system.

Our first result considers the abstract case.

Lemma 11.1. *Let $(X', \mathcal{F}')$ be a set system and k a positive integer such that $\mathcal{F}'^k$ shatters X'. Then for any integer $d \geq 1$, there exists a set system $(X, \mathcal{F})$ with $|X| = d \cdot |X'|$ such that*

1. $\text{VC-dim}(\mathcal{F}) \leq d \cdot \text{VC-dim}(\mathcal{F}')$, *and*
2. $\mathcal{F}^k$ *shatters X; that is,* $\text{VC-dim}\left(\mathcal{F}^k\right) \geq d \cdot |X'|$.

This immediately implies the following.

Corollary 11.2. *For any integer $d \geq 1$ and $k \geq C_9$, where C_9 is a sufficiently large integer, there exists a set system $(X, \mathcal{F})$ with $|X| = d \cdot \lfloor k \log k \rfloor$ such that*

1. $\text{VC-dim}(\mathcal{F}) \leq 5d$, *and*
2. $\mathcal{F}^{3k}$ *shatters X; that is,* $\text{VC-dim}\left(\mathcal{F}^{3k}\right) \geq d \cdot \lfloor k \log k \rfloor$.

Proof. Apply Theorem 10.2 to get a set system $(X', \mathcal{F}')$ such that $|X'| = \lfloor k \log k \rfloor$, $\text{VC-dim}(\mathcal{F}') \leq 5$ and $\mathcal{F}'^{3k}$ shatters X'. Then Lemma 11.1 applied to $(X', \mathcal{F}')$ gives the required set system $(X, \mathcal{F})$, with $|X| = d \cdot \lfloor k \log k \rfloor$, $\text{VC-dim}(\mathcal{F}) \leq 5d$ and $\text{VC-dim}\left(\mathcal{F}^{3k}\right) \geq d \cdot \lfloor k \log k \rfloor$. □

We remark here that the lower bound for the VC-dimension of $\mathcal{F}^{3k}$ given in Corollary 11.2 is tight within constant factors, as it matches the upper bound of Lemma 10.3.

Our second result shows that the construction of Lemma 11.1 can be 'realized geometrically' to extend the lower bound for the VC-dimension of the k-fold union of axis-aligned rectangles in $\mathbb{R}^2$ to that of boxes in $\mathbb{R}^d$.

THEOREM 11.3. *For any integer $r \geq 3$ and $d \geq 2$, there exists a set $\mathcal{B}$ of $\lfloor \frac{d}{2} \rfloor \cdot (r-1)\, 2^{r-2}$ axis-aligned boxes in $\mathbb{R}^d$ such that*

for any $\mathcal{S} \subseteq \mathcal{B}$, there exists a set $Q \subseteq \mathbb{R}^d$ of size 2^{r-1} such that $\mathcal{S} = \bigcup_{q \in Q} \mathcal{B}_q$,

where $\mathcal{B}_q$ is the set of boxes of $\mathcal{B}$ containing q.
In particular, since for any integer $k \geq 4$ there exists an integer $r \geq 3$ such that $2^{r-1} \leq k < 2^r$, the above gives a set $\mathcal{B}$ of size $\lfloor \frac{d}{2} \rfloor (r-1)\, 2^{r-2} \geq \lfloor \frac{d}{2} \rfloor \frac{k}{4} \log \frac{k}{2}$ and Q of size $2^{r-1} \leq k$.
In other words, if $\mathcal{F}$ is the dual set system induced by axis-aligned boxes in $\mathbb{R}^d$, then

$$\text{VC-dim}\left(\mathcal{F}^k\right) \geq \left\lfloor \frac{d}{2} \right\rfloor \cdot \frac{k}{4} \log \frac{k}{2}.$$

PROOF OF LEMMA 11.1. Let $(X_1, \mathcal{F}_1), \ldots, (X_d, \mathcal{F}_d)$ be d *distinct* copies of $(X', \mathcal{F}')$; in particular, for each $i = 1, \ldots, d$,

$$|X_i| = |X'|, \qquad \text{VC-dim}\left(\mathcal{F}_i\right) = \text{VC-dim}\left(\mathcal{F}'\right), \qquad \text{and} \qquad \mathcal{F}_i^k \text{ shatters } X_i.$$

Then the required set system $(X, \mathcal{F})$ is:

$$X = X_1 \cup \cdots \cup X_d,$$
$$\mathcal{F} = \Big\{ F_1 \cup \cdots \cup F_d \ : \ F_i \in \mathcal{F}_i \text{ for } i = 1, \ldots, d \Big\}.$$

Clearly $|X| = d \cdot |X'|$. It remains to show the following.

$\text{VC-dim}\,(\mathcal{F}) \leq d \cdot \text{VC-dim}\,(\mathcal{F}')$: Assume otherwise and let $Y \subseteq X$ be a set of size

$$d \cdot \text{VC-dim}\left(\mathcal{F}'\right) + 1$$

shattered by $\mathcal{F}$. By the pigeonhole principle, there exists an index $i \in \{1, \ldots, d\}$ such that Y contains at least $\text{VC-dim}\left(\mathcal{F}'\right) + 1$ points, denoted by the set Y_i, from X_i. But then Y_i is shattered by $\mathcal{F}_i$, a contradiction to the fact that $\text{VC-dim}\left(\mathcal{F}_i\right) = \text{VC-dim}\left(\mathcal{F}'\right)$.

$\mathcal{F}^k$ shatters X: Let $Y \subseteq X$. Then $Y = Y_1 \cup \cdots \cup Y_d$, where $Y_i \subseteq X_i$. Furthermore, for each $i = 1, \ldots, d$, as $\mathcal{F}_i^k$ shatters X_i, we have $Y_i = S_{i,1} \cup S_{i,2} \cup \cdots \cup S_{i,k}$, where $S_{i,j} \in \mathcal{F}_i$. Thus

$$Y = \bigcup_{i=1}^{d} Y_i = \bigcup_{i=1}^{d} \bigcup_{j=1}^{k} S_{i,j} = \bigcup_{j=1}^{k} \bigcup_{i=1}^{d} S_{i,j} = \bigcup_{j=1}^{k} F_j,$$

where $F_j = S_{1,j} \cup S_{2,j} \cup \cdots \cup S_{d,j} \in \mathcal{F}$.

This completes the proof. $\square$

We remark here that this idea of constructing 'product set systems' also works if they have different sizes (see discussion).

PROOF OF THEOREM 11.3. We prove the theorem for the case when the dimension d is even. The case where d is odd follows by applying the bound for $d-1$.

Apply Theorem 10.12 with parameter r to get a set $\mathcal{A} = \{A_1, \dots, A_n\}$ of axis-aligned rectangles in $\mathbb{R}^2$, where

$$n = (r-1)\,2^{r-2}.$$

Note that an axis-aligned box B in $\mathbb{R}^d$ can be seen as the product of d intervals, where the i-th interval is the projection of B onto the i-th axis. In particular, each A_i can be seen as a product of two intervals I_i^x and I_i^y, with $A_i = I_i^x \times I_i^y$. By scaling, we can assume that each interval lies in $[0,1]$.

We now 'lift' the rectangles of $\mathcal{A}$ to boxes in $\mathbb{R}^d$. Each such box will be the product of d intervals—two intervals from a rectangle of $\mathcal{A}$ and the other $d-2$ intervals set to $[0,1]$. Each rectangle A_i is lifted $\frac{d}{2}$ times, as follows.

Say $A_i = I_i^x \times I_i^y$. Then the j-th lift of A_i, for $j = 1, \dots, \frac{d}{2}$, will be the box

$$B_{i,j} = \underbrace{[0,1] \times \cdots \times [0,1]}_{2j-2 \text{ intervals}} \times \underbrace{I_i^x \times I_i^y}_{A_i = I_i^x \times I_i^y} \times \underbrace{[0,1] \times \cdots \times [0,1]}_{d-2j \text{ intervals}}.$$

The rectangle A_i is called the *corresponding rectangle* of the box $B_{i,j}$.

Let $\mathcal{B}_j$ be the set of $|\mathcal{A}|$ boxes formed by the j-th lift of all the rectangles of $\mathcal{A}$,

$$\mathcal{B}_j = \Big\{ B_{i,j} \;:\; A_i \in \mathcal{A} \Big\}.$$

The required set of boxes is then

$$\mathcal{B} = \bigcup_{j=1}^{\frac{d}{2}} \mathcal{B}_j, \qquad \text{with} \qquad |\mathcal{B}| = \frac{d}{2} \cdot |\mathcal{A}| = \frac{d}{2} \cdot (r-1)\,2^{r-2}.$$

It remains to show that for any set $\mathcal{B}' \subseteq \mathcal{B}$, there exists a set Q of 2^{r-1} points in $\mathbb{R}^d$ such that each box of $\mathcal{B}'$ contains at least one point of Q and no box of $\mathcal{B} \setminus \mathcal{B}'$ contains any point of Q.

Partition $\mathcal{B}'$ into disjoint sets $\mathcal{B}'_j = \mathcal{B}' \cap \mathcal{B}_j$, with

$$\mathcal{B}' = \mathcal{B}'_1 \cup \cdots \cup \mathcal{B}'_{\frac{d}{2}}.$$

For a fixed $j \in \left[1, \frac{d}{2}\right]$, by Theorem 10.12 applied to the corresponding rectangles of the boxes of $\mathcal{B}'_j$, we get a set Q_j of 2^{r-1} points lying in $[0,1]^2$ such that

- Q_j hits all the corresponding rectangles of the boxes of $\mathcal{B}'_j$.
- Q_j hits none of the corresponding rectangles of the boxes of $\mathcal{B}_j \setminus \mathcal{B}'_j$.

Now pick an arbitrary point $q_j \in Q_j$ from each $j = 1, \dots, \frac{d}{2}$, and consider the point $q \in \mathbb{R}^d$ obtained by 'packing' these $\frac{d}{2}$ points, each in $[0,1]^2$, into one point in $[0,1]^d$. That is,

$$q = \left(q_1^x, q_1^y, \dots, q_j^x, q_j^y, \dots, q_{\frac{d}{2}}^x, q_{\frac{d}{2}}^y \right).$$

Here is the key claim regarding the point q constructed above.

CLAIM 11.4. For each index $j \in \{1, \ldots, \frac{d}{2}\}$, the point q hits all the boxes in $\mathcal{B}'_j$ whose corresponding rectangles were hit by the point q_j. Furthermore, q hits no box of $\mathcal{B}_j \setminus \mathcal{B}'_j$.

PROOF. Take any $B \in \mathcal{B}'_j$ whose corresponding rectangle contains q_j—that is, the $(2j-1)$-th interval of B contains q_j^x and the $(2j)$-th interval contains q_j^y. Then q lies inside B, as the remaining coordinates of q are irrelevant since all the other intervals of B are $[0, 1]$.
For the second statement, assume for contradiction that q lies in a box $B \in \mathcal{B}_j \setminus \mathcal{B}'_j$. But then the point q_j lies in the corresponding rectangle of B, a contradiction to the property that Q_j hits none of the corresponding rectangles of $\mathcal{B}_j \setminus \mathcal{B}'_j$. □

Now given $Q_1, \ldots, Q_{\frac{d}{2}}$, each containing 2^{r-1} points in $\mathbb{R}^2$, pick an arbitrary point from each Q_j and pack these $\frac{d}{2}$ points into a point in $\mathbb{R}^d$, and add it to Q. Repeat the above step—that is, pick an arbitrary point from each of the remaining sets Q_j—and construct a second point in $\mathbb{R}^d$ and add to Q. This process goes on for 2^{r-1} steps, and thus Q consists of 2^{r-1} points.
Claim 11.4 implies that for each $j \in \{1, \ldots, \frac{d}{2}\}$, Q hits all boxes of $\mathcal{B}'_j$ and no box of $\mathcal{B}_j \setminus \mathcal{B}'_j$. This concludes the proof. □

Bibliography and discussion. The idea of the proof of Lemma 11.1 is from [**EA07**] which states it in a more general setting. The proof of Theorem 11.3 is from [**KMP17**].

[EA07] D. Eisenstat and D. Angluin, *The VC dimension of k-fold union*, Inform. Process. Lett. **101** (2007), no. 5, 181–184, DOI 10.1016/j.ipl.2006.10.004. MR2291190

[KMP17] A. Kupavskii, N. H. Mustafa, and J. Pach, *Near-optimal lower bounds for ϵ-nets for half-spaces and low complexity set systems*, A journey through discrete mathematics, Springer, Cham, 2017, pp. 527–541. MR3726612

2. Geometric Set Systems in $\mathbb{R}^d$

When scientists attempt to explain their work to the general public they are urged to simplify the ideas with which they work, to remove unnecessary technical language ... to explain how elementary particles of matter interact with one another we might describe them in terms of billiard balls colliding with each other. One Hungarian physicist once remarked in the course of writing a textbook that, although he would often be referring to the motions and collisions of billiard balls to illustrate the laws of mechanics, he had neither seen nor played this game and his knowledge of it was derived entirely from the study of physics books.

John Barrow

In this section we present lower bounds on the VC-dimension of k-fold unions of a variety of basic geometric set systems. These follow by applying various geometric transformations—lifting, duality, 'stretching'—to the dual set system induced on axis-aligned boxes in $\mathbb{R}^d$.

Lifting axis-aligned boxes in $\mathbb{R}^d$ to points in $\mathbb{R}^{2d}$: The positive orthant of $\mathbb{R}^d$ consists of points with all positive coordinates.

THEOREM 11.5. *For any integer $r \geq 3$ and $d \geq 4$, there exists a set P of $\lfloor \frac{d}{4} \rfloor \cdot (r-1)\, 2^{r-2}$ points lying in the positive orthant of $\mathbb{R}^d$ such that*

> *for any $P' \subseteq P$, there exists a set $\mathcal{B}$ of 2^{r-1} boxes in $\mathbb{R}^d$, each containing the origin, such that $P' = \bigcup_{B \in \mathcal{B}} (B \cap P)$.*

In particular, since for any integer $k \geq 4$ there exists an integer $r \geq 3$ such that $2^{r-1} \leq k < 2^r$, the above gives a set P of size $\lfloor \frac{d}{4} \rfloor \cdot (r-1)\, 2^{r-2} \geq \frac{(d-3)k}{16} \log \frac{k}{2}$ and $\mathcal{B}$ of size $2^{r-1} \leq k$.

In other words, if $\mathcal{F}$ is the primal set system induced on a set of points lying in the positive orthant of $\mathbb{R}^d$ by boxes containing the origin, then

$$\text{VC-dim}\left(\mathcal{F}^k\right) \geq \frac{(d-3)\,k}{16} \log \frac{k}{2}.$$

'Stretching' points in $\mathbb{R}^d$: Considering the x_d-axis as vertical, any half-space containing the point that is the 'minus infinity' of the x_d axis is called a downward-facing half-space.

THEOREM 11.6. *For any integer $r \geq 3$ and $d \geq 4$, there exists a set P of $\lfloor \frac{d}{4} \rfloor \cdot (r-1)\, 2^{r-2}$ points lying in the positive orthant of $\mathbb{R}^d$ such that*

> *for any $P' \subseteq P$, there exists a set $\mathcal{H}'$ of 2^{r-1} downward-facing half-spaces in $\mathbb{R}^d$ such that $P' = \bigcup_{h \in \mathcal{H}'} (h \cap P)$.*

In particular, since for any integer $k \geq 4$ there exists an integer $r \geq 3$ such that $2^{r-1} \leq k < 2^r$, the above gives a set P of size $\lfloor \frac{d}{4} \rfloor \cdot (r-1)\, 2^{r-2} \geq \frac{(d-3)k}{16} \log \frac{k}{2}$ and $\mathcal{H}'$ of size $2^{r-1} \leq k$.

In other words, if $\mathcal{F}$ is the primal set system induced on a set of points by half-spaces in $\mathbb{R}^d$, then

$$\text{VC-dim}\left(\mathcal{F}^k\right) \geq \frac{(d-3)\,k}{16} \log \frac{k}{2}.$$

Dualizing points in $\mathbb{R}^d$ to hyperplanes in $\mathbb{R}^d$: We now turn to the set system at the heart of Theorem 3.9. Given a set $\mathcal{H}$ of hyperplanes in $\mathbb{R}^d$ and an integer $k \geq 1$, let $\Delta_k(\mathcal{H})$ be the set system induced on $\mathcal{H}$ by intersection with k-dimensional simplices. That is,

$$\Delta_k(\mathcal{H}) = \left\{\mathcal{H}_\Delta \,:\, \Delta \text{ is a } k\text{-dimensional simplex in } \mathbb{R}^d\right\},$$

where $\mathcal{H}_\Delta$ is the set of hyperplanes of $\mathcal{H}$ intersecting Δ:

$$\mathcal{H}_\Delta = \left\{h \in \mathcal{H} \,:\, h \cap \Delta \neq \emptyset\right\}.$$

THEOREM 11.7. *For any integer $r \geq 3$ and $d \geq 4$, there exists a set $\mathcal{H}$ of $\lfloor \frac{d}{4} \rfloor \cdot (r-1)\,2^{r-2}$ hyperplanes in $\mathbb{R}^d$ such that*

> *for any $\mathcal{H}' \subseteq \mathcal{H}$, there exists a 2^{r-1}-dimensional simplex Δ in $\mathbb{R}^d$ such that $\mathcal{H}' = \mathcal{H}_\Delta$.*

In particular, since for any integer $k \geq 4$ there exists an integer $r \geq 3$ such that $2^{r-1} \leq k < 2^r$, the above gives a set $\mathcal{H}$ of size $\lfloor \frac{d}{4} \rfloor \cdot (r-1)\,2^{r-2} \geq \frac{(d-3)k}{16} \log \frac{k}{2}$ and Δ of dimension $2^{r-1} \leq k$.
In other words, if Δ_k is the set system induced on hyperplanes by intersection with k-dimensional simplices in $\mathbb{R}^d$, then

$$\text{VC-dim}(\Delta_k) \geq \frac{(d-3)\,k}{16} \log \frac{k}{2}.$$

PROOF OF THEOREM 11.5. Let $d' = \lfloor \frac{d}{2} \rfloor$. Apply Theorem 11.3 in $\mathbb{R}^{d'}$ with parameter r to get a set $\mathcal{B}$ of $\left\lfloor \frac{d'}{2} \right\rfloor \cdot (r-1)\,2^{r-2}$ axis-aligned boxes in $\mathbb{R}^{d'}$.
It suffices to show that one can map each $B \in \mathcal{B}$ to a point $\pi(B)$ lying in the positive orthant of $\mathbb{R}^{2d'}$ and each $p \in \mathbb{R}^{d'}$ to an axis-aligned box orthant(p) in $\mathbb{R}^{2d'}$ containing the origin, such that

$$p \in B \qquad \Longleftrightarrow \qquad \pi(B) \in \text{orthant}(p).$$

Then $\pi(\mathcal{B}) = \left\{\pi(B)\colon B \in \mathcal{B}\right\}$ gives the required set of points in $\mathbb{R}^{2d'} \subseteq \mathbb{R}^d$, of size

$$\left\lfloor \frac{d'}{2} \right\rfloor \cdot (r-1)\,2^{r-2} = \left\lfloor \frac{\lfloor \frac{d}{2} \rfloor}{2} \right\rfloor \cdot (r-1)\,2^{r-2} = \left\lfloor \frac{d}{4} \right\rfloor \cdot (r-1)\,2^{r-2}.$$

The map $\pi(\cdot)$: By translation we can assume that each $B \in \mathcal{B}$ is of the form

$$B = [s_1, e_1] \times [s_2, e_2] \times \cdots \times [s_{d'}, e_{d'}], \text{ with } e_i > s_i > 0 \text{ for all } i \in \{1, \ldots, d'\}.$$

Define the lifted point as

$$\pi(B) = \left(s_1, \frac{1}{e_1}, s_2, \frac{1}{e_2}, \ldots, s_{d'}, \frac{1}{e_{d'}}\right) \in \mathbb{R}^{2d'}.$$

Note that $\pi(B)$ lies in the positive orthant of $\mathbb{R}^{2d'}$.

The map orthant $(\cdot)$**:** Map each point $p = \Big(x_1(p), x_2(p), \ldots, x_{d'}(p)\Big) \in \mathbb{R}^{d'}$, with each $x_i(p) > 0$, to an axis-aligned box in $\mathbb{R}^{2d'}$ as

$$\text{orthant}(p) = \Big[0, x_1(p)\Big] \times \Big[0, \frac{1}{x_1(p)}\Big] \times \cdots \times \Big[0, x_{d'}(p)\Big] \times \Big[0, \frac{1}{x_{d'}(p)}\Big].$$

Now it can be verified that

$$p \in B$$
$$\Updownarrow$$
$$s_1 \le x_1(p) \le e_1, \quad \ldots, \quad s_{d'} \le x_{d'}(p) \le e_{d'}$$
$$\Updownarrow$$
$$0 < s_1 \le x_1(p), \quad 0 < \frac{1}{e_1} \le \frac{1}{x_1(p)}, \quad \ldots, \quad 0 < s_{d'} \le x_{d'}(p), \quad 0 < \frac{1}{e_{d'}} \le \frac{1}{x_{d'}(p)}$$
$$\Updownarrow$$
$$\pi(B) \in \text{orthant}(p).$$

□

PROOF OF THEOREM 11.6. Apply Theorem 11.5 in $\mathbb{R}^d$ and with parameter r to get a set P of $\lfloor \frac{d}{4} \rfloor \cdot (r-1)\, 2^{r-2}$ points lying in the positive orthant of $\mathbb{R}^d$ such that for any $Q \subseteq P$, there exists a set $\mathcal{B}$ of 2^{r-1} axis-aligned boxes, each containing the origin, such that $Q = \bigcup_{B \in \mathcal{B}} (B \cap P)$.
We now 'stretch' the points of P to get the required point set.

CLAIM 11.8. Given a set P of points lying in the positive orthant of $\mathbb{R}^d$, there exists a function $f\colon P \to \mathbb{R}^d$ such that the following holds: for any axis-aligned box $B \subset \mathbb{R}^d$ that contains the origin, there is a downward-facing half-space $h_B \subset \mathbb{R}^d$ such that for all $p \in P$,

$$p \in B \qquad \Longleftrightarrow \qquad f(p) \in h_B.$$

PROOF. For each $i \in \{1, \ldots, d\}$ do the following:
let $0 < \xi_{i,1} < \xi_{i,2} < \xi_{i,3} < \cdots$ denote the sorted sequence of the distinct x_i-coordinates of all the points of P. Now stretch these coordinates—while maintaining their relative order—to ensure that

$$\frac{\xi_{i,j+1}}{\xi_{i,j}} > d \quad \text{holds for every } j. \tag{11.9}$$

Let $f(p)$ be the stretched copy of the point p with these stretched coordinates[1] and set $f(P) = \{f(p) \,:\, p \in P\}$.
Observe that from the point of view of intersections with axis-aligned boxes, the actual values of the coordinates of the points of P do not matter—only their *relative ordering* does. Thus an axis-aligned box contains $Q \subseteq P$ if and only if there is a scaled axis-aligned copy of this box containing $f(Q)$. Thus it suffices to show that

[1]Note that the map $f(\cdot)$ depends on the points of P, and is constructed specifically from the given point set.

for each $p \in P$, $f(p) \in B$ if and only if $f(p) \in h_B$, where h_B is a downward-facing half-space that will be constructed from B.

Let B be an axis-aligned box containing the origin. We can shrink B, without changing its intersection with $f(P)$, so that B is of the form

$$B = [0, b_1] \times [0, b_2] \times \cdots \times [0, b_d],$$

where each $b_i > 0$ is equal to ξ_{i,j_i} for some j_i.

Define the downward-facing half-space h_B as

$$h_B \quad : \quad \frac{x_1}{b_1} + \frac{x_2}{b_2} + \cdots + \frac{x_d}{b_d} \leq d. \tag{11.10}$$

Now for any point $p \in P$:

$f(p) \in B$: That is, $x_i(f(p)) \in (0, b_i]$ for all $i = 1, \ldots, d$. Then each term of the L.H.S. of Equation (11.10) lies in $(0, 1]$ and so $f(p) \in h_B$.

$f(p) \notin B$: Then there exists a coordinate index $i \in \{1, \ldots, d\}$ such that $x_i(f(p)) > b_i$. Crucially, due to our stretching, Equation (11.9) implies that $x_i(f(p)) > d \cdot b_i$. But then the i-th term in the L.H.S. of Equation (11.10) is already larger than d (and all the other terms are non-negative), implying that $f(p) \notin h_B$.

□

We return the set $f(P)$, concluding the proof. □

PROOF OF THEOREM 11.7. Apply Theorem 11.6 in dimension d to get a set P of $\lfloor \frac{d}{4} \rfloor \cdot (r-1)\, 2^{r-2}$ points in $\mathbb{R}^d$. Let $\mathcal{H}$ be the set of hyperplanes dual to the points of P.

Theorem 11.6 implies the following fact for P:

> for any $P' \subseteq P$, there exists a set $\mathcal{H}'$ of 2^{r-1} downward-facing half-spaces in $\mathbb{R}^d$ such that the union of the half-spaces of $\mathcal{H}'$ contains precisely P'.

By duality, this translates to the following fact for $\mathcal{H}$:

> for any $\mathcal{H}' \subseteq \mathcal{H}$, there exists a set Q' of 2^{r-1} points in $\mathbb{R}^d$ such that $\mathcal{H}' = \bigcup_{q \in Q'} l_q$, where l_q is the set of hyperplanes of $\mathcal{H}$ lying below (w.r.t. the x_d-axis) the point q.

Now we complete the proof by showing that for any $\mathcal{H}' \subseteq \mathcal{H}$, there exists a 2^{r-1}-dimensional simplex Δ such that $\mathcal{H}' = \mathcal{H}_\Delta$; that is, Δ intersects precisely the hyperplanes of $\mathcal{H}'$.

Given any $\mathcal{H}' \subseteq \mathcal{H}$, let Q' be a set of 2^{r-1} points such that $\mathcal{H}' = \bigcup_{q \in Q'} l_q$. Let q' be a point of $\mathbb{R}^d$ lying below all the hyperplanes of $\mathcal{H}'$. Now $\Delta = \operatorname{conv}(Q' \cup \{q'\})$ intersects precisely the hyperplanes of $\mathcal{H}'$.

□

Bibliography and discussion. The ideas of lifting and stretching (stated for $\mathbb{R}^2$) are from [**PT13**], where they are used to prove lower bounds on sizes of ϵ-nets of the primal set system induced by half-spaces in $\mathbb{R}^4$. The relation to the VC-dimension of the k-fold union, as well as the use of duality, is from [**CMK19**], whose presentation we have essentially followed here.

[CMK19] M. Csikós, N. H. Mustafa, and A. Kupavskii, *Tight lower bounds on the VC-dimension of geometric set systems*, J. Mach. Learn. Res. **20** (2019), Paper No. 81, 8. MR3960935

[PT13] J. Pach and G. Tardos, *Tight lower bounds for the size of epsilon-nets*, J. Amer. Math. Soc. **26** (2013), no. 3, 645–658, DOI 10.1090/S0894-0347-2012-00759-0. MR3037784

3. Lower bounds on Epsilon-Nets

There is hope, but not for us.

Franz Kafka

This section presents consequences of lower bounds on the VC-dimension of k-fold unions to the sizes of ϵ-nets. This connection is the basis of the next statement.

CLAIM 11.11. Let $(X, \mathcal{F})$ be a set system and k a positive integer such that X is shattered by $\mathcal{F}^k$; that is, VC-dim $(\mathcal{F}^k) \geq |X|$. Then any $\frac{1}{2k}$-net of $\mathcal{F}$ must have size at least $\frac{|X|}{2}$.

PROOF. Let N be a $\frac{1}{2k}$-net of $\mathcal{F}$ and suppose that $N < \frac{|X|}{2}$. As X is shattered by $\mathcal{F}^k$, there exists a set $S \in \mathcal{F}^k$ containing precisely the points of $X \setminus N$. That is, there exist k sets $S_1, \ldots, S_k \in \mathcal{F}$, not necessarily distinct, such that $S_1 \cup \cdots \cup S_k = X \setminus N$. By the pigeonhole principle there exists a set $S \in \{S_1, \ldots, S_k\}$ with

$$|S| \geq \frac{|X \setminus N|}{k} \geq \frac{|X|}{2k}.$$

But S contains no point of N, a contradiction to the fact that N is a $\frac{1}{2k}$-net of $\mathcal{F}$. □

We next turn to deriving lower bounds for ϵ-net sizes using the above statement.

Abstract set systems: We now get a lower bound matching the upper bound in Theorem 6.1.

LEMMA 11.12. *There exists a small-enough positive constant δ such that given a parameter $\epsilon \in [0, \delta]$ and integers n and $d \geq 1$, there exists a set system $(X, \mathcal{F})$ with $|X| \geq n$, VC-dim $(\mathcal{F}) \leq 5d$, such that any ϵ-net of $\mathcal{F}$ has size at least $\left\lfloor \frac{d}{12\,\epsilon} \log \frac{1}{6\epsilon} \right\rfloor$.*

PROOF. Apply Corollary 11.2 with $k = \left\lfloor \frac{1}{6\epsilon} \right\rfloor$ (here we need $\delta \leq \frac{1}{6C_9}$) to get a set system $(X', \mathcal{F}')$ with $|X'| \geq \left\lfloor \frac{d}{6\epsilon} \log \frac{1}{6\epsilon} \right\rfloor$ such that

(1) VC-dim $(\mathcal{F}') \leq 5d$, and
(2) $\mathcal{F}'^{3k}$ shatters X'.

Now Claim 11.11 implies that any $\left(\frac{1}{2 \cdot 3 \left\lfloor \frac{1}{6\epsilon} \right\rfloor} \right)$-net of $\mathcal{F}'$ must have size at least $\frac{|X'|}{2}$. As $\left(\frac{1}{2 \cdot 3 \left\lfloor \frac{1}{6\epsilon} \right\rfloor} \right) \geq \epsilon$, any ϵ-net of $\mathcal{F}'$ has size at least

$$\frac{|X'|}{2} = \left\lfloor \frac{d}{12\,\epsilon} \log \frac{1}{6\epsilon} \right\rfloor.$$

We are almost done, except that we also want $|X'| \geq n$. That is easy—just duplicate elements of X' to get the required set X of size n, as follows.

for each $p \in X'$, let X_p be a set of $\left\lceil \frac{n}{|X'|} \right\rceil$ new elements and let

$$X = \bigcup_{p \in X'} X_p,$$

$$\mathcal{F} = \left\{ \bigcup_{p \in F'} X_p \ : \ F' \in \mathcal{F}' \right\}.$$

It can be verified that $|X| \geq n$, VC-dim $(\mathcal{F}) \leq 5d$ and $(X, \mathcal{F})$ has an ϵ-net of size t if and only if $(X', \mathcal{F}')$ has an ϵ-net of size t.

□

We remark that by a direct probabilistic construction, one can improve the lower bound of Lemma 11.12 by constant factors.

THEOREM 11.13. *Given any $\epsilon > 0$ and integer $d \geq 2$, there exists a set system $(X, \mathcal{F})$ such that* VC-dim $(\mathcal{F}) \leq d$ *and any ϵ-net of $\mathcal{F}$ has size at least* $\left(1 - \frac{2}{d} + \frac{1}{d(d+2)} + o(1)\right) \frac{d}{\epsilon} \log \frac{1}{\epsilon}$.

Axis-aligned boxes in $\mathbb{R}^d$: The proof of the next statement follows along the same lines, combining Theorem 11.3 with Claim 11.11.

LEMMA 11.14. *Given any $\epsilon \in \left[0, \frac{1}{8}\right]$ and integers n and $d \geq 2$, there exists a set $\mathcal{B}$ of n axis-aligned boxes in $\mathbb{R}^d$ such that any ϵ-net of the dual set system induced on $\mathcal{B}$ has size at least $\left\lfloor \frac{d}{2} \right\rfloor \frac{1}{16\epsilon} \log \frac{1}{4\epsilon}$.*

PROOF. Set r to be the largest integer satisfying $2^r \leq \frac{1}{2\epsilon}$; i.e., $r = \left\lfloor \log \frac{1}{2\epsilon} \right\rfloor \geq 2$. Apply Theorem 11.3 with d and $(r+1)$ to get a set $\mathcal{B}$ of $\left\lfloor \frac{d}{2} \right\rfloor 2^{r-1} r$ boxes in $\mathbb{R}^d$ such that $\mathcal{B}$ is shattered by the 2^r-fold union of the dual set system induced on $\mathcal{B}$.
Then Claim 11.11 implies that any $\frac{1}{2 \cdot 2^r}$-net of the dual set system induced on $\mathcal{B}$ has size at least $\frac{|\mathcal{B}|}{2}$. As $\frac{1}{2 \cdot 2^r} \geq \epsilon$, any ϵ-net of this set system has size at least

$$\frac{|\mathcal{B}|}{2} \geq \frac{\left\lfloor \frac{d}{2} \right\rfloor 2^r r}{4} \geq \left\lfloor \frac{d}{2} \right\rfloor \frac{1}{16\epsilon} \left(\log \frac{1}{2\epsilon} - 1 \right) = \left\lfloor \frac{d}{2} \right\rfloor \frac{1}{16\epsilon} \log \frac{1}{4\epsilon},$$

where the second step used the fact that $2^r = 2^{\left\lfloor \log \frac{1}{2\epsilon} \right\rfloor} \geq 2^{\log \frac{1}{2\epsilon} - 1} = \frac{1}{4\epsilon}$. Finally, similar to the proof of Lemma 11.12, one can 'duplicate' rectangles in the construction of Theorem 11.3 to increase the size of $\mathcal{B}$ to n (one has to be careful with the geometric placement of the duplicates). □

Half-spaces in $\mathbb{R}^d$: Theorem 11.6 together with Claim 11.11 implies the following.

LEMMA 11.15. *Given any $\epsilon \in \left[0, \frac{1}{8}\right]$ and integers n and $d \geq 4$, there exists a set P of n points in $\mathbb{R}^d$ such that any ϵ-net of the set system induced on P by half-spaces has size at least $\left\lfloor \frac{d}{4} \right\rfloor \frac{1}{16\epsilon} \log \frac{1}{4\epsilon}$.*

Bibliography and discussion. We presented the lower bound for ϵ-nets of abstract set systems first as its probabilistic proof is of independent interest. Technically that is not necessary, as it is implied by the lower bound for half-spaces in $\mathbb{R}^d$. The proof of Theorem 11.13 is given in [**KPW92**] (see also an exposition in [**PA95**]). The relation between lower bounds on the VC-dimension of k-fold unions and sizes of ϵ-nets was implicit in [**PT13**], and formulated explicitly by M. Csikós [**Csi16**].

[Csi16] M. Csikós. Personal communication. 2016.

[KPW92] J. Komlós, J. Pach, and G. Woeginger, *Almost tight bounds for ϵ-nets*, Discrete Comput. Geom. **7** (1992), no. 2, 163–173, DOI 10.1007/BF02187833. MR1139078

[PA95] J. Pach and P. K. Agarwal, *Combinatorial geometry*, Wiley-Interscience Series in Discrete Mathematics and Optimization, John Wiley & Sons, Inc., New York, 1995. A Wiley-Interscience Publication, DOI 10.1002/9781118033203. MR1354145

[PT13] J. Pach and G. Tardos, *Tight lower bounds for the size of epsilon-nets*, J. Amer. Math. Soc. **26** (2013), no. 3, 645–658, DOI 10.1090/S0894-0347-2012-00759-0. MR3037784

CHAPTER 12

Epsilon-Approximations: First Bounds

An ϵ-net N of a set system $(X, \mathcal{F})$ with n elements is a 'threshold' structure: N contains at least one element from each $S \in \mathcal{F}$ with $|S| \geq \epsilon n$. It does not matter if $|S| = \epsilon n$ or $|S| = n$—in both cases N is only required to contain at least one element of S.

In this chapter we consider a finer notion, where an approximation $A \subseteq X$ not only contains at least one element from each $S \in \mathcal{F}$ with $|S| \geq \epsilon n$, but the number of elements of A from S increases proportionally with the size of S. To see what notion of 'proportional' can be expected, consider the following sequence of approximations of $\mathcal{F}$ (these approximations are generally not possible; they should be seen as an 'ideal'):

Let $R_1 \subseteq X$ be such that $|R_1| = \frac{n}{2}$, and

$$\text{for all } S \in \mathcal{F}, \qquad |S \cap R_1| = \tfrac{|S|}{2}.$$

Repeating the same on the set system $(R_1, \mathcal{F}|_{R_1})$, let $R_2 \subseteq R_1$ be a set of size $\frac{|R_1|}{2} = \frac{n}{4}$ such that $|S \cap R_2| = \frac{|S \cap R_1|}{2} = \frac{|S|}{4}$. See the figure for an example where X is a set of points in $\mathbb{R}^2$ and $\mathcal{F}$ is induced by three disks. Continuing in this manner, let R_i be the set after i iterations. Then we have

$$|R_i| = \frac{n}{2^i} \qquad \text{and} \qquad |S \cap R_i| = \frac{|S|}{2^i} \quad \text{for all } S \in \mathcal{F}.$$

Each R_i is 'perfectly representative' of $\mathcal{F}$, in the sense that for each $S \in \mathcal{F}$ the proportion of points of S in R_i is equal to the proportion of points of S in X:

$$\left| \frac{|S|}{|X|} - \frac{|S \cap R_i|}{|R_i|} \right| = \left| \frac{|S|}{n} - \frac{|S|/2^i}{n/2^i} \right| = 0.$$

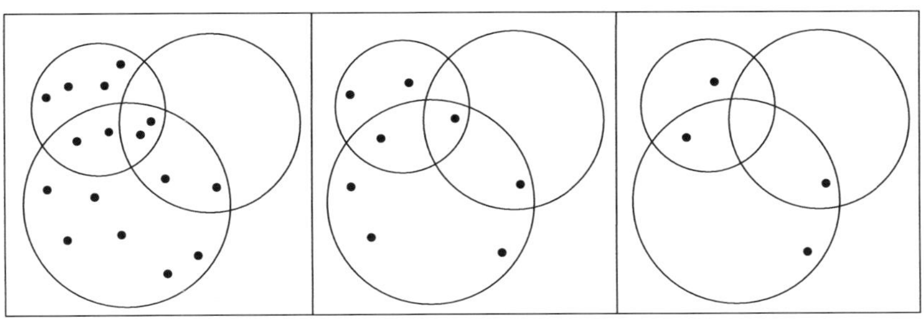

X $\qquad$ $R_1 \subseteq X$ $\qquad$ $R_2 \subseteq R_1$

How closely the above ideal can be realized is captured by the notion of an ϵ-approximation.

We use the notation $A = B \pm C$ as a shorthand for $A \in [B - C, B + C]$, where $A, B \in \mathbb{R}$ and $C \in \mathbb{R}^+$.

DEFINITION 12.1. Given a finite set system $(X, \mathcal{F})$ and a parameter $\epsilon > 0$, a set $A \subseteq X$ is an ϵ-approximation of $\mathcal{F}$ if for each $S \in \mathcal{F}$,

$$|S \cap A| = \frac{|S| \cdot |A|}{|X|} \pm \epsilon \cdot |A|.$$

Equivalently, dividing by $|A|$ gives

$$\left| \frac{|S|}{|X|} - \frac{|S \cap A|}{|A|} \right| \leq \epsilon.$$

The main topic of this chapter is the existence and construction of ϵ-approximations of small size. To see what bounds one can expect, it is instructive to consider the behavior of a fixed set $S \in \mathcal{F}$ with respect to a random sample A of size t picked uniformly out of all $\binom{n}{t}$ possible subsets of X.

For any $S \in \mathcal{F}$,

$$\mathrm{E}\left[|S \cap A|\right] = \sum_{p \in S} \Pr\left[p \in A\right] = |S| \cdot \frac{t}{n}.$$

We compute the probability that A is an ϵ-approximation of the set S. That is, A satisfies

$$|S \cap A| = \frac{|S|\, t}{n} \pm \epsilon\, t = \frac{|S|\, t}{n} \left(1 \pm \frac{\epsilon n}{|S|}\right).$$

Applying Chernoff's bound (Corollary 1.24) with $\mathrm{E}\left[|S \cap A|\right] = \frac{|S|t}{n}$ and $\delta = \frac{\epsilon n}{|S|}$,

$$\Pr\left[|S \cap A| > \frac{|S|t}{n}\left(1 + \frac{\epsilon n}{|S|}\right)\right] \leq \exp\left(-\frac{\frac{\epsilon^2 n^2}{|S|^2}}{2 + \frac{\epsilon n}{|S|}} \cdot \frac{|S|t}{n}\right)$$
$$= \exp\left(-\frac{\epsilon^2 n\, t}{2|S| + \epsilon n}\right) \leq e^{-\frac{\epsilon^2 t}{3}},$$

since $2|S| + \epsilon n \leq 3n$. The other direction is similar:

$$\Pr\left[|S \cap A| < \frac{|S|t}{n}\left(1 - \frac{\epsilon n}{|S|}\right)\right] \leq \exp\left(-\frac{\epsilon^2 n^2}{2|S|^2} \cdot \frac{|S|t}{n}\right)$$
$$= \exp\left(-\frac{\epsilon^2 n t}{2\,|S|}\right) \leq e^{-\frac{\epsilon^2 t}{2}}.$$

Thus to get even a constant probability, via Chernoff's bound, of A being an ϵ-approximation with respect to a *fixed* $S \in \mathcal{F}$, we need to set $t = \Omega\left(\frac{1}{\epsilon^2}\right)$. Furthermore, using the union bound over all sets of $\mathcal{F}$, we get

$$\Pr\left[A \text{ fails to be an } \epsilon\text{-approximation of } \mathcal{F}\right] \leq |\mathcal{F}|\, 2e^{-\frac{\epsilon^2 t}{3}}.$$

This immediately gives a first upper bound on the sizes of ϵ-approximations.

THEOREM 12.2. *Let $(X, \mathcal{F})$ be a finite set system and $\epsilon, \gamma > 0$ be given parameters. Then a uniform random sample $A \subseteq X$ of size at least $\frac{3}{\epsilon^2} \ln \frac{2|\mathcal{F}|}{\gamma}$ is an ϵ-approximation of $\mathcal{F}$ with probability at least $1 - \gamma$.*

Note that the bound of Theorem 12.2 holds for any set system. It can be seen as the ϵ-approximation analog of Theorem 2.3. Theorem 2.3 also describes a greedy deterministic $O(|X||\mathcal{F}|)$-time algorithm for the construction of ϵ-nets of size $O\left(\frac{1}{\epsilon}\log|\mathcal{F}|\right)$. This was generalized by Chazelle to a deterministic greedy $O(|X||\mathcal{F}|)$-time algorithm to compute an ϵ-approximation of size $O\left(\frac{1}{\epsilon^2}\log|\mathcal{F}|\right)$ (see Chapter 4 in 'The Discrepancy Method').

The main goal of this chapter is to show the existence of ϵ-approximations of size $O\left(\frac{d}{\epsilon^2}\log\frac{1}{\epsilon}\right)$ for set systems of VC-dimension d.

1. Proof using Induction

Theoretical physics is not just doing calculations. It's setting up the problem so that any fool could do the calculation.

Philip Anderson

The main theorem we will prove in this section is the following.

THEOREM 12.3. *There exists a positive constant C_1 such that the following is true. Let $(X, \mathcal{F})$ be a finite set system with* VC-dim $(\mathcal{F}) \leq d$, *and let $0 < \epsilon, \gamma < \frac{1}{2}$ be given parameters. Let A be a uniform random subset of X of size at least*

$$\frac{C_1}{\epsilon^2} \cdot \left(d \ln \frac{1}{\epsilon} + \ln \frac{1}{\gamma} \right).$$

Then A is an ϵ-approximation of $\mathcal{F}$ with probability at least $1 - \gamma$.

Overview of ideas. Given $(X, \mathcal{F})$ with $n = |X|$, the earlier bound of Theorem 12.2 depends on $|\mathcal{F}|$:

> a random sample $A_1 \subseteq X$ of size $\Omega\left(\frac{1}{\epsilon^2} \ln |\mathcal{F}|\right)$ is an ϵ-approximation of $\mathcal{F}$ with constant probability.

As $|\mathcal{F}| = O\left(n^d\right)$ if VC-dim $(\mathcal{F}) \leq d$ (Lemma 4.3), we have

$$|A_1| = O\left(\frac{1}{\epsilon^2} \ln n^d\right).$$

Since the size of A_1 is much smaller than that of X, the size of $\mathcal{F}|_{A_1}$ is much smaller than that of $\mathcal{F}$. Thus applying Theorem 12.2 again, this time to $\mathcal{F}|_{A_1}$, gives an ϵ-approximation $A_2 \subseteq A_1$ of $\mathcal{F}|_{A_1}$, with

$$|A_2| = O\left(\frac{1}{\epsilon^2} \ln |A_1|^d\right) = O\left(\frac{d}{\epsilon^2} \ln \left(\frac{1}{\epsilon^2} \ln n^d\right)\right) = O\left(\frac{d}{\epsilon^2} \ln \frac{1}{\epsilon^2} + \frac{d}{\epsilon^2} \ln \ln n^d\right). \tag{12.4}$$

Now observe that A_2 is a (2ϵ)-approximation of $\mathcal{F}$:

LEMMA 12.5. *Given a set system $(X, \mathcal{F})$, let $A \subseteq X$ be an ϵ_1-approximation of $\mathcal{F}$ and let $B \subseteq A$ be an ϵ_2-approximation of $\mathcal{F}|_A$. Then B is a $(\epsilon_1 + \epsilon_2)$-approximation of $\mathcal{F}$.*

PROOF. Fix any $S \in \mathcal{F}$. Then,

A is an ϵ_1-approximation of $\mathcal{F}$: $\quad |S \cap A| = \dfrac{|S| \cdot |A|}{|X|} \pm \epsilon_1 \cdot |A|.$

B is an ϵ_2-approximation of $\mathcal{F}|_A$: $\quad |(S \cap A) \cap B| = \dfrac{|S \cap A| \cdot |B|}{|A|} \pm \epsilon_2 \cdot |B|.$

As $B \subseteq A$,

$$|S \cap B| = |(S \cap A) \cap B| = \frac{\left(\frac{|S||A|}{|X|} \pm \epsilon_1 \cdot |A|\right) \cdot |B|}{|A|} \pm \epsilon_2 \cdot |B| = \frac{|S| \cdot |B|}{|X|} \pm (\epsilon_1 + \epsilon_2) \cdot |B|.$$

□

Thus if we take A_1 to be an $\frac{\epsilon}{2}$-approximation of $\mathcal{F}$ and A_2 a $\frac{\epsilon}{2}$-approximation of $\mathcal{F}|_{A_1}$, then A_2 is an ϵ-approximation of $\mathcal{F}$, of size $O\left(\frac{d}{\epsilon^2}\ln\frac{1}{\epsilon}+\frac{d}{\epsilon^2}\ln\ln n^d\right)$ (Equation (12.4)). This already improves upon the bound of Theorem 12.2!

In fact, we could keep applying Theorem 12.2 to get smaller and smaller approximations, with the error of approximation increasing with each application. Each successive iteration gives an improvement, as the size of each successive approximation is decreasing much faster than the increase in its error, as the formal proof of Theorem 12.3 now demonstrates.

This insight can be implemented in a proof in two ways: the straightforward method is to do the calculations explicitly over all iterations to prove that the last iteration produces an ϵ-approximation of the desired size with the desired probability. Alternatively, one can just do the calculations for a single step and let induction 'take care' of the remaining iterations. We give the second type of proof, though both are equivalent; the proof of the next section will follow the first method.

PROOF OF THEOREM 12.3. Given parameters ϵ and γ, let $T(\epsilon,\gamma)$ be a positive integer such that a uniform random sample of size at least $T(\epsilon,\gamma)$ from X is an ϵ-approximation of $\mathcal{F}$ with probability at least $1-\gamma$. Then our goal is to show that

$$T(\epsilon,\gamma)\leq\frac{C_1}{\epsilon^2}\cdot\left(d\ln\frac{1}{\epsilon}+\ln\frac{1}{\gamma}\right),$$

where C_1 is a sufficiently large constant.

The proof will proceed by induction on ϵ. When $\epsilon\leq 1/\sqrt{|X|}$, then as X itself is an ϵ-approximation of $\mathcal{F}$, we have $T(\epsilon,\gamma)=|X|\leq\frac{1}{\epsilon^2}$. Otherwise we construct an ϵ-approximation of $\mathcal{F}$ in two steps:

Step 1: take a uniform random sample A' from X of size large-enough so that A' is an $\frac{\epsilon}{2}$-approximation of $\mathcal{F}$ with probability at least $1-\frac{\gamma}{2}$. By inductive hypothesis, it suffices to set $|A'|\geq T\left(\frac{\epsilon}{2},\frac{\gamma}{2}\right)$.

Step 2: take a uniform random sample A from A' of size large-enough so that A is an $\frac{\epsilon}{2}$-approximation of $(A',\mathcal{F}|_{A'})$ with probability at least $1-\frac{\gamma}{2}$. By Theorem 12.2, it suffices to set

$$|A|\geq\frac{3}{(\epsilon/2)^2}\ln\frac{2\,|\mathcal{F}|_{A'}|}{(\gamma/2)}.$$

Thus the set A is a uniform random sample of X, and an ϵ-approximation of $\mathcal{F}$ (Lemma 12.5) with probability at least $1-\gamma$. So we can set

$$T(\epsilon,\gamma)=\frac{3}{(\epsilon/2)^2}\ln\frac{2\,|\mathcal{F}|_{A'}|}{(\gamma/2)},\qquad\text{where } |A'|=T\left(\frac{\epsilon}{2},\frac{\gamma}{2}\right).$$

Now one can verify that $T(\epsilon,\gamma)\leq\frac{C_1}{\epsilon^2}\left(d\ln\frac{1}{\epsilon}+\ln\frac{1}{\gamma}\right)$ for a sufficiently large absolute constant C_1, completing the proof.

The inductive step calculations are as follows:

$$
\begin{aligned}
T(\epsilon,\gamma) &= \frac{3}{(\epsilon/2)^2}\ln\frac{2\,|\mathcal{F}|_{A'}|}{(\gamma/2)} \le \frac{12}{\epsilon^2}\ln\left(\frac{4}{\gamma}\left(\frac{e\,T\left(\frac{\epsilon}{2},\frac{\gamma}{2}\right)}{d}\right)^d\right) \quad \left(\text{Lemma 4.3}\right)\\
&\le \frac{12}{\epsilon^2}\ln\left(\frac{4}{\gamma}\left(\frac{e\,\frac{4\,C_1}{\epsilon^2}\left(d\ln\frac{2}{\epsilon}+\ln\frac{2}{\gamma}\right)}{d}\right)^d\right) \quad \left(\text{inductive hypothesis}\right)\\
&\le \frac{12}{\epsilon^2}\ln\left(\frac{4}{\gamma}\left(\frac{e\frac{4\,C_1}{\epsilon^2}\frac{2}{\epsilon}\left(d+\ln\frac{2}{\gamma}\right)}{d}\right)^d\right)\\
&\le \frac{12}{\epsilon^2}\ln\left(\frac{4}{\gamma}\left(\frac{e\,8\,C_1}{\epsilon^3}\left(1+\frac{1}{d}\ln\frac{2}{\gamma}\right)\right)^d\right)\\
&\le \frac{12}{\epsilon^2}\ln\left(\frac{4}{\gamma}\left(\frac{e\,8\,C_1}{\epsilon^3}\exp\left(\frac{1}{d}\ln\frac{2}{\gamma}\right)\right)^d\right) = \frac{12}{\epsilon^2}\ln\left(\frac{4}{\gamma}\left(\frac{e\,8\,C_1}{\epsilon^3}\right)^d\frac{2}{\gamma}\right)\\
&= \frac{12}{\epsilon^2}\left(d\ln\frac{e\,8\,C_1}{\epsilon^3}+\ln\frac{8}{\gamma^2}\right) \le \frac{C_1}{\epsilon^2}\left(d\ln\frac{1}{\epsilon}+\ln\frac{1}{\gamma}\right),
\end{aligned}
$$

for a large-enough constant C_1.

□

We conclude by showing that Theorem 12.3 together with Theorem 2.3 implies bounds on ϵ-nets.

THEOREM 12.6. *Let $(X,\mathcal{F})$ be a finite set system with* VC-dim$(\mathcal{F}) \le d$ *for an integer $d \ge 1$, and let $\epsilon \in \left(0,\frac{1}{2}\right)$ be a given parameter. Then there exists an absolute constant $C_8 > 0$ such that a uniform random sample N of X of size $\frac{C_8}{\epsilon}\ln\frac{1}{\epsilon^d\gamma}$ is an ϵ-net of $\mathcal{F}$ with probability at least $1-\gamma$.*

PROOF. Let A be a uniform random sample of X of size $\frac{4C_1}{\epsilon^2}\left(d\ln\frac{2}{\epsilon}+\ln\frac{2}{\gamma}\right)$ and let N be a uniform random sample of A of size $\frac{2}{\epsilon}\ln\frac{2|\mathcal{F}|_A|}{\gamma}$. By Theorem 12.3, A is an $\frac{\epsilon}{2}$-approximation of $\mathcal{F}$ with probability $1-\frac{\gamma}{2}$, and by Theorem 2.3, N is an $\frac{\epsilon}{2}$-net of $\mathcal{F}|_A$ with probability at least $1-\frac{\gamma}{2}$. We first observe that N is an ϵ-net of $\mathcal{F}$ with probability at least $1-\gamma$:

If A is an $\frac{\epsilon}{2}$-approximation of $\mathcal{F}$, then for each $S \in \mathcal{F}$ with $|S| \ge \epsilon\,|X|$ we have

$$|S \cap A| \ge \frac{|S||A|}{|X|} - \frac{\epsilon}{2}|A| \ge \frac{\epsilon}{2}|A|.$$

Thus if N is an $\frac{\epsilon}{2}$-net of $\mathcal{F}|_A$, then N hits $S \cap A$ and so hits S.

Finally,

$$\begin{aligned}
|N| &\leq \frac{2}{\epsilon} \ln \frac{2|\mathcal{F}|_A|}{\gamma} \leq \frac{2}{\epsilon} \ln \left(\frac{2}{\gamma} \left(\frac{e|A|}{d} \right)^d \right) \\
&\leq \frac{2}{\epsilon} \ln \left(\frac{2}{\gamma} \left(\frac{e4C_1}{\epsilon^2} \right)^d \left(\ln \frac{2}{\epsilon} + \frac{1}{d} \ln \frac{2}{\gamma} \right)^d \right) \\
&\leq \frac{2}{\epsilon} \ln \left(\frac{2}{\gamma} \left(\frac{e8C_1}{\epsilon^3} \right)^d \left(1 + \frac{1}{d} \ln \frac{2}{\gamma} \right)^d \right) \\
&\leq \frac{2}{\epsilon} \ln \left(\frac{2}{\gamma} \left(\frac{e8C_1}{\epsilon^3} \right)^d \frac{2}{\gamma} \right) \\
&\leq \frac{C_8}{\epsilon} \ln \frac{1}{\epsilon^d \gamma},
\end{aligned}$$

by setting C_8 to be a sufficiently large constant depending only on C_1. □

Bibliography and discussion. The original proof of Theorem 12.3 is from [**VC71**]. The proof given in this section is folklore; see [**Mat95b, CM96**] for other more sophisticated proofs which also give efficient algorithms. The argument we gave can be 'compressed' into a direct combinatorial argument resembling the proof of Theorem 6.1. On the other hand, 'decompressing' the construction gives the discrepancy proof presented in the next section.

[CM96] B. Chazelle and J. Matoušek, *On linear-time deterministic algorithms for optimization problems in fixed dimension*, J. Algorithms **21** (1996), no. 3, 579–597, DOI 10.1006/jagm.1996.0060. MR1417665

[Mat95b] J. Matoušek, *Approximations and optimal geometric divide-and-conquer*, J. Comput. System Sci. **50** (1995), no. 2, 203–208, DOI 10.1006/jcss.1995.1018. 23rd Symposium on the Theory of Computing (New Orleans, LA, 1991). MR1330253

[VC71] V. N. Vapnik and A. Ja. Červonenkis, *The uniform convergence of frequencies of the appearance of events to their probabilities* (Russian, with English summary), Teor. Verojatnost. i Primenen. **16** (1971), 264–279. MR0288823

2. Proof using Discrepancy

The point is, don't let your ego disempower you. Ask the dumb question. And if they try to make you feel dumb, they are making a power move.

Matt Stoller

In this section we will see a relation between ϵ-approximations and another fundamental notion called combinatorial discrepancy, defined as follows.

DEFINITION 12.7. Given a set system $(X, \mathcal{F})$ and a two-coloring $\chi\colon X \to \{-1, 1\}$, define the discrepancy of a set $R \subseteq X$ with respect to χ as

$$\operatorname{disc}_\chi(R) = \left| \sum_{p \in R} \chi(p) \right|,$$

and the discrepancy of $\mathcal{F}$ with respect to χ as

$$\operatorname{disc}_\chi(\mathcal{F}) = \max_{R \in \mathcal{F}} \operatorname{disc}_\chi(R).$$

The discrepancy of $\mathcal{F}$ is then defined to be

$$\operatorname{disc}(\mathcal{F}) = \min_{\chi: X \to \{-1,1\}} \operatorname{disc}_\chi(\mathcal{F}).$$

As an illustration of the connection between approximations and discrepancy, we will first reprove the result of the previous section.

THEOREM 12.8. *Let $(X, \mathcal{F})$ be a finite set system with $\text{VC-dim}(\mathcal{F}) \leq d$ and let $\epsilon \in \left(0, \frac{1}{2}\right]$ be a given parameter. Then there exists an ϵ-approximation of $\mathcal{F}$ of size $O\left(\frac{d}{\epsilon^2} \log \frac{1}{\epsilon}\right)$.*

The following more general theorem establishes a relation between the size of ϵ-approximations and the combinatorial discrepancy of a set system.

THEOREM 12.9. *Let $(X, \mathcal{F})$ be a finite set system with $X \in \mathcal{F}$ and let $f(\cdot)$ be a function such that $\operatorname{disc}(\mathcal{F}|_Y) \leq f(|Y|)$ for all $Y \subseteq X$. Then for every integer $t \geq 0$, there exists a set $A \subseteq X$ of size $\left\lceil \frac{n}{2^t} \right\rceil$ such that A is an ϵ-approximation of $\mathcal{F}$, where*

$$\epsilon = \frac{2}{n} \left(f(n) + 2f\left(\left\lceil \frac{n}{2} \right\rceil\right) + \cdots + 2^{t-1} f\left(\left\lceil \frac{n}{2^{t-1}} \right\rceil\right) \right).$$

In particular, if there exists a constant $c > 1$ such that $f(2m) \leq \frac{2}{c} f(m)$ for all $m \geq \left\lceil \frac{n}{2^t} \right\rceil$, then there exists an ϵ-approximation of $\mathcal{F}$ of size $\left\lceil \frac{n}{2^t} \right\rceil$, and where

$$\epsilon = \Theta\left(\frac{2^t}{n} f\left(\left\lceil \frac{n}{2^t} \right\rceil \right) \right).$$

For technical simplicity we will assume that n is a power of 2.

Overview of ideas. The proof follows by iteratively selecting a set R_1 of half of the elements of X while also ensuring that R_1 contains close to half of the elements from each set of $\mathcal{F}$[1]. Of course a set $R_1 \subseteq X$ that contains $\frac{|X|}{2}$ points and also

[1] This idea is similar to the proof in Chapter 8.

exactly halves every set of $\mathcal{F}$ is not always possible. The accuracy of R_1 will have to depend on the number of sets in $\mathcal{F}$—e.g., for the complete set system $\mathcal{F} = 2^X$, for any choice of R_1 with $\frac{n}{2}$ elements, there will be a set of $\frac{n}{2}$ elements in $\mathcal{F}$—namely the set $X \setminus R_1$—that will not contain *any* point of R_1. Thus one asks for a set $R_1 \subseteq X$, $|R_1| = \frac{n}{2}$, such that for all $S \in \mathcal{F}$, $|S \cap R_1|$ is as close as possible to $\frac{|S|}{2}$. This is essentially the notion of combinatorial discrepancy. In fact, discrepancy can be seen as a 'first step' towards ϵ-approximations, as follows.

> Let χ be a two-coloring of X realizing disc $(\mathcal{F})$ and let X^+ be the elements of X with color '1'. Assume for simplicity that $|X^+| = \frac{|X|}{2}$. Then for any $S \in \mathcal{F}$,
>
> $$\big||S \cap X^+| - \left(|S| - |S \cap X^+|\right)\big| \leq \operatorname{disc}(\mathcal{F}),$$
>
> implying that
>
> $$\left||S \cap X^+| - \frac{|S|}{2}\right| \leq \frac{\operatorname{disc}(\mathcal{F})}{2}.$$
>
> In other words,
>
> $$|S \cap X^+| = \frac{|S| \cdot |X^+|}{|X|} \pm \frac{\operatorname{disc}(\mathcal{F})}{n}|X^+|,$$
>
> and so the set X^+ is an $\left(\frac{\operatorname{disc}(\mathcal{F})}{n}\right)$-approximation of $\mathcal{F}$.

To prove Theorem 12.8, we will first derive an upper bound on the discrepancy of set systems with VC-dimension at most d, such that exactly half the elements are colored '1'. Denote the set of these elements by R_1. Now iteratively repeat this halving process on $\mathcal{F}|_{R_1}$ to get a set $R_2 \subseteq R_1$ and so on. The proof will show that after t iterations, where t depends on $\frac{1}{\epsilon}$, the set R_t of $|X|/2^t$ elements is an ϵ-approximation of $\mathcal{F}$, satisfying the required bounds.
The same method, using the function $f(\cdot)$ to upper bound discrepancy at each iteration, gives the proof of Theorem 12.9.

Our first statement, following the above discussion, gives an upper bound on combinatorial discrepancy of set systems. We will state and prove it in a slightly different setting where, instead of assigning colors, the goal is to choose a subset of X of size $|X|/2$ that contains close to $\frac{|S|}{2}$ elements from each $S \in \mathcal{F}$.

LEMMA 12.10. *Given a set system $(X, \mathcal{F})$, there exists a set $R_1 \subseteq X$ with $|R_1| = \frac{n}{2}$ such that*

$$\text{for all } S \in \mathcal{F}, \qquad |S \cap R_1| = \frac{|S|}{2} \pm \sqrt{2|S| \ln(2|\mathcal{F}|)}.$$

PROOF. Let R_1 be a uniform random subset of X of size $\frac{n}{2}$. Note that for any $p \in X$, we have $\Pr[p \in R_1] = \frac{1}{2}$ and so for any $S \in \mathcal{F}$ we have $\mathrm{E}[|S \cap R_1|] = \frac{|S|}{2}$. For a positive integer Δ_S denoting the error term for $S \in \mathcal{F}$, Chernoff's bound (Corollary 1.24), applied to the indicator random variables for the elements of S, gives

$$\Pr\left[|S \cap R_1| > \frac{|S|}{2} + \Delta_S\right] = \Pr\left[|S \cap R_1| > \frac{|S|}{2}\left(1 + \frac{2\Delta_S}{|S|}\right)\right]$$
$$< \exp\left(-\frac{\left(\frac{2\Delta_S}{|S|}\right)^2}{2 + \frac{2\Delta_S}{|S|}} \cdot \frac{|S|}{2}\right)$$
$$= \exp\left(-\frac{\Delta_S^2}{|S| + \Delta_S}\right) < e^{-\frac{\Delta_S^2}{2|S|}},$$

where the last step follows from the assumption that $\Delta_S < \frac{|S|}{2}$ (otherwise the above probability is 0). Setting $\Delta_S = \sqrt{2|S|\ln(2|\mathcal{F}|)}$, the probability that there exists a set $S \in \mathcal{F}$ with $|S \cap R_1| > \frac{|S|}{2} + \Delta_S$ can be upper bounded, using the union bound, as

$$\sum_{S\in\mathcal{F}} \Pr\left[|S \cap R_1| > \frac{|S|}{2} + \Delta_S\right] < \sum_{S\in\mathcal{F}} e^{-\frac{\Delta_S^2}{2|S|}} = |\mathcal{F}| \cdot \frac{1}{2|\mathcal{F}|} = \frac{1}{2}.$$

By symmetry, the probability that there exists a set $S \in \mathcal{F}$ with $|S \cap R_1| < \frac{|S|}{2} - \Delta_S$ is also less than $\frac{1}{2}$. This implies the existence of the required set R_1. □

Note that Lemma 12.10 does not require $\mathcal{F}$ to have bounded VC-dimension and holds for any set system. The fact that $\mathcal{F}$ has VC-dimension at most d will only be used to plug in an upper bound on $|\mathcal{F}|$ via Lemma 4.3.

PROOF OF THEOREM 12.8. We will compute the sets $R_1, R_2, \ldots$ iteratively. Set $R_0 = X$. Apply Lemma 12.10 to $(X, \mathcal{F})$ to get a set $R_1 \subset X$, $|R_1| = \frac{n}{2}$, such that

$$\text{for all } S \in \mathcal{F}, \quad |S \cap R_1| = \frac{|S|}{2} \pm 2\sqrt{|R_0| \ln |\mathcal{F}|_{R_0}|}. \tag{12.11}$$

In fact Lemma 12.10 gives a stronger statement where the error for each set S is relative to the size of S—that is, $|S \cap R_1| = \frac{|S|}{2} \pm 2\sqrt{|S| \ln |\mathcal{F}|_{R_0}|}$. For ϵ-approximations the weaker statement suffices and simplifies subsequent calculations. The stronger statement yields a sensitive approximation; see Definition 14.7.

Consider the set system derived from projecting $\mathcal{F}$ onto R_1:

$$\mathcal{F}|_{R_1} = \{S \cap R_1 \colon S \in \mathcal{F}\}.$$

Now applying Lemma 12.10 on $(R_1, \mathcal{F}|_{R_1})$ gives a set $R_2 \subset R_1$, $|R_2| = \frac{|R_1|}{2} = \frac{n}{4}$, such that for all $S \in \mathcal{F}$,

$$|S \cap R_2| = \frac{|S \cap R_1|}{2} \pm 2\sqrt{|R_1| \ln |\mathcal{F}|_{R_1}|}$$
$$= \frac{\frac{|S|}{2} \pm 2\sqrt{|R_0| \ln |\mathcal{F}|_{R_0}|}}{2} \pm 2\sqrt{|R_1| \ln |\mathcal{F}|_{R_1}|}$$
$$= \frac{|S|}{4} \pm 2\sum_{j=0}^{1} \frac{\sqrt{|R_j| \ln |\mathcal{F}|_{R_j}|}}{2^{1-j}}.$$

Continuing on, after i iterations, we have the set R_i, $|R_i| = \frac{n}{2^i}$, such that for every $S \in \mathcal{F}$,

$$\begin{aligned}
|S \cap R_i| &= \frac{|S|}{2^i} \pm 2 \sum_{j=0}^{i-1} \frac{\sqrt{|R_j| \ln |\mathcal{F}|_{R_j}|}}{2^{(i-1)-j}} \\
&= \frac{|S|\,|R_i|}{n} \pm \left(\frac{2^{i+1}}{n} \sum_{j=0}^{i-1} \frac{\sqrt{|R_j| \ln |\mathcal{F}|_{R_j}|}}{2^{(i-1)-j}} \right) |R_i| \\
&= \frac{|S|\,|R_i|}{n} \pm \left(\frac{4}{\sqrt{n}} \sum_{j=0}^{i-1} \sqrt{2^j \ln |\mathcal{F}|_{R_j}|} \right) |R_i| \\
&= \frac{|S|\,|R_i|}{n} \pm \left(\frac{4}{\sqrt{n}} \sum_{j=0}^{i-1} \sqrt{2^j \ln \left(\frac{en}{2^j d} \right)^d} \right) |R_i|.
\end{aligned}$$

The last step uses Lemma 4.3. The above summation can be upper bounded by an increasing geometric series, since

$$\frac{2^{j+1} \ln \left(\frac{en}{2^{j+1} d} \right)^d}{2^j \ln \left(\frac{en}{2^j d} \right)^d} = \frac{2 \left(\ln \left(\frac{en}{2^j d} \right) - \ln 2 \right)}{\ln \left(\frac{en}{2^j d} \right)} = 2 - \frac{2 \ln 2}{\ln \left(\frac{en}{2^j d} \right)} \geq 1.1,$$

as $\frac{en}{2^j d} \geq 6$ for $j \leq i \leq \log \frac{en}{6d}$, which will be the case. Therefore there exists an absolute constant $c' \geq 1$ such that

$$|S \cap R_i| = \frac{|S|\,|R_i|}{n} \pm \left(c' \frac{4}{\sqrt{n}} \sqrt{2^i \ln \left(\frac{en}{2^i d} \right)^d} \right) |R_i|.$$

We set the number of iterations, i, to a value t such that the second term of the above expression is at most $\epsilon\,|R_i|$.

We verify that $t = \log \left(\frac{\epsilon^2 n}{dc \log \frac{1}{\epsilon}} \right)$ works, for a sufficiently large constant $c \geq 6$ depending only on c':

$$\begin{aligned}
c' \frac{4}{\sqrt{n}} \sqrt{2^t \ln \left(\frac{en}{2^t d} \right)^d} &= c' \frac{4}{\sqrt{n}} \sqrt{\frac{\epsilon^2 n}{dc \log \frac{1}{\epsilon}} \cdot d \cdot \ln \frac{en}{d} \frac{dc \log \frac{1}{\epsilon}}{\epsilon^2 n}} \\
&\leq c'\, 4 \epsilon \sqrt{\frac{1}{c \log \frac{1}{\epsilon}} \ln \frac{ec \log \frac{1}{\epsilon}}{\epsilon^2}} \\
&\leq c'\, 4 \epsilon \sqrt{\frac{3e \ln c}{c}} \leq \epsilon, \qquad \text{for sufficiently large } c.
\end{aligned}$$

Thus R_t is an ϵ-approximation of $\mathcal{F}$, with

$$|R_t| = \frac{n}{2^t} = n \frac{dc \log \frac{1}{\epsilon}}{\epsilon^2 n} = O\left(\frac{d}{\epsilon^2} \log \frac{1}{\epsilon} \right).$$

This completes the proof of Theorem 12.8. □

PROOF OF THEOREM 12.9. The proof follows by re-working the calculation of Theorem 12.8 using the discrepancy function $f(\cdot)$ instead of Lemma 12.10. Set $R_0 = X$ and for $i = 0, \ldots, t-1$, construct the set R_{i+1} of size $\frac{n}{2^{i+1}}$ as follows. Assume for simplicity that $n = |X|$ is a power of two.

Let χ_i be a two-coloring of $\mathcal{F}|_{R_i}$ such that $\operatorname{disc}_{\chi_i}(\mathcal{F}|_{R_i}) \le f(|R_i|)$ and let R_i^+ and R_i^- be the two color classes induced by χ_i. Say $|R_i^+| = \frac{|R_i|}{2} + \kappa$ and $|R_i^-| = \frac{|R_i|}{2} - \kappa$, for some integer $\kappa \ge 0$.

By the assumption that $X \in \mathcal{F}$ and hence $(X \cap R_i) = R_i \in \mathcal{F}|_{R_i}$, we have $\left| |R_i^+| - \left(|R_i| - |R_i^+|\right) \right| \le f(|R_i|)$, implying that $|R_i^+| = \frac{|R_i|}{2} \pm \frac{f(|R_i|)}{2}$, and so

$$\kappa \le \frac{f(|R_i|)}{2}.$$

Set $R_{i+1} \subseteq R_i^+$ to be any set of size $\frac{|R_i|}{2} = \frac{n}{2^{i+1}}$. Then for any set $S \in \mathcal{F}$, as $\left| |S \cap R_i^+| - \left(|S \cap R_i| - |S \cap R_i^+|\right) \right| \le f(|R_i|)$, we have $|S \cap R_i^+| = \frac{|S \cap R_i|}{2} \pm \frac{f(|R_i|)}{2}$ and thus

$$|S \cap R_{i+1}| = |S \cap R_i^+| \pm \kappa = \left(\frac{|S \cap R_i|}{2} \pm \frac{f(|R_i|)}{2}\right) \pm \kappa = \frac{|S \cap R_i|}{2} \pm f(|R_i|).$$

After t iterations we have computed a set R_t, $|R_t| = \frac{n}{2^t}$, such that for all $S \in \mathcal{F}$,

$$|S \cap R_t| = \frac{|S|}{2^t} \pm \sum_{j=0}^{t-1} \frac{f\left(\frac{n}{2^j}\right)}{2^{(t-1)-j}} = \frac{|S||R_t|}{n} \pm \left(\frac{2}{n}\sum_{j=0}^{t-1} 2^j f\left(\frac{n}{2^j}\right)\right)|R_t|.$$

This proves the first part.

For the second part, observe that if there exists a constant $c > 1$ such that $f(2m) \le \frac{2}{c} f(m)$ for all $m \ge \left\lceil \frac{n}{2^t} \right\rceil$, then $f\left(\frac{n}{2^j}\right) \le \frac{2}{c} f\left(\frac{n}{2^{j+1}}\right)$ or equivalently, $2^{j+1} f\left(\frac{n}{2^{j+1}}\right) \ge c \cdot 2^j f\left(\frac{n}{2^j}\right)$, for all $j \le t-1$. That is, the terms are increasing by at least a multiplicative factor of $c > 1$ and thus are upper bounded, within a constant factor depending on c, by the last term of the summation.

Thus we get a set R_t such that $|R_t| = \frac{n}{2^t}$ and for each $S \in \mathcal{F}$,

$$\begin{aligned} |S \cap R_t| &= \frac{|S||R_t|}{n} \pm \left(\frac{2}{n}\sum_{j=0}^{t-1} 2^j f\left(\frac{n}{2^j}\right)\right)|R_t| \\ &= \frac{|S||R_t|}{n} \pm \left(O\left(\frac{2^t f\left(\left\lceil \frac{n}{2^t}\right\rceil\right)}{n}\right)\right)|R_t|. \end{aligned}$$

□

Many of the best upper bounds on ϵ-approximations are derived from Theorem 12.9 together with bounds on combinatorial discrepancy for the corresponding set systems. For example, the primal set system $\mathcal{R}$ induced on a set P of n points by half-spaces in $\mathbb{R}^d$ has combinatorial discrepancy $O\left(n^{\frac{1}{2}-\frac{1}{2d}}\right)$. Thus Theorem 12.9, with

$$f(n) = O\left(n^{\frac{1}{2}-\frac{1}{2d}}\right) \quad \text{and} \quad 2^t = \Theta\left(\epsilon^{\frac{2d}{d+1}}\, n\right),$$

implies the existence of an ϵ-approximation of size $O\left(\frac{n}{2^t}\right) = O\left(\frac{1}{\epsilon^{\frac{2d}{d+1}}}\right)$ of $\mathcal{R}$.

Chapter 16 lists many other bounds on the sizes of ϵ-approximations derived via Theorem 12.9.

Bibliography and discussion. The connection between combinatorial discrepancy and ϵ-approximations was discovered in the important paper [**MWW9312b**]. An optimal bound on the discrepancy of half-spaces was given by Matoušek [**Mat95a**]. The proof of Theorem 12.8 gives a method for constructing an ϵ-approximation of size $O\left(\frac{d}{\epsilon^2}\log\frac{1}{\epsilon}\right)$, though it is randomized. A divide-and-conquer 'merge and reduce' procedure—which starts with an arbitrary partition of X into small-sized sets, which are then hierarchically merged—was used in [**CM9612b, Mat95b**] to derive a deterministic algorithm to compute an ϵ-approximation of size $O\left(\frac{d}{\epsilon^2}\log\frac{1}{\epsilon}\right)$ in time $O(d)^{3d}\frac{1}{\epsilon^{2d}}\left(\log\frac{d}{\epsilon}\right)^d n$.
Many other beautiful results in the general area of discrepancy theory have been nicely presented in the texts on the subject [**BC8712b, Mat99, Cha00, CST1412b**].

[BC8712b] J. Beck and W. W. L. Chen, *Irregularities of distribution*, Cambridge Tracts in Mathematics, vol. 89, Cambridge University Press, Cambridge, 1987, DOI 10.1017/CBO9780511565984. MR903025

[CM9612b] B. Chazelle and J. Matoušek, *On linear-time deterministic algorithms for optimization problems in fixed dimension*, J. Algorithms **21** (1996), no. 3, 579–597, DOI 10.1006/jagm.1996.0060. MR1417665

[CST1412b] N. Srivastava, *Discrepancy, graphs, and the Kadison-Singer problem*, Asia Pac. Math. Newsl. **3** (2013), no. 4, 15–20. MR3156130

[Cha00] B. Chazelle, *The discrepancy method*, Cambridge University Press, Cambridge, 2000. Randomness and complexity, DOI 10.1017/CBO9780511626371. MR1779341

[MWW9312b] J. Matoušek, E. Welzl, and L. Wernisch, *Discrepancy and approximations for bounded VC-dimension*, Combinatorica **13** (1993), no. 4, 455–466, DOI 10.1007/BF01303517. MR1262921

[Mat95a] J. Matoušek, *Tight upper bounds for the discrepancy of half-spaces*, Discrete Comput. Geom. **13** (1995), no. 3-4, 593–601, DOI 10.1007/BF02574066. MR1318799

[Mat95b] J. Matoušek, *Approximations and optimal geometric divide-and-conquer*, J. Comput. System Sci. **50** (1995), no. 2, 203–208, DOI 10.1006/jcss.1995.1018. 23rd Symposium on the Theory of Computing (New Orleans, LA, 1991). MR1330253

[Mat99] J. Matoušek. *Geometric Discrepancy: An Illustrated Guide*. Springer, 1999.

CHAPTER 13

Epsilon-Approximations: Improved Bounds

Recall the notion of an ϵ-approximation of a set system $(X, \mathcal{F})$.

DEFINITION 12.1. Given a finite set system $(X, \mathcal{F})$ and a parameter $\epsilon > 0$, a set $A \subseteq X$ is an ϵ-approximation of $\mathcal{F}$ if for each $S \in \mathcal{F}$,

$$|S \cap A| = \frac{|S| \cdot |A|}{|X|} \pm \epsilon \cdot |A|.$$

Equivalently, dividing by $|A|$ gives

$$\left| \frac{|S|}{|X|} - \frac{|S \cap A|}{|A|} \right| \leq \epsilon.$$

This chapter presents constructions of ϵ-approximations that improve the bounds of Chapter 12. In the first result, the algorithm is unchanged—we will take a single uniform random sample A from X and, improving upon previous analysis, show that if A has size $\Omega\left(\frac{d}{\epsilon^2}\right)$, then A is an ϵ-approximation of $\mathcal{F}$ with constant probability. The key tool used is *chaining*.

Chaining is essentially the idea of doing the analysis by partitioning each $S \in \mathcal{F}$ into a number of smaller sets (usually logarithmically many), each belonging to a distinct 'level'. The number of sets increases with increasing level while the size of each set decreases. The error of approximation for S can be upper bounded by the sum of approximation errors across levels, which turns out to be a geometric series and gives the improved bounds.

If one is constrained to taking a single uniform random sample, then the above bound of $\Omega\left(\frac{d}{\epsilon^2}\right)$ cannot be improved.

For the second result we turn to a geometric set system and derive improved bounds by taking a *non-uniform* random sample. In particular, we will prove that there exists an ϵ-approximation of size $o\left(\frac{1}{\epsilon^2}\right)$ for the primal set system induced by half-spaces in $\mathbb{R}^d$:

THEOREM 13.1. *Let P be a set of n points in $\mathbb{R}^d$ and $\epsilon \in \left(0, \frac{1}{2}\right]$ a given parameter. Then there exists an ϵ-approximation of size $C_3 \left(\frac{1}{\epsilon^2} \log \frac{1}{\epsilon}\right)^{\frac{d}{d+1}}$ of the primal set system $\mathcal{R}$ induced on P by half-spaces in $\mathbb{R}^d$, where C_3 is a constant depending on d.*

1. Improved Analysis via Chaining

Some people may sit back and say, I want to solve this problem and they sit down and say, 'How do I solve this problem?' I don't. I just move around in the mathematical waters, thinking about things, being curious, interested, talking to people, stirring up ideas; things emerge and I follow them up. Or I see something which connects up with something else I know about, and I try to put them together and things develop. I have practically never started off with any idea of what I'm going to be doing or where it's going to go. I'm interested in mathematics; I talk, I learn, I discuss and then interesting questions simply emerge. I have never started off with a particular goal, except the goal of understanding mathematics.

Michael Atiyah

The main theorem of this section improves the bound of Theorem 12.8 by a logarithmic factor.

THEOREM 13.2. *Let $(X, \mathcal{F})$, $|X| = n$, be a finite set system with* VC-dim$(\mathcal{F}) \leq d$ *and $\epsilon \in \left(0, \frac{1}{2}\right]$ a given parameter. Then a uniform random sample $A \subseteq X$ of size $C_2 \cdot \frac{d}{\epsilon^2}$ is an ϵ-approximation of $\mathcal{F}$ with constant probability, where $C_2 \geq 1$ is an absolute constant independent of X, $\mathcal{F}$ and d.*

The key point to note here is that the bound of $O\left(\frac{d}{\epsilon^2}\right)$ is achieved by taking a single *uniform* random sample. Furthermore if we want a random sample A to be an ϵ-approximation of $\mathcal{F}$ with, say probability at least $\frac{1}{2}$, then the above bound is asymptotically tight.

As seen earlier, a uniform random sample $A \subseteq X$ of size $\Theta\left(\frac{d}{\epsilon^2}\right)$ fails to be an ϵ-approximation for a fixed set $S \subseteq X$ with constant probability if $|S| = \Omega(n)$. As with all failures, this gives us useful information: to show that a uniform set $A \subseteq X$ of size $\Theta\left(\frac{d}{\epsilon^2}\right)$ fails to be an ϵ-approximation of $\mathcal{F}$ with probability less than 1, we can only afford to use the union bound over $O(1)$ sets of size $\Omega(n)$. Indeed this is precisely what we will be able to do.

Overview of ideas. For the moment we treat the size of A as a parameter t whose value will be set when needed. There are three additional ideas in the proof.

1. **Independence from $|\mathcal{F}|$:** To upper bound the probability that A fails to be an ϵ-approximation of $\mathcal{F}$, one upper bounds the probability that A fails to be an ϵ-approximation of a fixed set $S \in \mathcal{F}$ and then union bound is applied over all sets of $\mathcal{F}$. This adds a multiplicative factor of $\Theta(\log|\mathcal{F}|)$ in the sample size. An easy pre-processing step allows us to replace $|\mathcal{F}|$ by a function of $\frac{1}{\epsilon}$: we will first take a large-enough uniform random sample $A' \subseteq X$ such that A' is an $\frac{\epsilon}{2}$-approximation of $\mathcal{F}$ with constant probability. By Theorem 12.3 we can take $|A'| = \Theta\left(\frac{d}{\epsilon^2}\log\frac{1}{\epsilon}\right)$. Next recall the following lemma.

   LEMMA 12.5. *Given a set system $(X, \mathcal{F})$, let $A \subseteq X$ be an ϵ_1-approximation of $\mathcal{F}$ and let $B \subseteq A$ be an ϵ_2-approximation of $\mathcal{F}|_A$. Then B is a $(\epsilon_1 + \epsilon_2)$-approximation of $\mathcal{F}$.*

   Thus it suffices to prove that a uniform random sample $A \subseteq A'$ of the desired size is an $\frac{\epsilon}{2}$-approximation of $\mathcal{F}|_{A'}$. To minimize notations we will continue to work with $(X, \mathcal{F})$ and just assume that $n = |X| = \Theta\left(\frac{d}{\epsilon^2}\log\frac{1}{\epsilon}\right)$. Lemma 4.3 then implies that

$$|\mathcal{F}| \leq \left(\frac{e|X|}{d}\right)^d = \left(O\left(\frac{1}{\epsilon^2}\log\frac{1}{\epsilon}\right)\right)^d. \tag{13.3}$$

We emphasize that the construction itself consists of taking a single uniform random sample of the size indicated by Theorem 13.2. The step above is only a proof technique, where we *think* of taking the random sample in two iterations for analysis purposes.

2. **Small sets of $\mathcal{F}$:** A random sample A of size t fails to be an ϵ-approximation of a set $S \in \mathcal{F}$ if the random variable $|S \cap A|$ differs by greater than ϵt from its expectation $\frac{|S|t}{n}$. Note that this interval ϵt is *independent* of the size of S. If $|S|$ is large, the deviation of $|S \cap A|$ from its expectation will be larger (in absolute terms) and so there will be a higher probability that $|S \cap A|$ falls outside this ϵt 'margin of error'. Specifically, as we want

$$|S \cap A| = \frac{|S| \cdot t}{n} \pm \epsilon t = \frac{|S| \cdot t}{n}\left(1 \pm \frac{\epsilon n}{|S|}\right) = \mathrm{E}\left[|S \cap A|\right] \cdot \left(1 \pm \frac{\epsilon n}{|S|}\right),$$

applying Chernoff's bound (Corollary 1.24) with $\delta = \frac{\epsilon n}{|S|}$, the probability that A *fails* to be an ϵ-approximation of S is

$$\Pr\left[|S \cap A| > \left(1 + \frac{\epsilon n}{|S|}\right)\mathrm{E}\left[|S \cap A|\right]\right] + \Pr\left[|S \cap A| < \left(1 - \frac{\epsilon n}{|S|}\right)\mathrm{E}\left[|S \cap A|\right]\right]$$

$$\leq 2\exp\left(-\frac{\left(\frac{\epsilon n}{|S|}\right)^2}{2 + \frac{\epsilon n}{|S|}} \cdot \frac{|S|t}{n}\right) = 2\exp\left(-\frac{\epsilon^2 tn}{2|S| + \epsilon n}\right). \tag{13.4}$$

This is as expected: as the size of S increases, so does the above upper bound on the probability that A fails to be an ϵ-approximation of S. Indeed, the next claim shows that a random set A of size $t = c\frac{d}{\epsilon^2}$, for a sufficiently large constant c, is an ϵ-approximation of all sets $S \in \mathcal{F}$ with $|S| \leq \frac{n}{\log\frac{1}{\epsilon}}$ with high probability.

CLAIM 13.5. There exists a sufficiently large constant c such that with probability at least $1 - e^{-100d}$, a uniform random set $A \subseteq X$ of size $c\frac{d}{\epsilon^2}$ is an ϵ-approximation, for $\epsilon \leq \frac{1}{2}$, of all sets of $\mathcal{F}$ of size at most $\frac{n}{\log\frac{1}{\epsilon}}$.

PROOF. The union bound together with Equation (13.4) implies that A fails to be such an ϵ-approximation with probability at most

$$|\mathcal{F}| \cdot 2\exp\left(-\frac{\epsilon^2 tn}{2\frac{n}{\log\frac{1}{\epsilon}} + \epsilon n}\right) \leq |\mathcal{F}| \cdot 2\exp\left(-\frac{\epsilon^2 t\log\frac{1}{\epsilon}}{2 + \epsilon\log\frac{1}{\epsilon}}\right)$$

$$\leq |\mathcal{F}| \cdot 2\exp\left(-\frac{1}{3}\epsilon^2 t\log\frac{1}{\epsilon}\right) = \left(O\left(\frac{1}{\epsilon^2}\log\frac{1}{\epsilon}\right)\right)^d \cdot \epsilon^{c\,d/3} \leq e^{-100\,d},$$

for $t = c\frac{d}{\epsilon^2}$, where c is a sufficiently large constant, and $\epsilon \leq \frac{1}{2}$. □

3. Large sets of $\mathcal{F}$: It remains to deal with the sets of $\mathcal{F}$ of size greater than $\frac{n}{\log \frac{1}{\epsilon}}$.

For a set $S \in \mathcal{F}$ with $|S| = \Omega(n)$, Equation (13.4) implies that

$$\Pr[A \text{ fails to be an } \epsilon\text{-approximation of } S] \leq 2 \exp\left(-\frac{\epsilon^2 tn}{2|S| + \epsilon n}\right) = e^{-\Theta(\epsilon^2 t)}.$$

Having to use the union bound over all $\left(O\left(\frac{1}{\epsilon^2}\log\frac{1}{\epsilon}\right)\right)^d$ sets of $\mathcal{F}$ forces us to set $t = \Omega\left(\frac{d}{\epsilon^2}\log\frac{1}{\epsilon}\right)$ and the additional logarithmic term reappears in the size of A.

However, in this worst-case scenario—when considering sets of size $\Omega(n)$ in $\mathcal{F}$—it is clear that these sets have a lot of elements of X in common. Towards this we recall the following packing statement:

THEOREM 5.8. *Let $\mathcal{F} = \{S_1, \ldots, S_m\}$ be a set system on a set X of n elements, with* VC-dim$(\mathcal{F}) \leq d$. *Let $\delta \in [n]$ be an integer such that for every $1 \leq i < j \leq m$, we have $|\Delta(S_i, S_j)| \geq \delta$ (that is, the size of the symmetric difference between every pair of sets of $\mathcal{F}$ is at least δ). Then $|\mathcal{F}| \leq 2\left(\frac{8en}{\delta}\right)^d$.*

We will use Theorem 5.8 to show that there are few sets that are 'basic' and that all other sets can be derived, with small modifications, from these basic sets. Thus we will apply the union bound in two separate scenarios: on the basic sets which could have large sizes but are few in number, and the other 'modification' sets, which could be many but will have small size.

Before we present the proof of Theorem 13.2, we illustrate the above ideas by a short proof showing that a uniform random sample A of size $\Theta\left(\frac{d}{\epsilon^2}\log\log\frac{1}{\epsilon}\right)$ is an ϵ-approximation of $\mathcal{F}$ with constant probability.

For a parameter $\eta \geq \epsilon n$ to be fixed later, let $\mathcal{P}$ be a *maximal* subset of $\mathcal{F}$ such that the size of the symmetric difference between every pair of sets of $\mathcal{P}$ is at least η. Theorem 5.8 implies that $|\mathcal{P}| = O\left(\left(\frac{8en}{\eta}\right)^d\right)$. By the maximality of $\mathcal{P}$, for any $S \in \mathcal{F} \setminus \mathcal{P}$ there exists a set $F_S \in \mathcal{P}$ such that $|\Delta(S, F_S)| < \eta$. Define

$$\mathcal{G} = \{S \setminus F_S : S \in \mathcal{F} \setminus \mathcal{P}\} \bigcup \{F_S \setminus S : S \in \mathcal{F} \setminus \mathcal{P}\}.$$

Note that $|\mathcal{G}| \leq 2|\mathcal{F}| = \left(O\left(\frac{1}{\epsilon^2}\log\frac{1}{\epsilon}\right)\right)^d$ by Equation (13.3) and each set of $\mathcal{G}$ has size less than η. Each $S \in \mathcal{F}$ can be 'derived' from a set of $\mathcal{P}$ together with two sets of $\mathcal{G}$—that is, $S = F_S - (F_S \setminus S) + (S \setminus F_S)$. Therefore if A is a $\frac{\epsilon}{3}$-approximation of both $\mathcal{P}$ and $\mathcal{G}$, then A is an ϵ-approximation of $\mathcal{F}$ (see Claim 13.6). Thus the probability that a uniform random sample A of size t fails to be an ϵ-approximation of $\mathcal{F}$ can be upper bounded, using Equation (13.4), by

$$\Pr\left[A \text{ is not an } \frac{\epsilon}{3}\text{-approx. of } \mathcal{P}\right] + \Pr\left[A \text{ is not an } \frac{\epsilon}{3}\text{-approx. of } \mathcal{G}\right]$$
$$\leq |\mathcal{P}| \cdot 2\exp\left(-\frac{\epsilon^2 tn}{9\,(2n + (\epsilon/3)n)}\right) \quad + \quad |\mathcal{G}| \cdot 2\exp\left(-\frac{\epsilon^2 tn}{9\,(2\eta + (\epsilon/3)n)}\right)$$
$$\leq O\left(\left(\frac{8en}{\eta}\right)^d\right) \cdot 2\exp\left(-\frac{\epsilon^2 t}{21}\right) \quad + \quad \left(O\left(\frac{1}{\epsilon^2}\log\frac{1}{\epsilon}\right)\right)^d \cdot 2\exp\left(-\frac{\epsilon^2 tn}{21\,\eta}\right)$$

Setting $\eta = \frac{n}{\log \frac{1}{\epsilon}}$ and $t = c \cdot \frac{d}{\epsilon^2} \log\log \frac{1}{\epsilon}$ for a sufficiently large constant c,

$$= O\left(\left(8e \log \frac{1}{\epsilon}\right)^d\right) \cdot 2\exp\left(-\frac{c\, d \log\log \frac{1}{\epsilon}}{21}\right)$$
$$+ \left(O\left(\frac{1}{\epsilon^2} \log \frac{1}{\epsilon}\right)\right)^d \cdot 2\exp\left(-\frac{c\, d(\log \frac{1}{\epsilon})(\log\log \frac{1}{\epsilon})}{21}\right)$$
$$< \frac{1}{2}.$$

Thus A is an ϵ-approximation of size $\Theta\left(\frac{d}{\epsilon^2} \log\log \frac{1}{\epsilon}\right)$ with probability at least $\frac{1}{2}$.

The proof of the main theorem improves the above bound by using the thematic idea of partitioning $\mathcal{F}$ into a logarithmic number of levels instead of just two; then the error becomes a geometric series, and we get rid of the logarithmic term completely. We will need the following decomposition claim whose proof follows immediately from the definition of ϵ-approximations (and is similar to the proof of Lemma 12.5).

CLAIM 13.6. Let X be a set of n elements and $A \subseteq X$ such that A is an ϵ_1-approximation of S_1 and an ϵ_2-approximation of S_2. Then

(1) If $S_1 \cap S_2 = \emptyset$, then A is a $(\epsilon_1 + \epsilon_2)$-approximation of $S_1 \cup S_2$.
(2) If $S_1 \subseteq S_2$, then A is a $(\epsilon_1 + \epsilon_2)$-approximation of $S_2 \setminus S_1$.

PROOF OF THEOREM 13.2. The proof is divided into three stages.

Initial random sample: Given $(X, \mathcal{F})$, we first take a uniform random sample $A' \subseteq X$ of sufficient size so that A' is an $\frac{\epsilon}{2}$-approximation of $\mathcal{F}$ with constant probability. Theorem 12.3 implies that we can take $|A'| = \Theta\left(\frac{d}{\epsilon^2} \log \frac{1}{\epsilon}\right)$. Then it suffices to show that a uniform random sample $A \subseteq A'$ of size $t = C_2 \frac{d}{\epsilon^2}$ is an $\frac{\epsilon}{2}$-approximation of $\mathcal{F}|_{A'}$ with constant probability (see Lemma 12.5). For notational convenience we simply assume that $X = A'$, $n = |A'|$ and so by Lemma 4.3,

$$|\mathcal{F}| \leq \left(\frac{e\,|X|}{d}\right)^d = \left(O\left(\frac{1}{\epsilon^2} \log \frac{1}{\epsilon}\right)\right)^d.$$

Partitioning $\mathcal{F}$ into levels: We construct a series of maximal packings, as follows. Let k be a parameter to be fixed later. For each $i = 1, \ldots, k$, let

$\mathcal{P}_i$: a maximal subset of $\mathcal{F}$ such that the size of the symmetric difference between every pair of sets of $\mathcal{P}_i$ is at least $\frac{n}{2^i}$.

By first constructing $\mathcal{P}_1$, then $\mathcal{P}_2$ by starting with $\mathcal{P}_1$ and adding sets to it and so on, we can assume that $\mathcal{P}_1 \subseteq \mathcal{P}_2 \subseteq \cdots \subseteq \mathcal{P}_k$[1]. Theorem 5.8 implies that

$$|\mathcal{P}_i| = O\left(\left(\frac{8en}{n/2^i}\right)^d\right) = O\left((8e)^d\, 2^{di}\right).$$

Set $\mathcal{P}_{k+1} = \mathcal{F}$.

[1]Forcing inclusion is done to minimize the number of cases one has to consider in the proof. The whole argument works even when $\mathcal{P}_i$ is not a subset of $\mathcal{P}_{i+1}$.

For each $i \in [1,k]$, the maximality of $\mathcal{P}_i$ implies that for each $S \in \mathcal{P}_{i+1}$ there exists a set $F_S \in \mathcal{P}_i$ such that $|\Delta(S, F_S)| < \frac{n}{2^i}$. For $i = 1, \ldots, k$ define the sets

$$\mathcal{A}_i = \{S \setminus F_S \colon S \in \mathcal{P}_{i-1}\} \qquad \text{and} \qquad \mathcal{B}_i = \{F_S \setminus S \colon S \in \mathcal{P}_{i+1}\}.$$

Note that each $S \in \mathcal{A}_i \cup \mathcal{B}_i$ satisfies $|S| < \frac{n}{2^i}$. Furthermore,

$$|\mathcal{A}_i|, |\mathcal{B}_i| \leq |\mathcal{P}_{i+1}| = O\left((8e)^d \, 2^{d(i+1)}\right) \quad \text{for} \quad i \in [1, k-1],$$

while $|\mathcal{A}_k|, |\mathcal{B}_k| \leq |\mathcal{F}|$.

Here one can see the tradeoff clearly: as i increases, the number of sets in $\mathcal{A}_i, \mathcal{B}_i$ increase, while the size of each set decreases.

We set the parameter k such that the sets in $\mathcal{A}_k \cup \mathcal{B}_k$ have size at most ϵn. That is,

$$k = \left\lceil \log \frac{1}{\epsilon} \right\rceil.$$

Define

$$\epsilon_i = \frac{\epsilon}{30}\sqrt{\frac{i}{2^i}}.$$

Note that

$$\sum_{j=1}^{\infty} \sqrt{\frac{j}{2^j}} = \frac{2}{\sqrt{2}} + \sum_{j=3}^{\infty} \sqrt{\frac{j}{2^j}} \leq \sqrt{2} + \sqrt{\frac{3}{8}} \sum_{i=0}^{\infty} \left(\sqrt{\frac{2}{3}}\right)^i \leq 5, \tag{13.7}$$

where the second step uses that for $j \geq 3$, $\frac{\sqrt{(j+1)/2^{j+1}}}{\sqrt{j/2^j}} = \sqrt{\frac{1}{2} + \frac{1}{2j}} \leq \sqrt{\frac{2}{3}}$.

LEMMA 13.8. *For $C_2 \geq 1$ sufficiently large, A is simultaneously an*
(i) *$\frac{\epsilon}{20}$-approximation of all sets of $\mathcal{P}_1$, and*
(ii) *ϵ_i-approximation of all the sets in $\mathcal{A}_i \cup \mathcal{B}_i$, for $i = 1 \ldots k-1$, and*
(iii) *$\frac{\epsilon}{20}$-approximation of all sets in $\mathcal{A}_k \cup \mathcal{B}_k$,*
with probability at least $1 - 3e^{-100\,d}$.

PROOF. (*i*) By the union bound and Equation (13.4), the probability that A fails to be an $\frac{\epsilon}{20}$-approximation of $\mathcal{P}_1$ can be upper bounded as

$$\begin{aligned} |\mathcal{P}_1| \cdot 2\exp\left(-\frac{(\epsilon/20)^2 nt}{2\,n + (\epsilon/20)n}\right) &= O\left((16\,e)^d\right) \cdot \exp\left(-\frac{\epsilon^2 t}{400 \cdot 3}\right) \\ &= O\left((16\,e)^d\right) \cdot e^{-\Omega(C_2 d)} \leq e^{-100\,d}, \end{aligned}$$

for C_2 sufficiently large.

(*ii*) As each set of $\mathcal{A}_i \cup \mathcal{B}_i$ has size less than $\frac{n}{2^i}$, A fails to be an ϵ_i-approximation of a fixed $S \in \mathcal{A}_i \cup \mathcal{B}_i$ with probability at most

$$\begin{aligned} 2\exp\left(-\frac{\epsilon_i^2 nt}{2|S| + \epsilon_i n}\right) &\leq 2\exp\left(-\frac{\frac{i}{900\,2^i}\epsilon^2 nt}{\frac{2\,n}{2^i} + \frac{\epsilon}{30}\sqrt{\frac{i}{2^i}}n}\right) = \exp\left(-\Omega\left(\frac{i\epsilon^2 t}{1 + \sqrt{2^i\, i}\,\epsilon}\right)\right) \\ &= \exp\left(-\Omega\left(\frac{i\epsilon^2 t}{1 + \sqrt{(1/\epsilon)\log(1/\epsilon)}\,\epsilon}\right)\right) = \exp\left(-\Omega\left(i\epsilon^2 t\right)\right). \end{aligned}$$

Using the union bound, the probability of failure can be upper bounded as

$$\sum_{i=1}^{k-1} |\mathcal{A}_i \cup \mathcal{B}_i| \cdot \exp\left(-\Omega\left(i\epsilon^2 t\right)\right) = \sum_{i=1}^{k-1} O\left((8e)^d\ 2^{d(i+1)}\right) \cdot e^{-\Omega(C_2 di)} \leq e^{-100\,d},$$

for a sufficiently large C_2, as the summation is upper bounded by a geometric series.
(iii) As $|\mathcal{A}_k \cup \mathcal{B}_k| \leq 2|\mathcal{F}|$ and each set has size less than $\frac{n}{2^i} \leq \epsilon n$, this follows from Claim 13.5, for C_2 sufficiently large. □

Chaining: Next we show that each set $S \in \mathcal{F}$ can be written in terms of sets in $\mathcal{P}_1$, $\mathcal{A}_1, \ldots, \mathcal{A}_k$, $\mathcal{B}_1, \ldots, \mathcal{B}_k$.

LEMMA 13.9. *For each $j \in [1,k]$ and $S \in \mathcal{P}_{j+1}$, there exist a set $I \in \mathcal{P}_1$ and sets*

$$A_1 \in \mathcal{A}_1,\ A_2 \in \mathcal{A}_2,\ \ldots,\ A_j \in \mathcal{A}_j \qquad \text{and} \qquad B_1 \in \mathcal{B}_1,\ B_2 \in \mathcal{B}_2,\ \ldots,\ B_j \in \mathcal{B}_j$$

such that $A_i \cap B_i = \emptyset$ for all $i = 1, \ldots, j$ and

$$S = \left(\cdots\left(\left((I - B_1 + A_1) - B_2 + A_2\right) - B_3 + A_3\right)\cdots - B_j + A_j\right). \tag{13.10}$$

Furthermore, if A is an ϵ_0-approximation of all the sets of $\mathcal{P}_1$ and simultaneously an ϵ_i-approximation of $\mathcal{A}_i \cup \mathcal{B}_i$ for all $i = 1, \ldots, k$, then A is also an $(\epsilon_0 + 2\epsilon_1 + 2\epsilon_2 + \cdots + 2\epsilon_j)$-approximation of all the sets of $\mathcal{P}_{j+1}$.

PROOF. As $S \in \mathcal{P}_{j+1}$, there exists $F_S \in \mathcal{P}_j$ is such that $|\Delta(S, F_S)| < \frac{n}{2^j}$, and further we can write

$$S = F_S + (S \setminus F_S) - (F_S \setminus S) = F_S + A_j - B_j,$$

where $A_j \in \mathcal{A}_j$ and $B_j \in \mathcal{B}_j$. We can again write $F_S \in \mathcal{P}_j$ as an addition/subtraction of sets in $\mathcal{A}_{j-1}$ and $\mathcal{B}_{j-1}$. Continuing like this recursively, any set $S \in \mathcal{P}_{j+1}$ can be written as addition of j sets of $\mathcal{A}_i$ and subtraction of j sets of $\mathcal{B}_i$, ending up at some set $I \in \mathcal{P}_1$. This gives Equation (13.10).
The proof of the second part proceeds by induction on j and Claim 13.6. □

We now conclude the proof of Theorem 13.2: Lemma 13.8 and Lemma 13.9 together imply that with probability at least $1 - 3e^{-100\,d}$, A is an ϵ'-approximation of each $S \in \mathcal{F} = \mathcal{P}_{k+1}$, where

$$\epsilon' \leq \frac{\epsilon}{20} + \left(2\sum_{i=1}^{k-1} \frac{\epsilon}{30}\sqrt{\frac{i}{2^i}}\right) + \frac{\epsilon}{20} \leq \frac{\epsilon}{10} + \left(\frac{\epsilon}{15} \cdot \sum_{i=1}^{\infty} \sqrt{\frac{i}{2^i}}\right) \leq \frac{\epsilon}{2},$$

where the last step follows from Equation (13.7). □

In the proof of Theorem 13.2 we omitted the precise relation between the size of the random sample and the probability that it is an ϵ-approximation—we simply wanted a uniform random sample to be an ϵ-approximation with constant probability. By introducing this probability as a parameter in the sample size and re-working the above proof, we arrive at the following theorem (we omit its proof as we will give the proof of a more general result, Theorem 14.3, later).

THEOREM 13.11. *Let $(X, \mathcal{F})$ be a set system with* VC-dim$(\mathcal{F}) \leq d$ *and $\epsilon, \gamma > 0$ be two given parameters. Then a uniform random sample $A \subseteq X$ of size*

$$c \cdot \left(\frac{d}{\epsilon^2} + \frac{1}{\epsilon^2} \log \frac{1}{\gamma} \right)$$

is an ϵ-approximation of $\mathcal{F}$ with probability at least $1-\gamma$, where $c \geq 1$ is an absolute constant independent of X, $\mathcal{F}$ and d.

Bibliography and discussion. The presentation in the text follows that of [**CM21**], which is a simplification of the influential work in [**LLS01**] and [**Tal94**]. Chaining is also used for the related problem of upper bounding combinatorial discrepancy of set systems with finite VC-dimension (see [**Mat95a**]). If one is restricted to taking a single random sample, then the bound of Theorem 13.11 is tight (see [**LLS01**]).

[CM21] M. Csikós and N. H. Mustafa. *Optimal approximations made easy*. *CoRR* abs/2008.08970, 2021. arXiv:2008.08970. URL: https://arxiv.org/abs/2008.08970.

[LLS01] Y. Li, P. M. Long, and A. Srinivasan, *Improved bounds on the sample complexity of learning*, J. Comput. System Sci. **62** (2001), no. 3, 516–527, DOI 10.1006/jcss.2000.1741. MR1824457

[Mat95a] J. Matoušek, *Tight upper bounds for the discrepancy of half-spaces*, Discrete Comput. Geom. **13** (1995), no. 3-4, 593–601, DOI 10.1007/BF02574066. MR1318799

[Tal94] M. Talagrand, *Sharper bounds for Gaussian and empirical processes*, Ann. Probab. **22** (1994), no. 1, 28–76. MR1258865

2. Near-Optimal Bounds via Partitioning

From him [Alexandre Grothendieck] and his example, I have also learned not to take glory in the difficulty of a proof: difficulty means we have not understood. The ideal is to be able to paint a landscape in which the proof is obvious. I admire how often he succeeded in reaching this ideal.

Pierre Deligne

We now improve the upper bound on the size of ϵ-approximations of the primal set system induced by half-spaces in $\mathbb{R}^d$.

THEOREM 13.1. *Let P be a set of n points in $\mathbb{R}^d$ and $\epsilon \in \left(0, \frac{1}{2}\right]$ a given parameter. Then there exists an ϵ-approximation of size $C_3 \left(\frac{1}{\epsilon^2} \log \frac{1}{\epsilon}\right)^{\frac{d}{d+1}}$ of the primal set system $\mathcal{R}$ induced on P by half-spaces in $\mathbb{R}^d$, where C_3 is a constant depending on d.*

With additional work, the logarithmic term can also be removed (see discussion).

This improves the previous upper bound of $O\left(\frac{d}{\epsilon^2}\right)$ of Theorem 13.2. However the price we pay for this improvement is that the ϵ-approximation, while still a random sample of X, will no longer be a *uniform* random sample. This is necessary—as stated in the previous section, the bound of Theorem 13.2 is optimal if one is restricted to taking a uniform random sample.

Overview of ideas. For simplicity consider the case when X is a unit cube in $\mathbb{R}^d$, say $X = [0,1]^d$. Then our goal is to construct a small set $A \subseteq [0,1]^d$ such that for any half-space h^+,

$$|h^+ \cap A| = \text{vol}\left(h^+ \cap X\right) \cdot t \,\pm\, \epsilon t,$$

where $t = |A|$.

A natural first try is to partition $[0,1]^d$ into t smaller cubes by a uniform $t^{\frac{1}{d}} \times \cdots \times t^{\frac{1}{d}}$ grid. Note that each small cube has side-length $\frac{1}{t^{1/d}}$ and volume $\frac{1}{t}$. Now let A be the t centers of these cubes (in fact, pick any one point from each of the t small cubes).

For any fixed half-space h^+, the 'error' of approximating $\text{vol}\left(h^+ \cap X\right) \cdot t$ by $|h^+ \cap A|$ comes from the cubes intersecting the bounding hyperplane h of h^+. Since any hyperplane can only intersect $O\left(t^{1-\frac{1}{d}}\right)$ cubes in a grid, we have

$$|h^+ \cap A| = \text{vol}\left(h^+ \cap X\right) \cdot t \,\pm\, O\left(t^{1-\frac{1}{d}}\right) = \text{vol}\left(h^+ \cap X\right) \cdot t \,\pm\, O\left(\frac{1}{t^{\frac{1}{d}}}\right) t.$$

Setting $t = \Theta\left(\frac{1}{\epsilon^d}\right)$ implies that A is an ϵ-approximation, of size $\Theta\left(\frac{1}{\epsilon^d}\right)$. Of course, the size of A is too big.

The above bound can be improved by the following idea (called *jittered sampling* in computer graphics literature):

> instead of the centers, let A be a set constructed by picking a point uniformly at random from each small cube.

Then Chernoff's bound implies that, with high probability, each half-space h^+ has error $\tilde{O}\left(\sqrt{t^{1-1/d}}\right)$ from the cubes intersecting h, giving

$$|h^+ \cap A| = \mathrm{vol}\left(h^+ \cap X\right) \cdot t \pm \tilde{O}\left(\sqrt{t^{1-\frac{1}{d}}}\right) = \mathrm{vol}\left(h^+ \cap X\right) \cdot t \pm \tilde{O}\left(\frac{1}{t^{\frac{d+1}{2d}}}\right) t.$$

Setting $t = \tilde{\Theta}\left(\frac{1}{\epsilon^2}\right)^{\frac{d}{d+1}}$ implies that A is an ϵ-approximation of the desired size.

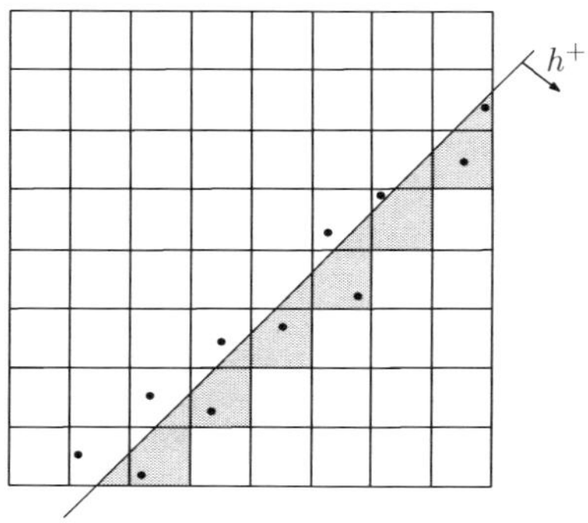

We outline the Chernoff bound calculation. Let $\mathrm{vol}'\left(h^+ \cap X\right)$ be the volume of $h^+ \cap X$ from the cubes which intersect h and let $A' \subseteq A$ be the points chosen from these $O\left(t^{1-\frac{1}{d}}\right)$ cubes. See figure where the points of A' are drawn. Let Δ be the error of approximation due to these cubes; that is,

$$\begin{aligned} |h^+ \cap A'| &= \mathrm{vol}'\left(h^+ \cap X\right) \cdot |A'| \pm \Delta \\ &= \mathrm{vol}'\left(h^+ \cap X\right) \cdot |A'| \left(1 \pm \frac{\Delta}{\mathrm{vol}'\left(h^+ \cap X\right) \cdot |A'|}\right). \end{aligned}$$

Since $\mathrm{E}\left[|h^+ \cap A'|\right] = \mathrm{vol}'\left(h^+ \cap X\right) \cdot |A'|$, the probability that the above fails for a fixed h^+ can be upper bounded using Chernoff's bound by

$$\exp\left(-\Omega\left(\frac{\Delta^2}{|A'|}\right)\right) = \exp\left(-\Omega\left(\frac{\Delta^2}{t^{1-1/d}}\right)\right).$$

Setting $\Delta = \Theta\left(\sqrt{t^{1-1/d}}\right)$ implies that a fixed h^+ has error $O\left(\sqrt{t^{1-1/d}}\right)$ with positive probability; increasing t by a multiplicative logarithmic term and using the union bound gets this probability small-enough so that *each possible* half-space has error $\tilde{O}\left(\sqrt{t^{1-1/d}}\right)$, completing the outline.

To extend this idea for an arbitrary point set in $\mathbb{R}^d$, we will need a partitioning analog of the uniform grid. That is the wonderful Matoušek's simplicial partition theorem, whose statement we recall.

THEOREM 9.5. *Let P be a set of n points in $\mathbb{R}^d$ and $t \in \left[2, \frac{n}{2}\right]$ a given integer parameter. Then there exists a partition $\{P_1, \ldots, P_t\}$ of P with $|P_i| = \left\lfloor \frac{n}{t} \right\rfloor$ for $i = 1, \ldots, t-1$, such that any hyperplane in $\mathbb{R}^d$ intersects the convex-hull of at most $C_5 \cdot t^{1-\frac{1}{d}}$ sets of this partition. Here C_5 is a constant depending only on d.*

The proof of Theorem 13.1, which we now present in detail, follows pretty much along the lines of the proof for $[0,1]^d$ that is outlined above.

Proof of Theorem 13.1. Let $\mathcal{R}$ be the primal set system induced on P by half-spaces in $\mathbb{R}^d$. As before[2] it suffices to first take an $\frac{\epsilon}{2}$-approximation A' of size $O\left(\frac{d}{\epsilon^2}\right)$ of $\mathcal{R}$—say using Theorem 13.2—and then return an $\frac{\epsilon}{2}$-approximation of the set system $(A', \mathcal{R}|_{A'})$. Thus one can assume that $|P| = n = \Theta\left(\frac{d}{\epsilon^2}\right)$. We will also assume that $d \geq 2$.

> When $d = 1$, a set of $\Theta\left(\frac{1}{\epsilon}\right)$ points of P, constructed by picking every ϵn-th point of P as ordered from left to right, is an ϵ-approximation for the set system induced by intervals.

Let $t = C_3\left(\frac{d}{\epsilon^2}\log\frac{1}{\epsilon}\right)^{\frac{d}{d+1}}$, with the constant C_3 to be fixed later; the choice of this value of t will become clear later in the proof. The algorithm to construct an ϵ-approximation $A \subseteq P$ is the following.

Epsilon-Approximation Algorithm for Half-spaces $\left(P \subset \mathbb{R}^d, \epsilon > 0\right)$.

Apply Theorem 9.5 to P with parameter t to get $\Xi = \left\{P_1, \ldots, P_t\right\}$.
$A = \emptyset$.
for $i = 1, \ldots, t$ **do**
 add a point chosen uniformly at random from P_i to A.
return A.

We will argue that the set A of size t is an ϵ-approximation of $\mathcal{R}$ with constant probability. For technical simplicity we will assume that t is an integer and n is a multiple of t; thus each set of Ξ has size precisely $\frac{n}{t}$.

Let h^+ be a half-space in $\mathbb{R}^d$, with bounding hyperplane h. For A to be an ϵ-approximation of $h^+ \cap P$, we require

$$|h^+ \cap A| = \frac{|h^+ \cap P|\,t}{n} \pm \epsilon t.$$

> A set $P_i \in \Xi$ that is completely contained in h^+ is 'perfectly represented' by A, in the sense that if h^+ only contained points of P from such sets, then $|h^+ \cap A| = \frac{|h^+ \cap P|}{(n/t)} = \frac{|h^+ \cap P|\,t}{n}$.
>
> The error is thus due to the sets of Ξ whose convex-hull intersects the hyperplane h. Denote these sets by $P_1, \ldots, P_m$ and let
> $$P^h = h^+ \cap (P_1 \cup \cdots \cup P_m).$$
> Let A^h be the m points of A chosen from the sets $P_1, \ldots, P_m$. See figure where the points of A^h are drawn in solid.

The following statement is intuitive; for completeness, we give the proof.

Lemma 13.12. *A is an ϵ-approximation of $h^+ \cap P$ if*

$$\left|h^+ \cap A^h\right| = \frac{|P^h|}{(n/t)}\left(1 \pm \frac{\epsilon n}{|P^h|}\right).$$

[2]See Lemma 12.5.

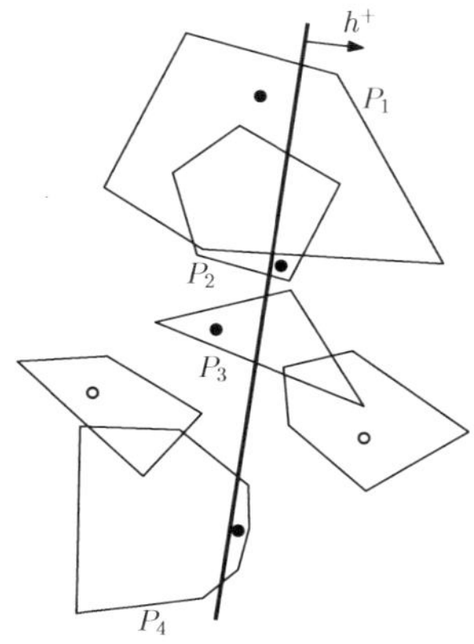

PROOF. Let t' be the number of sets of Ξ completely contained in h^+. Setting ζ_h to be a positive real such that $\left|h^+ \cap A^h\right| = \frac{|P^h|}{(n/t)}(1 \pm \zeta_h)$, we have

$$\begin{aligned}\left|h^+ \cap A\right| &= \left|h^+ \cap A^h\right| + t' = \frac{|P^h|}{(n/t)}(1 \pm \zeta_h) + t' = \left(\frac{|P^h|}{(n/t)} + t'\right) \pm \frac{|P^h|\zeta_h}{(n/t)}\\ &= \frac{\left(t'\frac{n}{t} + |P^h|\right)\cdot t}{n} \pm \frac{|P^h|\cdot\zeta_h\cdot t}{n} = \frac{|h^+ \cap P|\cdot t}{n} \pm \frac{|P^h|\cdot\zeta_h\cdot t}{n}.\end{aligned}$$

For A to be an ϵ-approximation of $h^+ \cap P$, we require

$$\frac{|P^h|\cdot\zeta_h\cdot t}{n} \le \epsilon\cdot t, \qquad \text{and hence} \qquad \zeta_h \le \frac{\epsilon n}{|P^h|}.$$

□

The key point here is that $\frac{\epsilon n}{|P^h|}$ is not too small, as $|P^h|$ is not too large:

$$|P^h| \le m\cdot\frac{n}{t} \le C_5\, t^{1-\frac{1}{d}}\cdot\frac{n}{t} = \frac{C_5\, n}{t^{\frac{1}{d}}}, \qquad \text{by Theorem 9.5.}$$

For the probability calculations, we will need the following tail bound.

LEMMA 13.13. *For any parameter* $\zeta > 0$,

$$\Pr\left[\left|h^+ \cap A^h\right| \ne \frac{|P^h|}{(n/t)}(1 \pm \zeta)\right] \le 2\exp\left(-\frac{\zeta^2}{2+\zeta}\frac{|P^h|}{n/t}\right).$$

PROOF. Let $p_i = \frac{|h^+ \cap P_i|}{|P_i|}$ be the probability that the point chosen from P_i lies in h^+. As

$$\mathrm{E}\left[\left|h^+ \cap A^h\right|\right] = \sum_{i=1}^m p_i = \sum_{i=1}^m \frac{|h^+ \cap P_i|}{|P_i|} = \sum_{i=1}^m \frac{|h^+ \cap P_i|}{(n/t)} = \frac{|P^h|}{(n/t)}, \tag{13.14}$$

and the m elements are chosen independently, the required statement follows from Chernoff's bound (Corollary 1.23). □

Lemmas 13.12 and 13.13 together imply an upper bound on the probability that A fails to be an ϵ-approximation of $h^+ \cap P$:

$$\begin{aligned}
\Pr\left[\left|h^+ \cap A^h\right| \neq \frac{|P^h|}{(n/t)}\left(1 \pm \frac{\epsilon n}{|P^h|}\right)\right] &\leq 2\exp\left(-\frac{\left(\frac{\epsilon n}{|P^h|}\right)^2}{2 + \frac{\epsilon n}{|P^h|}} \cdot \frac{|P^h|}{(n/t)}\right) \\
&= 2\exp\left(-\frac{\epsilon^2 n t}{2|P^h| + \epsilon n}\right) \\
&\leq 2\exp\left(-\frac{\epsilon^2 n t}{2\,\frac{C_5\, n}{t^{\frac{1}{d}}} + \epsilon n}\right) \\
&= 2\exp\left(-\frac{\epsilon^2 t^{\frac{d+1}{d}}}{2\,C_5 + \epsilon\, t^{\frac{1}{d}}}\right).
\end{aligned}$$

Recalling that $t = C_3\left(\frac{d}{\epsilon^2}\log\frac{1}{\epsilon}\right)^{\frac{d}{d+1}}$,

$$\begin{aligned}
&\leq 2\exp\left(-\frac{C_3^{\frac{d+1}{d}}\, d\log\frac{1}{\epsilon}}{2\,C_5 + C_3^{\frac{1}{d}}\left(d\,\epsilon^{d-1}\,\log\frac{1}{\epsilon}\right)^{\frac{1}{d+1}}}\right) \\
&\leq 2\exp\left(-\frac{C_3^{\frac{d+1}{d}}\, d\log\frac{1}{\epsilon}}{2\,C_5 + C_3^{\frac{1}{d}}}\right) \\
&\leq 2\exp\left(-\frac{C_3}{4\,C_5}d\log\frac{1}{\epsilon}\right),
\end{aligned}$$

where the second-to-last step uses the fact that $d\,\epsilon^{d-1}\log\frac{1}{\epsilon} \leq 1$ for $d \geq 2$ and $\epsilon \leq \frac{1}{2}$. On the other hand, the number of distinct subsets of X induced by half-spaces in $\mathbb{R}^d$ can be upper bounded using Equation (1.13) by

$$2\sum_{i=0}^{d}\binom{n}{i} \leq 2\left(\frac{e\,n}{d}\right)^d = 2\left(\frac{e\cdot\Theta\left(\frac{d}{\epsilon^2}\right)}{d}\right)^d = \left(O\left(\frac{1}{\epsilon}\right)\right)^{2d}.$$

Using the union bound over all distinct subsets induced by half-spaces,

$$\begin{aligned}
\Pr\left[A \text{ is not an } \epsilon\text{-approximation of } \mathcal{R}\,\right] &= \left(O\left(\frac{1}{\epsilon}\right)\right)^{2d} \cdot 2\exp\left(-\frac{C_3}{4\,C_5}d\log\frac{1}{\epsilon}\right) \\
&< \frac{1}{2},
\end{aligned}$$

for a sufficiently large constant C_3 depending only on C_5.

This concludes the proof of Theorem 13.1. □

Bibliography and discussion. The elegant proof presented here is due to Suri et al. [**STZ06**] and is related to the discrepancy-based constructions of ϵ-approximations we saw in Chapter 12 (see also Matheny and Phillips [**MP18**] for an experimental evaluation). Matoušek's simplicial partition theorem, a fundamental tool in combinatorial and computational geometry, is from [**Mat92**] (see [**Cha12**] for efficient algorithms for its construction). The optimal bound without the logarithmic factors is achieved by using Theorem 12.9 together with optimal bounds on discrepancy of set systems with bounded VC-dimension [**Mat95a**] (see Chapter 16).

[Cha12] T. M. Chan, *Optimal partition trees*, Discrete Comput. Geom. **47** (2012), no. 4, 661–690, DOI 10.1007/s00454-012-9410-z. MR2901245

[MP18] M. Matheny and J. M. Phillips, *Practical low-dimensional halfspace range space sampling*, 26th European Symposium on Algorithms, LIPIcs. Leibniz Int. Proc. Inform., vol. 112, Schloss Dagstuhl. Leibniz-Zent. Inform., Wadern, 2018, pp. Art. No. 62, 14. MR3847688

[Mat92] J. Matoušek, *Efficient partition trees*, Discrete Comput. Geom. **8** (1992), no. 3, 315–334, DOI 10.1007/BF02293051. ACM Symposium on Computational Geometry (North Conway, NH, 1991). MR1174360

[Mat95a] J. Matoušek, *Tight upper bounds for the discrepancy of half-spaces*, Discrete Comput. Geom. **13** (1995), no. 3-4, 593–601, DOI 10.1007/BF02574066. MR1318799

[STZ06] S. Suri, C. D. Tóth, and Y. Zhou, *Range counting over multidimensional data streams*, Discrete Comput. Geom. **36** (2006), no. 4, 633–655, DOI 10.1007/s00454-006-1269-4. MR2267550

CHAPTER 14

Epsilon-Approximations: Relative Case

Chapter 13 introduced the notion of an ϵ-approximation and presented upper bounds on sizes of ϵ-approximations of geometric and combinatorial set systems. We now consider other notions of approximations.

Given a set system $(X, \mathcal{F})$ and a target sample size t, our goal is to construct a set A of size t that minimizes the error of approximating each $S \in \mathcal{F}$. That is, we would like to prove upper bounds on the value of η for which the following inequality can be satisfied by A for *all* $S \in \mathcal{F}$:

$$|S \cap A| = \frac{|S|\,t}{n} \pm \eta = \frac{|S|t}{n}\left(1 \pm \eta \frac{n}{|S|t}\right).$$

For a uniform random A of size t, we have $\mathrm{E}\left[|S \cap A|\right] = \frac{|S|t}{n}$ and thus η is the additive error beyond the expectation. The value of η is guided by two technical considerations.

Tail bounds: One of the main methods of constructing approximations is via random sampling—if A is a uniform random sample of X of size t, then Chernoff's bound (Corollary 1.24) implies the following.

FACT 14.1. Let $S \subseteq [n]$ and A a uniform random subset of $[n]$ of size t. Then

$$\Pr\left[|S \cap A| \neq \frac{|S|\,t}{n} \pm \eta\right] \leq 2\exp\left(-\frac{\frac{\eta^2 n^2}{|S|^2 t^2}}{2 + \frac{\eta n}{|S|t}} \cdot \frac{|S|t}{n}\right)$$
$$= 2\exp\left(-\frac{\eta^2 n}{2|S|t + \eta n}\right).$$

Fact 14.1 thus prohibits η from being too small for uniformly sampled approximations.

Chaining: A key property used in the proof of Theorem 12.8 to get the bound of $O\left(\frac{d}{\epsilon^2}\right)$ for ϵ-approximations was that ϵ-approximations are additive—that is, if A is an ϵ_1-approximation of a set S_1 and an ϵ_2-approximation of S_2, then A is an $(\epsilon_1 + \epsilon_2)$-approximation of $S_1 \cup S_2$. This allowed us to 'decompose' the analysis into a logarithmic number of sub-problems (see the proof of Theorem 13.2).

In this chapter we will consider two other notions of approximations that are possible within these constraints, the main one being that of relative (ϵ, δ)-approximations.

To see the intuition behind relative (ϵ, δ)-approximations, recall that for A to be an ϵ-approximation of $\mathcal{F}$, each $S \in \mathcal{F}$ must satisfy the property $|S \cap A| = \frac{|S|\,|A|}{|X|} \pm \epsilon\,|A|$. As the error margin $\epsilon\,|A|$ is independent of the size of S, the probability that a uniform random sample A satisfies the above property for a set S decreases with the size of $|S|$. In fact, when $|S| = \Theta(\epsilon n)$ one can show that a random sample A of size $\tilde{\Theta}\left(\frac{d}{\epsilon}\right)$ suffices with high probability (see the proof of Claim 13.5). It is the other extreme—when $|S| = \Omega(n)$—that forces us to set $|A| = \Omega\left(\frac{1}{\epsilon^2}\right)$.

Therefore a more fine-tuned notion of approximation is to specify *two* error margins—an *absolute* error, and a *relative* error that increases with $|S|$ when $|S| = \Omega(\epsilon n)$.

A first try is to require, for each $S \in \mathcal{F}$, that

$$|S \cap A| = \frac{|S|\,|A|}{|X|} \pm \max\left\{\frac{|S|}{|X|},\ \epsilon\right\} |A|.$$

Working this out using Fact 14.1 indeed shows that a uniform random sample A of size $\tilde{\Theta}\left(\frac{d}{\epsilon}\right)$ satisfies the above for each $S \in \mathcal{F}$. However, note that when $\frac{|S|}{|X|} \geq \epsilon$, the error term $\frac{|S|}{|X|}|A|$ is as large as the main first term of the R.H.S. above! Thus we will scale the entire error term by another user-given parameter, denoted by $\delta > 0$.

This brings us to the notion of a relative (ϵ, δ)-approximation.

DEFINITION 14.2. Given a finite set system $(X, \mathcal{F})$ and parameters $0 < \delta, \epsilon \leq 1$, a set $A \subseteq X$ is a relative (ϵ, δ)-approximation of $\mathcal{F}$ if for each $S \in \mathcal{F}$,

$$|S \cap A| = \frac{|S|\,|A|}{|X|} \pm \delta \cdot \max\left\{\frac{|S|\,|A|}{|X|},\ \epsilon \cdot |A|\right\}.$$

Equivalently, dividing by $|A|$ gives

$$\left|\frac{|S|}{|X|} - \frac{|S \cap A|}{|A|}\right| \leq \delta \cdot \max\left\{\frac{|S|}{|X|},\ \epsilon\right\}.$$

In particular,

$$|S \cap A| = \begin{cases} (1 \pm \delta)\,\frac{|S|\,|A|}{|X|} & \text{if } |S| \geq \epsilon\,|X|, \\ \frac{|S|\,|A|}{|X|} \pm \delta\,\epsilon & \text{otherwise.} \end{cases}$$

The main theorem of this chapter, and the most general statement on approximations presented in this text, is the following.

THEOREM 14.3. *There exists a positive constant C_4 such that the following is true. Let $(X, \mathcal{F})$ be a finite set system with* VC-dim$(\mathcal{F}) \leq d$ *and let $0 < \delta, \epsilon, \gamma \leq \frac{1}{2}$ be given parameters. Then a uniform random sample $A \subseteq X$ of size at least*

$$C_4 \cdot \left(\frac{d}{\epsilon\,\delta^2} \ln\frac{1}{\epsilon} + \frac{1}{\epsilon\delta^2} \ln\frac{1}{\gamma}\right)$$

is a relative (ϵ, δ)-approximation of $\mathcal{F}$ with probability at least $1 - \gamma$.

We remark that Theorem 14.3 implies earlier bounds on ϵ-nets and ϵ-approximations for set systems with VC-dim$(\mathcal{F}) \leq d$:

ϵ-nets: Let N be a relative $\left(\epsilon, \frac{1}{2}\right)$-approximation, of size $O\left(\frac{d}{\epsilon}\log\frac{1}{\epsilon}\right)$ by Theorem 14.3. Then for any $S \in \mathcal{F}$ with $|S| \geq \epsilon n$,

$$|S \cap N| = \frac{|S||N|}{n} \pm \frac{1}{2}\max\left\{\frac{|S||N|}{n}, \epsilon \cdot |N|\right\} \geq \frac{|S||N|}{n} - \frac{|S||N|}{2\,n} > 0.$$

That is, N is an ϵ-net of $\mathcal{F}$.

ϵ-approximations: Let A be a relative $\left(\frac{1}{2}, \epsilon\right)$-approximation, of size $O\left(\frac{d}{\epsilon^2}\right)$ by Theorem 14.3. Then for any $S \in \mathcal{F}$,

$$|S \cap A| = \frac{|S||A|}{n} \pm \epsilon \max\left\{\frac{|S||A|}{n}, \frac{1}{2}|A|\right\} = \frac{|S||A|}{n} \pm \epsilon|A|.$$

That is, A is an ϵ-approximation of $\mathcal{F}$.

1. Relative and Sensitive Approximations

Yes, some people are brighter than others but I really believe that most people can really get to quite a good level in mathematics if they're prepared to deal with these more psychological issues of how to handle the situation of being stuck.

Andrew Wiles

Let $(X, \mathcal{F})$ be a finite set system, with $n = |X|$. Recall the notion of relative (ϵ, δ)-approximations.

DEFINITION 14.2. Given a finite set system $(X, \mathcal{F})$ and parameters $0 < \delta, \epsilon \leq 1$, a set $A \subseteq X$ is a relative (ϵ, δ)-approximation of $\mathcal{F}$ if for each $S \in \mathcal{F}$,

$$|S \cap A| = \frac{|S|\,|A|}{|X|} \pm \delta \cdot \max\left\{\frac{|S|\,|A|}{|X|},\ \epsilon \cdot |A|\right\}.$$

Equivalently, dividing by $|A|$ gives

$$\left|\frac{|S|}{|X|} - \frac{|S \cap A|}{|A|}\right| \leq \delta \cdot \max\left\{\frac{|S|}{|X|},\ \epsilon\right\}.$$

In particular,

$$|S \cap A| = \begin{cases} (1 \pm \delta)\,\frac{|S|\,|A|}{|X|} & \text{if } |S| \geq \epsilon\,|X|, \\ \frac{|S|\,|A|}{|X|} \pm \delta\,\epsilon & \text{otherwise.} \end{cases}$$

As an $(\epsilon\delta)$-approximation is a relative (ϵ, δ)-approximation, since

$$\epsilon\delta \leq \delta \max\left\{\frac{|S|}{|X|},\ \epsilon\right\},$$

Theorem 13.2 implies the existence of a relative (ϵ, δ)-approximation of size $O\left(\frac{d}{\epsilon^2\delta^2}\right)$. A direct calculation shows that this bound can be improved.
We will need the following statement.

LEMMA 14.4. *Given $(X, \mathcal{F})$, let $A' \subseteq X$ be a relative $\left(\epsilon, \frac{\delta}{3}\right)$-approximation of $\mathcal{F}$. Further let $A \subseteq A'$ be a relative $\left(\epsilon, \frac{\delta}{3}\right)$-approximation of $(A', \mathcal{F}|_{A'})$. Then A is a relative (ϵ, δ)-approximation of $(X, \mathcal{F})$.*

PROOF. Set $t' = |A'|$ and $t = |A|$. Then for each $S \in \mathcal{F}$,

$$|S \cap A'| = \frac{|S|t'}{n} \pm \frac{\delta}{3} \max\left\{\frac{|S|t'}{n}, \epsilon t'\right\}, \quad \text{and}$$

$$|(S \cap A') \cap A| = \frac{|S \cap A'|\,t}{t'} \pm \frac{\delta}{3} \max\left\{\frac{|S \cap A'|\,t}{t'}, \epsilon t\right\}.$$

Since $(S \cap A') \cap A = S \cap A$, the above two statements together imply that $|S \cap A|$ is

$$= \frac{\left(\frac{|S|t'}{n} \pm \frac{\delta}{3} \max\left\{\frac{|S|t'}{n}, \epsilon t'\right\}\right) t}{t'} \pm \frac{\delta}{3} \max\left\{\frac{\left(\frac{|S|t'}{n} \pm \frac{\delta}{3} \max\left\{\frac{|S|t'}{n}, \epsilon t'\right\}\right) t}{t'}, \epsilon t\right\}$$

$$= \frac{|S|t}{n} \pm \frac{\delta}{3} \max\left\{\frac{|S|t}{n}, \epsilon t\right\} \pm \frac{\delta}{3} \max\left\{\frac{|S|t}{n} \pm \frac{\delta}{3} \max\left\{\frac{|S|t}{n}, \epsilon t\right\}, \epsilon t\right\}$$

$$= \frac{|S|t}{n} \pm \frac{\delta}{3}\left(\max\left\{\frac{|S|t}{n}, \epsilon t\right\} \pm \max\left\{\left(1+\frac{\delta}{3}\right)\frac{|S|t}{n}, \left(1+\frac{\delta}{3}\right)\epsilon t\right\}\right)$$

$$= \frac{|S|t}{n} \pm \frac{\delta}{3}\left(2+\frac{\delta}{3}\right) \max\left\{\frac{|S|t}{n}, \epsilon t\right\}.$$

As $\frac{\delta}{3}\left(2+\frac{\delta}{3}\right) \leq \delta$, A is a relative (ϵ, δ)-approximation of $\mathcal{F}$. □

We now come to the first main theorem of this section.

THEOREM 14.5. *Let $(X, \mathcal{F})$ be a finite set system with* VC-dim$(\mathcal{F}) \leq d$ *and* $0 < \delta, \epsilon \leq \frac{1}{2}$ *be given parameters. Let $A \subseteq X$ be a uniform random subset of X of size*

$$c \cdot \frac{d}{\epsilon \delta^2} \log \frac{1}{\epsilon\, \delta},$$

where $c > 0$ is a sufficiently large constant. Then A is a relative (ϵ, δ)-approximation of $\mathcal{F}$ with probability at least $\frac{1}{2}$.

PROOF. The first step is to get rid of the dependence on n. By Theorem 13.11, a uniform random sample A' of size $\Theta\left(\frac{d}{\epsilon^2\delta^2}\right)$ is a $\frac{\epsilon\,\delta}{3}$-approximation of $\mathcal{F}$ with probability at least $\frac{3}{4}$. Thus it suffices to prove that given a set system $(X, \mathcal{F})$ with

$$|X| = \Theta\left(\frac{d}{\epsilon^2\delta^2}\right) \qquad \text{and} \qquad |\mathcal{F}| = O\left(\left(\frac{1}{\epsilon^2\delta^2}\right)^d\right) \qquad \big(\text{by Lemma 4.3}\big),$$

a uniform random sample $A \subseteq X$ of size

$$t = c \cdot \frac{d}{\epsilon \delta^2} \log \frac{1}{\epsilon\, \delta},$$

for a sufficiently large constant c, is a relative $\left(\epsilon, \frac{\delta}{3}\right)$-approximation of $\mathcal{F}$ with probability at least $\frac{3}{4}$. Then Lemma 14.4 would complete the proof of Theorem 14.5. We will need the following upper bound on the probability that A fails to be a relative (ϵ, δ)-approximation for a fixed set.

LEMMA 14.6. *For a fixed $S \in \mathcal{F}$ and a uniform random sample $A \subseteq X$ of size t,*

$$\Pr\big[A \textit{ is not a relative } (\epsilon, \delta)\textit{-approximation of } S\big] \leq 2 \exp\left(-\frac{\epsilon\delta^2 t}{3}\right).$$

PROOF. We consider two cases.

$|S| < \epsilon n$**:** Then $\delta \max\left\{\frac{|S|\,t}{n}, \epsilon t\right\} = \delta\epsilon t$. By Fact 14.1 with $\eta = \delta\epsilon t$,

$$\Pr\big[A \text{ is not a relative } (\epsilon, \delta)\text{-approximation of } S\big] \leq 2\exp\left(-\frac{\delta^2\epsilon^2 t^2\, n}{2|S|t + \epsilon\delta t n}\right)$$

$$\leq 2\exp\left(-\frac{\delta^2\epsilon^2 t^2\, n}{2\epsilon n t + \epsilon\delta t n}\right) = 2\exp\left(-\frac{\delta^2\epsilon t}{2+\delta}\right) \leq 2\exp\left(-\frac{\delta^2\epsilon t}{3}\right).$$

$|S| \geq \epsilon n$**:** In this case $\delta \max\left\{\frac{|S|\,t}{n}, \epsilon\, t\right\} = \delta \frac{|S|t}{n}$. By Fact 14.1 with $\eta = \delta \frac{|S|t}{n}$,

$$\Pr\left[A \text{ is not a relative } (\epsilon,\delta)\text{-approximation of } S\right] \leq 2\exp\left(-\frac{\delta^2 \frac{|S|^2 t^2}{n^2} n}{2|S|t + \delta \frac{|S|t}{n} n}\right)$$

$$\leq 2\exp\left(-\frac{\delta^2 \frac{|S|t}{n}}{2+\delta}\right) \leq 2\exp\left(-\frac{\delta^2 \epsilon t}{3}\right), \qquad \text{since } |S| \geq \epsilon n.$$

□

Finally, by the union bound and Lemma 14.6,

$$\Pr\left[A \text{ is not a relative } \left(\epsilon, \frac{\delta}{3}\right)\text{-approximation of } \mathcal{F}\right] \leq |\mathcal{F}| \cdot 2\exp\left(-\frac{\epsilon\delta^2 t}{27}\right)$$

$$= O\left(\left(\frac{1}{\epsilon^2\delta^2}\right)^d\right) \cdot 2\exp\left(-\frac{\epsilon\delta^2 t}{27}\right) \leq \frac{1}{4},$$

by setting $t = \frac{c}{\epsilon\delta^2} \log\left(\frac{1}{\epsilon^2\delta^2}\right)^d$ for a sufficiently large constant c.

Taking into account the earlier failure probability for A', we conclude that A is a relative (ϵ,δ)-approximation of $\mathcal{F}$ with probability at least $\frac{1}{2}$. □

Observe that Theorem 14.5 states the existence of a relative (ϵ,δ)-approximation whose size depends superlinearly on $\frac{1}{\delta}$. Thus it is reasonable to hope that the factor $\log\frac{1}{\delta}$ in it can be removed using chaining, in a manner similar to the proof of Theorem 13.2. Indeed this is the case, and will be the subject of the next section.

We conclude this chapter by presenting another notion of approximation, called *sensitive* ϵ*-approximations*.

Let A be a uniform random sample of X of size t. Recall the following.

FACT 14.1. *Let* $S \subseteq [n]$ *and* A *a uniform random subset of* $[n]$ *of size* t. *Then*

$$\Pr\left[|S \cap A| \neq \frac{|S|\,t}{n} \pm \eta\right] \leq 2\exp\left(-\frac{\frac{\eta^2 n^2}{|S|^2 t^2}}{2 + \frac{\eta n}{|S|t}} \cdot \frac{|S|t}{n}\right)$$

$$= 2\exp\left(-\frac{\eta^2 n}{2|S|t + \eta n}\right).$$

For this probability of failure to be less than 1, we need

$$\frac{\eta^2 n}{2|S|t + \eta n} = \frac{1}{\frac{2|S|t}{\eta^2 n} + \frac{1}{\eta}} = \Omega(1).$$

Or equivalently, $\dfrac{2|S|t}{\eta^2 n} + \dfrac{1}{\eta} = O(1)$,

which is satisfied if $\eta = \Omega\left(\max\left\{\sqrt{\frac{|S|t}{n}}, 1\right\}\right) = \Omega\left(\sqrt{\frac{|S|t}{n}} + 1\right).$

In particular, setting $\eta = \frac{1}{2}\left(\sqrt{\frac{|S|t}{n}} + 1\right)$ (the constant $\frac{1}{2}$ is somewhat arbitrary here), we arrive at the following constraint:

$$|S \cap A| = \frac{|S|\,t}{n} \pm \frac{1}{2}\left(\sqrt{\frac{|S|\,t}{n}} + 1\right) = \frac{|S|\,t}{n} \pm \frac{\sqrt{t}}{2}\left(\sqrt{\frac{|S|}{n}} + \frac{1}{\sqrt{t}}\right).$$

Equivalently, dividing by t gives

$$\left|\frac{|S|}{n} - \frac{|S \cap A|}{t}\right| \leq \frac{1}{2\sqrt{t}}\left(\sqrt{\frac{|S|}{n}} + \frac{1}{\sqrt{t}}\right).$$

Parameterizing this statement with respect to ϵ by setting $\epsilon = \frac{1}{\sqrt{t}}$, we arrive at the notion of a *sensitive ϵ-approximation.*

DEFINITION 14.7. Given a finite set system $(X, \mathcal{F})$ and a parameter $0 < \epsilon \leq 1$, a set $A \subseteq X$ is a sensitive ϵ-approximation of $\mathcal{F}$ if for each $S \in \mathcal{F}$,

$$\left|\frac{|S|}{|X|} - \frac{|S \cap A|}{|A|}\right| \leq \frac{\epsilon}{2}\left(\sqrt{\frac{|S|}{|X|}} + \epsilon\right).$$

Equivalently, multiplying by $|A|$ gives

$$|S \cap A| = \frac{|S|\,|A|}{|X|} \pm \frac{\epsilon\,|A|}{2}\left(\sqrt{\frac{|S|}{|X|}} + \epsilon\right).$$

Observe that a sensitive ϵ-approximation N of $\mathcal{F}$ is also an ϵ'-net, for any $\epsilon' > \epsilon^2$:

For any $S \in \mathcal{F}$ with $|S| > \epsilon^2 n$,

$$|S \cap N| = \frac{|S|\,|N|}{n} \pm \frac{\epsilon\,|N|}{2}\left(\sqrt{\frac{|S|}{n}} + \epsilon\right) > \frac{|S|\,|N|}{n} - \epsilon|N|\sqrt{\frac{|S|}{n}} > 0,$$

where the second step uses $\epsilon < \sqrt{\frac{|S|}{n}}$.

Thus the lower bound of $\Omega\left(\frac{d}{\epsilon^2}\log\frac{1}{\epsilon}\right)$ for the size of a ϵ^2-net (Lemma 11.12) implies a lower bound of $\Omega\left(\frac{d}{\epsilon^2}\log\frac{1}{\epsilon}\right)$ for the size of any sensitive ϵ-approximation. The following theorem gives the matching upper bound.

THEOREM 14.8. *There exists a positive constant C_{10} such that the following is true. Let $(X, \mathcal{F})$ be a finite system with* VC-dim$(\mathcal{F}) \leq d$. *For given parameters $0 < \epsilon \leq \frac{1}{2}$ and $\gamma \in (0,1)$, let $A \subseteq X$ be a subset of size at least*

$$\frac{C_{10}}{\epsilon^2}\left(d\log\frac{1}{\epsilon} + \log\frac{1}{\gamma}\right)$$

chosen uniformly at random. Then A is a sensitive ϵ-approximation of $\mathcal{F}$ with probability at least $1 - \gamma$.

PROOF. Let A be a random sample of X of size t. We first need an upper bound on the probability that A fails to be a sensitive ϵ-approximation for a fixed set.

LEMMA 14.9. *For a fixed $S \in \mathcal{F}$ and a uniform random sample $A \subseteq X$ of size t,*

$$\Pr\left[A \textit{ is not a sensitive } \epsilon\textit{-approximation of } S\right] \leq 2\exp\left(-\frac{\epsilon^2 t}{10}\right).$$

PROOF. We consider two cases.

$|S| < \epsilon^2 n$: Then $\frac{\epsilon t}{2}\left(\sqrt{\frac{|S|}{n}}+\epsilon\right) \geq \frac{\epsilon^2 t}{2}$ and so applying Fact 14.1 with $\eta = \frac{\epsilon^2 t}{2}$,

$$\Pr\left[A \text{ is not a sensitive } \epsilon\text{-approx. of } S\right] \leq 2\exp\left(-\frac{\epsilon^4 t^2\, n/4}{2\epsilon^2 nt + \epsilon^2 tn/2}\right) = 2\exp\left(-\frac{\epsilon^2 t}{10}\right).$$

$|S| \geq \epsilon^2 n$: In this case $\frac{\epsilon t}{2}\left(\sqrt{\frac{|S|}{n}}+\epsilon\right) \geq \frac{\epsilon t}{2}\sqrt{\frac{|S|}{n}}$ and so applying Fact 14.1 with $\eta = \frac{\epsilon t}{2}\sqrt{\frac{|S|}{n}}$,

$$\Pr\left[A \text{ not a sensitive } \epsilon\text{-approx. of } S\right] \leq 2\exp\left(-\frac{\epsilon^2 t^2 \frac{|S|}{n}\, n/4}{2|S|t + \epsilon t\sqrt{\frac{|S|}{n}}n/2}\right)$$
$$= 2\exp\left(-\frac{\epsilon^2 t/4}{2+\epsilon\sqrt{\frac{n}{|S|}}/2}\right) \leq 2\exp\left(-\frac{\epsilon^2 t}{10}\right).$$

□

Lemma 14.9 together with the union bound implies that a uniform random sample of size t fails to be a sensitive ϵ-approximation with probability at most $|\mathcal{F}| \cdot 2\exp\left(-\frac{\epsilon^2 t}{10}\right)$, which implies that

> a uniform random sample of size at least $\frac{10}{\epsilon^2}\ln\frac{2|\mathcal{F}|}{\gamma}$ is a ϵ-sensitive approximation of $\mathcal{F}$ with probability at least $1-\gamma$.

Now one can complete the proof via induction, along the lines of the proof of Theorem 12.3[1]. Let $T(\epsilon,\gamma)$ be a positive integer such that a uniform random sample of size at least $T(\epsilon,\gamma)$ is a sensitive ϵ-approximation of $\mathcal{F}$ with probability at least $1-\gamma$. We will prove that for a large-enough constant C_{10}, we have $T(\epsilon,\gamma) \leq \frac{C_{10}}{\epsilon^2}\left(d\log\frac{1}{\epsilon}+\log\frac{1}{\gamma}\right)$.

We will need the following fact that is easy to verify:

> if A' is a sensitive $\frac{\epsilon}{2}$-approximation of $\mathcal{F}$ and furthermore A is a sensitive $\frac{\epsilon}{4}$-approximation of $\mathcal{F}|_{A'}$, then A is a sensitive ϵ-approximation of $\mathcal{F}$.

As a uniform random sample $A' \subseteq X$ of size $T\left(\frac{\epsilon}{2},\frac{\gamma}{2}\right)$ is a sensitive $\frac{\epsilon}{2}$-approximation of $\mathcal{F}$ with probability at least $1-\frac{\gamma}{2}$ (induction hypothesis), and a uniform random sample $A \subseteq A'$ of size at least $\frac{10}{(\epsilon/4)^2}\ln\frac{2|\mathcal{F}|_{A'}|}{(\gamma/2)}$ is a sensitive $\frac{\epsilon}{4}$-approximation of $\mathcal{F}|_{A'}$

[1] Alternatively, similar to the proof of Theorem 14.5, we could also first take an $\frac{\epsilon}{2}$-approximation of $(X,\mathcal{F})$ to get rid of the dependence on $|X|$ and then apply the union bound.

with probability at least $1 - \frac{\gamma}{2}$, we have

$$T(\epsilon, \gamma) \leq |A| = \frac{10}{(\epsilon/4)^2} \ln \frac{2|\mathcal{F}|_{A'}|}{(\gamma/2)} \leq \frac{10}{(\epsilon/4)^2} \ln \frac{2\left(\frac{eT\left(\frac{\epsilon}{2},\frac{\gamma}{2}\right)}{d}\right)^d}{(\gamma/2)}$$

$$\leq \frac{C_{10}}{\epsilon^2}\left(d \log \frac{1}{\epsilon} + \log \frac{1}{\gamma}\right),$$

for a large-enough constant $C_{10} > 1$. □

Another proof is via relative (ϵ, δ)-approximations, for appropriate values of ϵ and δ, by applying the more general Theorem 14.3 (see discussion).

Bibliography and discussion. The notion of sensitive approximations together with Theorem 14.8 is from [**BCM99**]. Bounds on relative (ϵ, δ)-approximations were given in [**HPS11**], based on the fact that the notion of relative (ϵ, δ)-approximations is, within constants, equivalent to that of so-called (ν, α)-samples from the important paper [**LLS01**]. The precise results and proofs around (ν, α)-samples can be read in the text [**HP11**]. The paper [**HPS11**] also contains a study of various notions of approximations together with some improvements for geometric set systems; it also contains the construction of sensitive ϵ-approximations via relative (ϵ, δ)-approximations. Further improvements for geometric set systems are shown in [**Ezr16**], and is the key paper that revisited and strengthened Haussler's packing theorem for geometric set systems.

[BCM99] H. Brönnimann, B. Chazelle, and J. Matoušek, *Product range spaces, sensitive sampling, and derandomization*, SIAM J. Comput. **28** (1999), no. 5, 1552–1575, DOI 10.1137/S0097539796260321. MR1694192

[Ezr16] E. Ezra, *A size-sensitive discrepancy bound for set systems of bounded primal shatter dimension*, SIAM J. Comput. **45** (2016), no. 1, 84–101, DOI 10.1137/140977746. MR3452262

[HP11] S. Har-Peled, *Geometric approximation algorithms*, Mathematical Surveys and Monographs, vol. 173, American Mathematical Society, Providence, RI, 2011, DOI 10.1090/surv/173. MR2760023

[HPS11] S. Har-Peled and M. Sharir, *Relative (p, ϵ)-approximations in geometry*, Discrete Comput. Geom. **45** (2011), no. 3, 462–496, DOI 10.1007/s00454-010-9248-1. MR2770547

[LLS01] Y. Li, P. M. Long, and A. Srinivasan, *Improved bounds on the sample complexity of learning*, J. Comput. System Sci. **62** (2001), no. 3, 516–527, DOI 10.1006/jcss.2000.1741. MR1824457

2. Improved Bounds for Relative Approximations

Mathematics is like a flight of fancy, but one in which the fanciful turns out to be real and to have been present all along. Doing mathematics has the feel of fanciful invention, but it is really a process for sharpening our perception so that we discover patterns that are everywhere around... To share in the delight and the intellectual experience of mathematics—to fly where before we walked—that is the goal of mathematical education.

William Thurston

The main theorem of this section improves upon Theorem 14.5 by a multiplicative factor of $\log \frac{1}{\delta}$.

THEOREM 14.3. *There exists a positive constant C_4 such that the following is true. Let $(X, \mathcal{F})$ be a finite set system with* VC-dim $(\mathcal{F}) \leq d$ *and let $0 < \delta, \epsilon, \gamma \leq \frac{1}{2}$ be given parameters. Then a uniform random sample $A \subseteq X$ of size at least*

$$C_4 \cdot \left(\frac{d}{\epsilon\, \delta^2} \ln \frac{1}{\epsilon} + \frac{1}{\epsilon \delta^2} \ln \frac{1}{\gamma} \right)$$

is a relative (ϵ, δ)-approximation of $\mathcal{F}$ with probability at least $1 - \gamma$.

Overview of ideas. Recall the notion of relative (ϵ, δ)-approximations.

DEFINITION 14.2. Given a finite set system $(X, \mathcal{F})$ and parameters $0 < \delta, \epsilon \leq 1$, a set $A \subseteq X$ is a relative (ϵ, δ)-approximation of $\mathcal{F}$ if for each $S \in \mathcal{F}$,

$$|S \cap A| = \frac{|S|\,|A|}{|X|} \ \pm\ \delta \cdot \max \left\{ \frac{|S|\,|A|}{|X|},\ \epsilon \cdot |A| \right\}.$$

Equivalently, dividing by $|A|$ gives

$$\left| \frac{|S|}{|X|} - \frac{|S \cap A|}{|A|} \right| \leq \delta \cdot \max \left\{ \frac{|S|}{|X|},\ \epsilon \right\}.$$

In particular,

$$|S \cap A| = \begin{cases} (1 \pm \delta)\ \frac{|S|\,|A|}{|X|} & \text{if } |S| \geq \epsilon\, |X|, \\[2ex] \frac{|S|\,|A|}{|X|} \pm \delta\, \epsilon & \text{otherwise.} \end{cases}$$

The proof of Theorem 14.3 uses chaining, pretty much following the proof of Theorem 13.2—with one subtlety. As the goal is to remove a factor of $\log \frac{1}{\delta}$ from the bound of Theorem 14.5, a natural first try is to apply chaining on δ. That is, show the existence of a small number of set systems on X—say $\mathcal{F}_1, \ldots, \mathcal{F}_t$—such that each set of $\mathcal{F}$ can be 'derived' from a combination of t sets, one from each $\mathcal{F}_i$. Then it suffices to show, using an appropriate analog of Claim 13.6, that if a uniform random sample A is a relative (ϵ, δ_i)-approximation of each $(X, \mathcal{F}_i)$—the δ_i's being chosen so that $\delta_1 + \cdots + \delta_t$ is at most δ—then A is a relative (ϵ, δ)-approximation of $(X, \mathcal{F})$.

The problem with this approach is that Claim 13.6 does not hold for the error function $\delta \max \left\{ \frac{|S|}{|X|}, \epsilon \right\}$. That is, given $S_1 \subseteq S_2 \subseteq X$, if A is a relative (ϵ, δ_1)-approximation of S_1 and a relative (ϵ, δ_2)-approximation of S_2, then A need not be a relative $(\epsilon, O(\delta_1 + \delta_2))$-approximation of $S_2 \setminus S_1$. This happens when $|S_2 \setminus S_1|$ is much smaller than $\min \{|S_1|, |S_2|\}$.

The trick is to instead do chaining on the parameter ϵ, since the dependence on ϵ in the error function does not involve $|S|$. We then proceed precisely as in the proof of Theorem 13.2 and calculations will show that with probability at least $1-\gamma$, A has total error at most $\delta \max\left\{\frac{|S|}{n}+c_1\epsilon, c_2\epsilon\right\}$ for each $S \in \mathcal{F}^2$, for two absolute constants c_1 and c_2. Since

$$\delta \cdot \max\left\{\frac{|S|}{n}+c_1\epsilon,\ c_2\epsilon\right\} \le \delta \cdot \max\left\{\frac{2\,|S|}{n}, 2\,c_1\epsilon,\ c_2\epsilon\right\}$$
$$\le 2\delta \cdot \max\left\{\frac{|S|}{n}, \left(c_1+\frac{c_2}{2}\right)\epsilon\right\},$$

we arrive at a relative $\left(\left(c_1+\frac{c_2}{2}\right)\epsilon, 2\delta\right)$-approximation of $\mathcal{F}$.
Of course one can re-do the entire calculation with $\delta' = \frac{\delta}{2}$ and $\epsilon' = \frac{\epsilon}{\left(c_1+\frac{c_2}{2}\right)}$ to get a relative (ϵ,δ)-approximation. However, for notational convenience, we will simply do the calculations with parameters ϵ, δ and end up with a relative $(O(\epsilon), O(\delta))$-approximation.

We first recall the useful tail bound.

FACT 14.1. Let $S \subseteq [n]$ and A a uniform random subset of $[n]$ of size t. Then

$$\Pr\left[|S\cap A| \ne \frac{|S|\,t}{n} \pm \eta\right] \le 2\exp\left(-\frac{\frac{\eta^2 n^2}{|S|^2 t^2}}{2+\frac{\eta n}{|S|t}} \cdot \frac{|S|t}{n}\right)$$
$$= 2\exp\left(-\frac{\eta^2 n}{2|S|t+\eta n}\right).$$

PROOF OF THEOREM 14.3. The analysis can be divided into three stages.

1. From X to A'. Given $(X,\mathcal{F})$, let $A' \subseteq X$ be a uniform random sample of sufficient size such that A' is a $\frac{\epsilon\delta}{3}$-approximation of $\mathcal{F}$ with probability at least $1-\frac{\gamma}{2}$. Note that A' is also a relative $\left(\epsilon, \frac{\delta}{3}\right)$-approximation of $\mathcal{F}$. Recall the following:

LEMMA 14.4. *Given $(X,\mathcal{F})$, let $A' \subseteq X$ be a relative $\left(\epsilon,\frac{\delta}{3}\right)$-approximation of $\mathcal{F}$. Further let $A \subseteq A'$ be a relative $\left(\epsilon,\frac{\delta}{3}\right)$-approximation of $(A', \mathcal{F}|_{A'})$. Then A is a relative (ϵ,δ)-approximation of $(X,\mathcal{F})$.*

Thus it suffices to show that a uniform random sample $A \subseteq A'$ of size at least

$$t = C_4 \cdot \left(\frac{d}{\epsilon\delta^2}\ln\frac{1}{\epsilon} + \frac{1}{\epsilon\delta^2}\ln\frac{1}{\gamma}\right) = \frac{C_4}{\epsilon\delta^2}\ln\frac{1}{\epsilon^d\gamma},$$

where C_4 is a sufficiently large constant whose value will be fixed later, is a relative $\left(\epsilon,\frac{\delta}{3}\right)$-approximation of $\mathcal{F}|_{A'}$ with probability at least $1-\frac{\gamma}{2}$.

[2]That is, $\left|\frac{|S|}{|X|} - \frac{|S\cap A|}{|A|}\right| \le \delta\max\left\{\frac{|S|}{n}+c_1\epsilon, c_2\epsilon\right\}$.

To avoid additional notations we will simply assume that $X = A'$, and so by Theorem 12.3 we have

$$|X| = n = \frac{9C_1}{\epsilon^2\delta^2}\left(d\ln\frac{3}{\epsilon\delta} + \ln\frac{2}{\gamma}\right) \le \frac{27C_1 d}{\epsilon^3\delta^3}\left(1 + \frac{1}{d}\ln\frac{2}{\gamma}\right) \le \frac{27C_1 d}{\epsilon^3\delta^3}\exp\left(\frac{1}{d}\ln\frac{2}{\gamma}\right),$$

$$|\mathcal{F}| \le \left(\frac{e|X|}{d}\right)^d \le \left(\frac{27eC_1}{\epsilon^3\delta^3}\right)^d \exp\left(\ln\frac{2}{\gamma}\right) = \left(\frac{27eC_1}{\epsilon^3\delta^3}\right)^d \frac{2}{\gamma},$$

where the first step in the bound on $|\mathcal{F}|$ follows from Lemma 4.3.

2. Constructing small set systems. Set $k = \lceil \log \frac{1}{\delta} \rceil$ and for each $i \in [1, k]$, let

$$\mathcal{P}_i : \quad \text{a } \textit{maximal } \tfrac{\epsilon n}{2^i}\text{-packing of } (X, \mathcal{F}).$$

That is, the size of the set symmetric difference between any pair of sets in $\mathcal{P}_i$ is at least $\frac{\epsilon n}{2^i}$. Set $\mathcal{P}_{k+1} = \mathcal{F}$.

For any $i \in [1, k]$ and $S \in \mathcal{P}_{i+1}$, the maximality of $\mathcal{P}_i$ implies that there exists a set $F_S \in \mathcal{P}_i$ such that $|\Delta(S, F_S)| < \frac{\epsilon n}{2^i}$ (F_S could be the set S itself). Define

$$\mathcal{A}_i = \Big\{ S \setminus F_S \colon S \in \mathcal{P}_{i+1} \Big\} \qquad \text{and} \qquad \mathcal{B}_i = \Big\{ F_S \setminus S \colon S \in \mathcal{P}_{i+1} \Big\},$$

where each set in $\mathcal{A}_i \cup \mathcal{B}_i$ has size less than $\frac{\epsilon n}{2^i}$. Furthermore, by Theorem 5.8,

$$\text{for } i \in [1, k-1] \colon |\mathcal{A}_i|, |\mathcal{B}_i| \le |\mathcal{P}_{i+1}| = O\left((8e)^d \left(\frac{2^{i+1}}{\epsilon}\right)^d\right) \quad \text{and} \quad |\mathcal{A}_k|, |\mathcal{B}_k| \le |\mathcal{F}|.$$

Let

$$\epsilon_i = \sqrt{\frac{i}{2^i}}\,\epsilon.$$

Note that

$$\sum_{j=1}^{\infty} \sqrt{\frac{j}{2^j}} = \frac{2}{\sqrt{2}} + \sum_{j=3}^{\infty} \sqrt{\frac{j}{2^j}} \le \sqrt{2} + \sqrt{\frac{3}{8}} \sum_{i=0}^{\infty} \left(\sqrt{\frac{2}{3}}\right)^i \le 5, \tag{14.10}$$

where the second step uses that for $j \ge 3$, $\frac{\sqrt{(j+1)/2^{j+1}}}{\sqrt{j/2^j}} = \sqrt{\frac{1}{2} + \frac{1}{2j}} \le \sqrt{\frac{2}{3}}$.

CLAIM 14.11. *For a sufficiently large constant C_4, with probability at least $1 - \frac{\gamma}{2}$, A is simultaneously*

(i) *a relative (ϵ, δ)-approximation of $\mathcal{P}_1$, and*
(ii) *a relative (ϵ, δ)-approximation of $\mathcal{A}_k \cup \mathcal{B}_k$, and*
(iii) *a relative (ϵ_i, δ)-approximation of $\mathcal{A}_i \cup \mathcal{B}_i$, $i \in [1, k-1]$.*

PROOF. *(i)* Applying Lemma 14.6, the probability of failure is at most

$$\begin{aligned}
|\mathcal{P}_1| \cdot 2\exp\left(-\frac{\epsilon\delta^2 t}{3}\right) &= O\left(\left(\frac{32e}{\epsilon}\right)^d\right) \cdot 2\exp\left(-\frac{\epsilon\delta^2 t}{3}\right) \\
&= O\left(\left(\frac{32e}{\epsilon}\right)^d\right) \cdot 2\exp\left(-\frac{C_4 \ln\frac{1}{\epsilon^d\gamma}}{3}\right) \\
&= O\left(\left(\frac{32e}{\epsilon}\right)^d\right) \cdot 2\left(\epsilon^d\gamma\right)^{C_4/3} \le \frac{\gamma}{6},
\end{aligned}$$

for a sufficiently large constant C_4.

(ii) We have $|\mathcal{A}_k|, |\mathcal{B}_k| \leq |\mathcal{F}|$ and each set in $\mathcal{A}_k \cup \mathcal{B}_k$ has size at most $\frac{\epsilon n}{2^k} \leq \epsilon\delta n$. Thus $\delta \max\left\{\frac{|S|t}{n}, \epsilon t\right\} = \delta\epsilon t$ and by applying Fact 14.1 with $\eta = \delta\epsilon t$, the probability of failure is at most

$$2|\mathcal{F}| \cdot 2\exp\left(-\frac{\delta^2\epsilon^2 t^2 \cdot n}{2\epsilon\delta n \cdot t + \delta\epsilon t \cdot n}\right) = 4|\mathcal{F}| \cdot \exp\left(-\frac{\delta\epsilon t}{3}\right) = 4|\mathcal{F}| \cdot \exp\left(-\frac{C_4 \ln \frac{1}{\epsilon^d\gamma}}{3\delta}\right)$$

$$\leq 4\left(\frac{27\, e\, C_1}{\epsilon^3\delta^3}\right)^d \frac{2}{\gamma} \cdot \left(\epsilon^d\gamma\right)^{C_4/3\delta} \leq \frac{\gamma}{6},$$

for a sufficiently large constant C_4.

(iii) For any $i \in [1, k-1]$ and $S \in \mathcal{A}_i \cup \mathcal{B}_i$, we have $|S| < \frac{\epsilon n}{2^i} \leq \epsilon_i n$. Thus $\delta \max\left\{\frac{|S|t}{n}, \epsilon_i t\right\} = \delta\epsilon_i t$ and by applying Fact 14.1 with $\eta = \delta\epsilon_i t$, the probability of failure for a fixed $S \in \mathcal{A}_i \cup \mathcal{B}_i$ is at most

$$2\exp\left(-\frac{\delta^2\epsilon_i^2 t^2 \cdot n}{2|S| \cdot t + \delta\epsilon_i t \cdot n}\right) \leq 2\exp\left(-\frac{\delta^2\, t\, \frac{i}{2^i}\, \epsilon^2}{2\epsilon/2^i + \delta\,\sqrt{i/2^i}\,\epsilon}\right)$$

$$= 2\exp\left(-\frac{\delta^2\, t\, i\, \epsilon}{2 + \delta\,\sqrt{i 2^i}}\right) \leq 2\exp\left(-\frac{\delta^2\, t\, i\, \epsilon}{2 + \delta\,\sqrt{\log(1/\delta)\cdot(1/\delta)}}\right)$$

$$\leq 2\exp\left(-\frac{C_4\, i\, \ln\frac{1}{\epsilon^d\gamma}}{3}\right).$$

The probability that A fails to be a relative (ϵ_i, δ)-approximation of some $\mathcal{A}_i \cup \mathcal{B}_i$ is at most

$$\sum_{i=1}^{k-1} |\mathcal{A}_i \cup \mathcal{B}_i| \cdot 2\exp\left(-\frac{C_4\, i \ln\frac{1}{\epsilon^d\gamma}}{3}\right) \leq 4\sum_{i=1}^{k-1} O\left(\left(\frac{16\, e\, 2^i}{\epsilon}\right)^d\right)\left(\epsilon^d\gamma\right)^{C_4\, i/3}$$

$$\leq \frac{\gamma}{12}\sum_{i=1}^{k-1}(2\epsilon)^{id} \leq \frac{\gamma}{6},$$

for a sufficiently large constant C_4. □

3. Chaining. Let $S \in \mathcal{F}$. There exists a set $S_k \in \mathcal{P}_k$ with $A_k = S \setminus S_k \in \mathcal{A}_k$ and $B_k = S_k \setminus S \in \mathcal{B}_k$ such that $S = (S_k \setminus B_k) \cup A_k$. Thus

$$\left|\frac{|S|}{n} - \frac{|S \cap A|}{t}\right| = \left|\frac{|(S_k \setminus B_k) \cup A_k|}{n} - \frac{|\big((S_k \setminus B_k) \cup A_k\big) \cap A|}{t}\right|$$

Since $B_k \subseteq S_k$ and $S_k \cap A_k = \emptyset$,

$$= \left|\left(\frac{|S_k|}{n} - \frac{|B_k|}{n} + \frac{|A_k|}{n}\right) - \left(\frac{|S_k \cap A|}{t} - \frac{|B_k \cap A|}{t} + \frac{|A_k \cap A|}{t}\right)\right|$$

$$\leq \left|\frac{|S_k|}{n} - \frac{|S_k \cap A|}{t}\right| + \left|\frac{|A_k|}{n} - \frac{|A_k \cap A|}{t}\right| + \left|\frac{|B_k|}{n} - \frac{|B_k \cap A|}{t}\right|$$

Using Claim 14.11 (ii),

$$\leq \left| \frac{|S_k|}{n} - \frac{|S_k \cap A|}{t} \right| + \delta \max\left\{ \frac{|A_k|}{n}, \epsilon \right\} + \delta \max\left\{ \frac{|B_k|}{n}, \epsilon \right\}$$
$$= \left| \frac{|S_k|}{n} - \frac{|S_k \cap A|}{t} \right| + 2\,\delta\epsilon, \qquad \text{as} \quad |A_k|, |B_k| \leq \epsilon n.$$

The above argument can now be repeated to upper bound $\left| \frac{|S_k|}{n} - \frac{|S_k \cap A|}{t} \right|$. That is, one can write S_k in terms of a set $S_{k-1} \in \mathcal{P}_{k-1}$, $A_{k-1} \in \mathcal{A}_{k-1}$, $B_{k-1} \in \mathcal{B}_{k-1}$ and so on until we reach a set $S_1 \in \mathcal{P}_1$. Using Claim 14.11 (iii), we arrive at

$$\leq \left| \frac{|S_1|}{n} - \frac{|S_1 \cap A|}{t} \right| + 2\delta \sum_{j=1}^{k-1} \epsilon_j + 2\delta\epsilon \leq \left| \frac{|S_1|}{n} - \frac{|S_1 \cap A|}{t} \right| + 12\delta\epsilon,$$

where the last step follows from Equation (14.10). Using Claim 14.11 (i) as well as the fact that $|S_1| \leq |S| + \sum_{j=1}^{k} |B_i| \leq |S| + \sum_{j=1}^{\infty} \frac{\epsilon n}{2^j} \leq |S| + 2\epsilon n$,

$$\leq \delta \max\left\{ \frac{|S_1|}{n}, \epsilon \right\} + 12\delta\epsilon \leq \delta \left(\frac{|S|}{n} + 2\epsilon \right) + 12\delta\epsilon$$
$$\leq \delta \left(\frac{|S|}{n} + 14\epsilon \right) \leq 2\delta \max\left\{ \frac{|S|}{n}, 14\epsilon \right\}.$$

This concludes the proof. □

Bibliography and discussion. The basic idea of the proof follows [**LLS0114b**], with an earlier use of this idea in the important paper [**Mat95a**]. The presentation given in the text is from [**CM21**].

[CM21] M. Csikós and N. H. Mustafa. *Optimal approximations made easy. CoRR* abs/2008.08970, 2021. arXiv:2008.08970. URL: https://arxiv.org/abs/2008.08970.

[LLS0114b] Y. Li, P. M. Long, and A. Srinivasan, *Improved bounds on the sample complexity of learning*, J. Comput. System Sci. **62** (2001), no. 3, 516–527, DOI 10.1006/jcss.2000.1741. MR1824457

[Mat95a] J. Matoušek, *Tight upper bounds for the discrepancy of half-spaces*, Discrete Comput. Geom. **13** (1995), no. 3-4, 593–601, DOI 10.1007/BF02574066. MR1318799

CHAPTER 15

Epsilon-Approximations: Functional Case

Given a set P of n points in $\mathbb{R}^d$ and a parameter $\epsilon > 0$, let A be an ϵ-approximation of the primal set system induced on P by balls in $\mathbb{R}^d$. Then clearly A can be used to approximate P 'combinatorially' with respect to balls:

Let $P_{q,r} = \text{Ball}\,(q,r) \cap P$ be the set of points of P contained in the ball of radius r centered at q; similarly set $A_{q,r} = \text{Ball}\,(q,r) \cap A$.

Then for any $q \in \mathbb{R}^d$ and $r > 0$, $|P_{q,r}|$ can be approximated by $|A_{q,r}| \cdot \frac{|P|}{|A|}$, since by the definition of ϵ-approximations,

$$|A_{q,r}| = \frac{|P_{q,r}|\,|A|}{|P|} \pm \epsilon|A| \quad \text{or equivalently,}$$

$$|P_{q,r}| = |A_{q,r}| \cdot \frac{|P|}{|A|} \pm \epsilon|P|. \tag{15.1}$$

The new idea in this chapter is the observation that since for any $q \in \mathbb{R}^d$, Equation (15.1) holds for *every* radius r, the set A can also be used to approximate the sum of distances from q to the points of P. In particular, here is another property that holds for A: for any $q \in \mathbb{R}^d$ and $r > 0$,

$$\left| \frac{\sum_{p \in P_{q,r}} \text{dist}\,(p,q)}{n} - \frac{\sum_{p \in A_{q,r}} \text{dist}\,(p,q)}{|A|} \right| \leq 3\epsilon r. \tag{15.2}$$

To see the intuition for Equation (15.2), we sketch the proof for a weaker bound of $3\sqrt{\epsilon} r$.

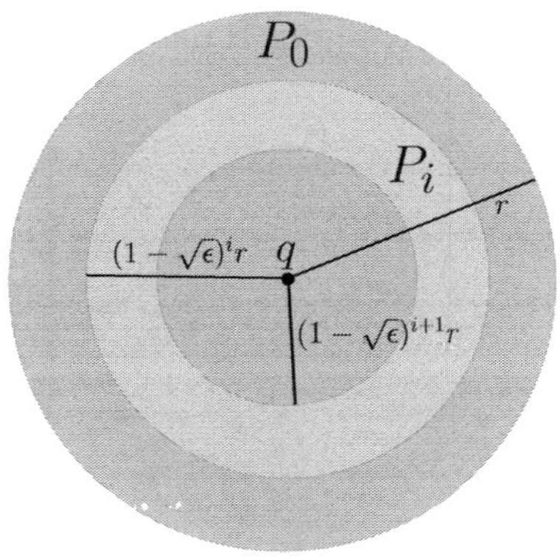

Recall that we say A is an ϵ-approximation of a set $P' \subseteq P$ if $|P' \cap A| = \frac{|P'|\,|A|}{|P|} \pm \epsilon|A|$.

Partition $P_{q,r}$ into disjoint sets $P_0, P_1, \ldots$, where $p \in P_i$ if and only if

$$\mathrm{dist}(p,q) \in \left(\left(1-\sqrt{\epsilon}\right)^{i+1} r, \left(1-\sqrt{\epsilon}\right)^{i} r \right].$$

That is, P_i is the set of points of P lying in the region

$$\mathrm{Ball}\left(q, \left(1-\sqrt{\epsilon}\right)^{i} r\right) \setminus \mathrm{Ball}\left(q, \left(1-\sqrt{\epsilon}\right)^{i+1} r\right).$$

Now the sum of distances of the points of A to q can be approximated by summing up over the P_i's:

$$\left(\sum_{i=0}^{\infty} \left(1-\sqrt{\epsilon}\right)^{i+1} r \cdot |P_i \cap A| \right) < \sum_{p \in A_{q,r}} \mathrm{dist}(p,q) \le \left(\sum_{i=0}^{\infty} \left(1-\sqrt{\epsilon}\right)^{i} r \cdot |P_i \cap A| \right).$$

We remark that we only need to do the above sum till index $i = \frac{1}{\sqrt{\epsilon}} \ln \frac{1}{\epsilon}$, as after that the average sum of distances is most ϵr in any case. Using the fact that A is a 2ϵ-approximation of each P_i (by Claim 13.6),

$$\begin{aligned}
\left(\sum_{i=0}^{\infty} \left(1-\sqrt{\epsilon}\right)^{i+1} r \left(\frac{|P_i|\,|A|}{n} - 2\epsilon|A| \right) \right) &< \sum_{p \in A_{q,r}} \mathrm{dist}(p,q) \le \\
&\left(\sum_{i=0}^{\infty} \left(1-\sqrt{\epsilon}\right)^{i} r \left(\frac{|P_i|\,|A|}{n} + 2\epsilon|A| \right) \right) \\
\left(1-\sqrt{\epsilon}\right) \left(\sum_{i=0}^{\infty} \left(1-\sqrt{\epsilon}\right)^{i} \left(\frac{|P_i|}{n} - 2\epsilon \right) \right) &< \frac{\sum_{p \in A_{q,r}} \mathrm{dist}(p,q)}{r\,|A|} \le \\
&\left(\sum_{i=0}^{\infty} \left(1-\sqrt{\epsilon}\right)^{i} \left(\frac{|P_i|}{n} + 2\epsilon \right) \right).
\end{aligned}$$

Using the fact that $2\epsilon \sum_{i=0}^{\infty} \left(1-\sqrt{\epsilon}\right)^{i} = 2\epsilon \cdot \frac{1}{1-(1-\sqrt{\epsilon})} = 2\sqrt{\epsilon}$,

$$\begin{aligned}
\left(1-\sqrt{\epsilon}\right) \left(\left(\frac{1}{n} \sum_{i=0}^{\infty} \left(1-\sqrt{\epsilon}\right)^{i} |P_i| \right) - 2\sqrt{\epsilon} \right) &< \frac{\sum_{p \in A_{q,r}} \mathrm{dist}(p,q)}{r\,|A|} \le \\
&\left(\frac{1}{n} \sum_{i=0}^{\infty} \left(1-\sqrt{\epsilon}\right)^{i} |P_i| \right) + 2\sqrt{\epsilon}.
\end{aligned}$$

Similarly approximating $\sum_{p \in P_{q,r}} \mathrm{dist}(p,q)$ over the P_i's gives

$$\begin{aligned}
\left(1-\sqrt{\epsilon}\right) \left(\sum_{i=0}^{\infty} \left(1-\sqrt{\epsilon}\right)^{i} r\,|P_i| \right) &< \sum_{p \in P_{q,r}} \mathrm{dist}(p,q) \le \\
&\left(\sum_{i=0}^{\infty} \left(1-\sqrt{\epsilon}\right)^{i} r\,|P_i| \right).
\end{aligned}$$

Dividing by rn,

$$\begin{aligned}
\left(1-\sqrt{\epsilon}\right) \left(\frac{1}{n} \sum_{i=0}^{\infty} \left(1-\sqrt{\epsilon}\right)^{i} |P_i| \right) &< \frac{\sum_{p \in P_{q,r}} \mathrm{dist}(p,q)}{rn} \le \\
&\left(\frac{1}{n} \sum_{i=0}^{\infty} \left(1-\sqrt{\epsilon}\right)^{i} |P_i| \right).
\end{aligned}$$

These together imply the desired bound:

$$\begin{aligned}
\left| \frac{\sum_{p \in P_{q,r}} \mathrm{dist}(p,q)}{r\,n} - \frac{\sum_{p \in A_{q,r}} \mathrm{dist}(p,q)}{r\,|A|} \right| &\le 2\sqrt{\epsilon} + \sqrt{\epsilon} \left(\frac{1}{n} \sum_{i=0}^{\infty} \left(1-\sqrt{\epsilon}\right)^{i} |P_i| \right) \\
&< 2\sqrt{\epsilon} + \sqrt{\epsilon} \left(\frac{1}{n} \sum_{i=0}^{\infty} |P_i| \right) \\
&\le 3\sqrt{\epsilon}.
\end{aligned}$$

As we will see later, the improvement to $3\epsilon r$ presented later in this chapter follows with more precise calculations.

Consider now another application of the same idea on a slightly more complicated distance function where the point q is replaced by a set of k points. For any set X of k points in $\mathbb{R}^d$ and $p \in P$, define

$$\operatorname{dist}(p, X) = \min_{q \in X} \operatorname{dist}(p, q)\,. \tag{15.3}$$

Further let

$$P_{X,r} = \{p \in P : \ \operatorname{dist}(p, X) \leq r\}\,.$$

Observe that $P_{X,r}$ is the set of points of P that lie in the union of the k balls of radius r centered at the points of X. Denote this union by $\operatorname{Ball}(X, r)$.

As earlier, our goal is to estimate, for any given $X \in \left(\mathbb{R}^d\right)^k$ and $r \geq 0$, the expression

$$\sum_{p \in P_{X,r}} \operatorname{dist}(p, X).$$

Not surprisingly, if A is an ϵ-approximation of the set system induced on P by the union of k balls, then one can show that

$$\left| \frac{\sum_{p \in P_{X,r}} \operatorname{dist}(p, X)}{n} - \frac{\sum_{p \in A_{X,r}} \operatorname{dist}(p, X)}{|A|} \right| \leq 3\epsilon r.$$

The first result of this chapter is a more general statement which implies both the above two instances.

The second result is its application to an algorithmic problem central to several domains: the k-median clustering problem, where given a set P of points in $\mathbb{R}^d$ and an integer parameter $k > 0$, the goal is to partition the points of P into k clusters based on certain geometric criteria.

1. A Functional View of Approximations

A common error of judgment among mathematicians is the confusion between telling the truth and giving a logically correct presentation. The two objectives are antithetical and hard to reconcile. Most presentations obeying the current Diktats of linear rigor are a long way from telling the truth; any reader of such a presentation is forced to start writing on the margin, or deciphering on a separate sheet of paper.

The truth of any piece of mathematical writing consists of realizing what the author is "up to"; it is the tradition of mathematics to do whatever it takes to avoid giving away this secret.

Gian-Carlo Rota

Recall the following statement.

Let P be a set of n points in $\mathbb{R}^d$. Each $p \in P$ defines the function

$$\mathrm{dist}\,(p, X) = \min_{q \in X} \mathrm{dist}\,(p, q)\,,$$

where X is a finite set of points in $\mathbb{R}^d$.

Then for any positive integer k and $\epsilon > 0$, there exists an ϵ-approximation $A \subseteq P$ such that for any set X of k points in $\mathbb{R}^d$ and $r \in \mathbb{R}^+$,

$$\left| \frac{\sum_{p \in P_{X,r}} \mathrm{dist}\,(p, X)}{n} - \frac{\sum_{p \in A_{X,r}} \mathrm{dist}\,(p, X)}{|A|} \right| \leq 3\epsilon r, \tag{15.4}$$

where $P_{X,r} = \{p \in P\colon \mathrm{dist}\,(p, X) \leq r\}$ and $A_{X,r} = A \cap P_{X,r}$.

Note that the role of each $p \in P$ is captured by the function $\mathrm{dist}\,(p, \cdot)$. We now prove Equation (15.4) in an abstract setting where $\mathrm{dist}\,(p, \cdot)$ is replaced by an arbitrary function $g_p\colon \mathcal{X} \to \mathbb{R}^+$, where $\mathcal{X}$ is a given domain. That is,

$$\begin{aligned} \text{set of all } k\text{-tuples of points in } \mathbb{R}^d &\longrightarrow \text{ a domain } \mathcal{X}, \\ \text{set of } n \text{ functions } \mathrm{dist}\,(p, \cdot),\ p \in P &\longrightarrow \text{ set } G \text{ of } n \text{ functions from } \mathcal{X} \text{ to } \mathbb{R}^+, \\ P_{X,r} = \{p \in P\colon \mathrm{dist}(p, X) \leq r\} &\longrightarrow G_{X,r} = \left\{g \in G\colon g\,(X) \leq r\right\}. \end{aligned}$$

The main theorem of this section states and proves the analog of Equation (15.4) in this abstract setting for a set G of n functions.

THEOREM 15.5. *Let $G = \{g_1, \ldots, g_n\}$ be a set of n functions over a domain $\mathcal{X}$*[1], *where $g_i\colon \mathcal{X} \to \mathbb{R}^+$. Define the set system $(G, \mathcal{F})$, with*

$$\mathcal{F} = \left\{G_{X,r}\colon X \in \mathcal{X} \text{ and } r \in \mathbb{R}^+\right\}, \quad \text{where} \quad G_{X,r} = \left\{g \in G\colon g\,(X) \leq r\right\}.$$

Let $A \subseteq G$ be an ϵ-approximation of $\mathcal{F}$, for a given parameter $\epsilon > 0$. Then for any $X \in \mathcal{X}$ and $r \in \mathbb{R}^+$, setting $A_{X,r} = A \cap G_{X,r}$, we have

$$\left| \frac{\sum_{g \in G_{X,r}} g\,(X)}{|G|} - \frac{\sum_{g \in A_{X,r}} g\,(X)}{|A|} \right| \leq 3\epsilon r.$$

To visualize Theorem 15.5, consider the case when $\mathcal{X} = \mathbb{R}$. The figure illustrates an example of five functions $g_1, \ldots, g_5$. In this example, the set $G_{X,r}$ is simply the set of functions lying below the point (X, r).

[1] $\mathcal{X}$ need not be finite. In the previous example, $\mathcal{X}$ was the set of all k-tuples of points in $\mathbb{R}^d$.

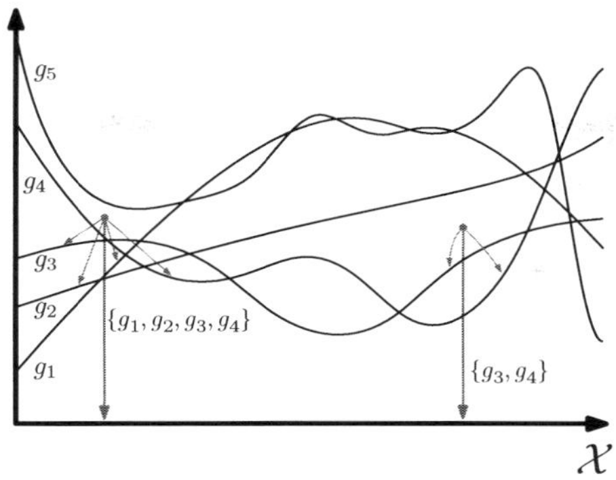

Before we proceed to the proof, we illustrate the versatility of Theorem 15.5 by showing a specific consequence.

Let P be a set of n points in $\mathbb{R}^d$ and $\mathcal{X}$ the set of all k-tuples of points in $\mathbb{R}^d$. Additionally, for each $p \in P$ we are given a function $f_p \colon \mathcal{X} \to \mathbb{R}^+$. Set

$$G = \{f_p \colon p \in P\}.$$

For each $X \in \mathcal{X}$, let r_X be the smallest value for which $G_{X,r} = G$. That is,

$$r_X = \max_{p \in P} f_p(X).$$

Applying Theorem 15.5 to P and G, we arrive at the following.

COROLLARY 15.6. *Let P be a set of n points in $\mathbb{R}^d$ and k a positive integer. Further each $p \in P$ has an associated function*

$$f_p \colon \left(\mathbb{R}^d\right)^k \to \mathbb{R}^+.$$

These functions define a set system $(P, \mathcal{R})$, with

$$\mathcal{R} = \left\{P_{X,r} \colon X \in \left(\mathbb{R}^d\right)^k \quad \textit{and} \quad r \in \mathbb{R}^+\right\},$$

$$\textit{where } P_{X,r} = \left\{p \in P \colon f_p(X) \le r\right\}.$$

Let A be an ϵ-approximation of $\mathcal{R}$. Then for any $X \subseteq \mathbb{R}^d$ with $|X| = k$, we have

$$\left| \frac{\sum_{p \in P} f_p(X)}{|P|} - \frac{\sum_{p \in A} f_p(X)}{|A|} \right| \le 3\epsilon r_X = 3\epsilon \max_{p \in P} f_p(X).$$

Note that for the case where $f_p(X) = \mathrm{dist}(p, X)$, $\mathcal{R}$ is precisely the primal set system induced on P by the union of k equal-radius balls in $\mathbb{R}^d$.

Overview of ideas. For a fixed $X \in \mathcal{X}$ and $r \in \mathbb{R}^+$, we need to relate the quantities

$$\sum_{g \in G_{X,r}} g(X) \qquad \text{and} \qquad \sum_{g \in A_{X,r}} g(X).$$

As earlier, one way to proceed is to partition $G_{X,r}$ into disjoint sets $G_0, G_1, \ldots$, where

$$G_i = \left\{ g \in G : g(X) \in \left((1-\epsilon)^{i+1} r, (1-\epsilon)^i r \right] \right\}.$$

Then all the functions in G_i have approximately the same value on X, and so one can approximately bound the summation of functions in G_i in terms of $|G_i|$, and the summation of functions in $A \cap G_i$ in terms of $|A \cap G_i|$.

However, we present a different proof based on an elegant trick: we sort the functions in $G_{X,r}$ by increasing $g(X)$ values, and rewrite each of the above two summations as sums of the interval lengths between two consecutive function values in the sorted order. Concretely,

let $G_{X,r} = \{g_1, \ldots, g_t\}$, sorted by increasing $g(X)$ values.

Then any interval

$$C_i = g_i(X) - g_{i-1}(X)$$

is 'contributed' in $\sum_{g \in G_{X,r}} g(X)$ by precisely the functions $\{g_i, \ldots, g_t\}$ (see fig-

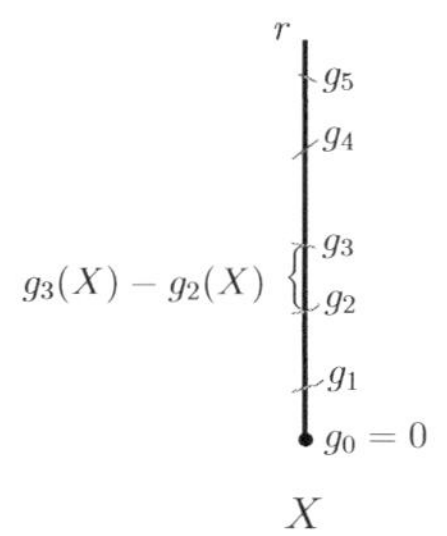

ure). Summing over all consecutive intervals, and using the fact that A is an 2ϵ-approximation of each $\{g_i, \ldots, g_t\}$, we get the required bound.

In the formal proof one has to be a little careful though, as multiple functions might have the same value on X.

PROOF OF THEOREM 15.5. Fix any $X \in \mathcal{X}$ and $r \in \mathbb{R}^+$. Sort the functions in $G_{X,r}$ by increasing $g(X)$ values, and partition $G_{X,r}$ into groups along this order:

$$G_{X,r} = G_1 \cup \cdots \cup G_m,$$

where all the functions in G_i, $i \in [m]$, have the same value on X. Note that $m \leq |G_{X,r}|$. Set

$$G_{\geq i} = G_i \cup \cdots \cup G_m.$$

As A is an ϵ-approximation of the sets $G_1 \cup \cdots \cup G_i$ for all $i \in [m]$, Claim 13.6 (2) implies the following.

CLAIM 15.7. For each $i \in [m]$, A is a 2ϵ-approximation of $G_{\geq i}$.

Now we sum up the functions in $G_{X,r}$ and $A_{X,r}$ by summing up over the differences between values of adjacent functions. For each $i \in [m]$, fix an arbitrary function $g_i \in G_i$ and let $g_0 = 0$. Then

$$\sum_{g \in G_{X,r}} g(X) = \sum_{i=1}^{m} \big(g_i(X) - g_{i-1}(X)\big) \cdot \big|G_{\geq i}\big|, \qquad \text{and}$$

$$\sum_{g \in A_{X,r}} g(X) = \sum_{i=1}^{m} \big(g_i(X) - g_{i-1}(X)\big) \cdot \big|A \cap G_{\geq i}\big|.$$

Thus the required expression

$$\left| \frac{\sum_{g \in G_{X,r}} g(X)}{n} - \frac{\sum_{g \in A_{X,r}} g(X)}{|A|} \right|$$

is upper bounded by

$$\begin{aligned} &\left| \left(\sum_{i=1}^{m} \big(g_i(X) - g_{i-1}(X)\big) \cdot \frac{|G_{\geq i}|}{n} \right) - \left(\sum_{i=1}^{m} \big(g_i(X) - g_{i-1}(X)\big) \cdot \frac{|A \cap G_{\geq i}|}{|A|} \right) \right| \\ &\leq \sum_{i=1}^{m} \big(g_i(X) - g_{i-1}(X)\big) \cdot \left| \frac{|G_{\geq i}|}{n} - \frac{|A \cap G_{\geq i}|}{|A|} \right| \\ &\leq \sum_{i=1}^{m} \big(g_i(X) - g_{i-1}(X)\big) \cdot 2\epsilon \leq 2\epsilon r, \end{aligned}$$

where the second-to-last step used Claim 15.7. □

Bibliography and discussion. The material in this section is from [**FL11**] (with some simplifications). The size of the ϵ-approximation in Theorem 15.5 depends on a parameter called the *pseudo-dimension*, which is a generalization of the notion of VC-dimension for general functions (see [**HP11**, Chapter 7] for details).

[FL11] D. Feldman and M. Langberg, *A unified framework for approximating and clustering data*, STOC'11—Proceedings of the 43rd ACM Symposium on Theory of Computing, ACM, New York, 2011, pp. 569–578, DOI 10.1145/1993636.1993712. MR2932007

[HP11] S. Har-Peled, *Geometric approximation algorithms*, Mathematical Surveys and Monographs, vol. 173, American Mathematical Society, Providence, RI, 2011, DOI 10.1090/surv/173. MR2760023

2. Application: Sensitivity and Coresets for Clustering

Elegant algorithms are easy to program correctly, as well as being efficient. A clever algorithm that is clean and elegant is much more likely to be used than a messy one. When people understand how an algorithm works, which is much more likely with an elegant algorithm, they are more likely to have confidence in the results it produces.

Also, elegant solutions are much easier to generalize, to extend to other problems. My goal is to find general approaches and solutions, not ad hoc tricks.

Robert Tarjan

Given a set P of n points in $\mathbb{R}^d$, the k-median problem asks to compute a set X of k points that minimizes the cost function[2]

$$\mathrm{Cost}(P,k) = \min_{\substack{X \subseteq \mathbb{R}^d \\ |X|=k}} \mathrm{Cost}(P,X), \quad \text{where} \quad \mathrm{Cost}(P,X) = \sum_{p\in P} \mathrm{dist}(p,X).$$

> One approach towards solving this problem is to first compute a smaller set A that 'approximates' P with respect to $\mathrm{Cost}(P,X)$. That is, for *every* set X of k points in $\mathbb{R}^d$ we would like $\mathrm{Cost}(P,X)$ to be approximately equal to $\mathrm{Cost}(A,X)$ (scaled up appropriately). Then the original problem on P is reduced to finding an X minimizing $\mathrm{Cost}(A,X)$—an easier problem if $|A|$ is much smaller than $|P|$.

This leads to the following definition.

DEFINITION 15.8. Given a set P of n points in $\mathbb{R}^d$ and a parameter $\epsilon > 0$, a set $A \subseteq \mathbb{R}^d$ together with a weight function $w\colon A \to \mathbb{R}$ is an ϵ-coreset for the k-median problem on P if for every $X \subseteq \mathbb{R}^d$ of k points,

$$\sum_{p\in A} \mathrm{dist}\,(p,X)\cdot w\,(p) = (1\pm\epsilon)\cdot \sum_{p\in P}\mathrm{dist}\,(p,X)\,. \tag{15.9}$$

Our goal then is to construct an ϵ-coreset for the k-median problem. We will prove two main theorems, the first of which is the following.

THEOREM 15.10. *Let P be a set of n points in $\mathbb{R}^d$ and $k \in \mathbb{Z}^+$, $\epsilon > 0$ be two given parameters. Define*

$$S = \sum_{p\in P} \sup_{\substack{Y\subseteq \mathbb{R}^d \\ |Y|=k}} \frac{\mathrm{dist}(p,Y)}{\sum_{q\in P}\mathrm{dist}(q,Y)}.$$

Then there exists an ϵ-coreset $A \subseteq P$ of size $O\left(\frac{S^2\, d\, k \log k}{\epsilon^2}\right)$ for the k-median problem on P.

The size of the ϵ-coreset in Theorem 15.10 relies on the seemingly mysterious quantity S; however its proof will demonstrate that S 'falls out' naturally when constructing coresets using ϵ-approximations. There do exist good upper bounds on S but using techniques and ideas outside the scope of this text (see discussion).

[2]Recall that $\mathrm{dist}\,(p,X) = \min_{q\in X}\mathrm{dist}\,(p,X)$.

Our second main result shows that the dependency on S can be removed if one is also given an approximate solution B.

THEOREM 15.11. *Let P be a set of n points in $\mathbb{R}^d$ and $k \in \mathbb{Z}^+, \epsilon > 0$ be two given parameters. Further let $B \subseteq \mathbb{R}^d$ be a set of points and $C \geq 1$ such that*

$$\text{Cost}(P, B) \leq C \cdot \text{Cost}(P, k).$$

Then there exists an ϵ-coreset for the k-median problem on P of size

$$O\left(\frac{C^2\, d\, k \log k}{\epsilon^2} + |B|\right).$$

Note that $|B|$ could be larger than k. Furthermore B is only a C-approximation to $\text{Cost}(P, k)$, where C can be large, even a function of k and n. Theorem 15.11 shows that this approximate solution is already sufficient to get a small ϵ-coreset for P.

Overview of ideas. The proof of Theorems 15.10 and 15.11 rely on the following two insights. Let X be any set of k points in $\mathbb{R}^d$.

Relation to ϵ-approximations: Rewrite Equation (15.9) to get

$$\left|\sum_{p\in P} \text{dist}\,(p, X) - \sum_{p\in A} \text{dist}\,(p, X) \cdot w\,(p)\right| \leq \epsilon \cdot \sum_{p\in P} \text{dist}\,(p, X)\,.$$

This resembles the notion of ϵ-approximations. Indeed, applying Corollary 15.6 to P with functions $f_p\,(X) = \text{dist}\,(p, X)$ for each $p \in P$, an ϵ-approximation $A \subseteq P$ of the set system induced on P by the union of k balls in $\mathbb{R}^d$ satisfies

$$\left|\sum_{p\in P} \text{dist}(p, X) - \sum_{p\in A} \text{dist}(p, X)\frac{|P|}{|A|}\right| \leq 3\,\epsilon \cdot |P| \cdot \max_{p\in P} \text{dist}(p, X).$$

The set A would be an $O\,(\epsilon)$-coreset, with weight function $w(p) = \frac{|P|}{|A|}$, if for each X,

$$|P| \cdot \max_{p\in P} \text{dist}\,(p, X) = O\left(\sum_{q\in P} \text{dist}\,(q, X)\right),$$

or equivalently, if for each p and each X,

$$\text{dist}\,(p, X) = O\left(\frac{\sum_{q\in P} \text{dist}\,(q, X)}{|P|}\right).$$

This is not the case, of course—each distance cannot be upper bounded by the average distance for *all* $X \subseteq \mathbb{R}^d$.

Weighted ϵ-approximations: The condition $\frac{\text{dist}(p,X)}{\sum_{q\in P} \text{dist}(q,X)} = O(\frac{1}{|P|})$ suggests that one should construct an ϵ-approximation according to a weight distribution, which can then be set depending on the relative values of dist $(p, \cdot)$. This idea, sometimes called *importance sampling*, is thematically very similar to the idea in Theorem 8.20. Specifically, consider the following weighted version of Corollary 15.6.

LEMMA 15.12. *Let P be a set of n points in $\mathbb{R}^d$ and k a positive integer. For each $p \in P$ we are given a rational weight m_p and a function $f_p\colon (\mathbb{R}^d)^k \to \mathbb{R}^+$. These functions define a set system $(P, \mathcal{R})$ with*

$$\mathcal{R} = \left\{ P_{X,r}\colon X \subseteq \mathbb{R}^d, |X| = k \text{ and } r \in \mathbb{R} \right\}, \text{ where } P_{X,r} = \left\{ p \in P\colon f_p(X) \le r \right\}.$$

Then given $\epsilon > 0$ there exists a multiset $A \subseteq P$ of size $O\left(\frac{\text{VC-dim}(\mathcal{R})}{\epsilon^2}\right)$ such that for any $X \in (\mathbb{R}^d)^k$,

$$\left| \frac{\sum_{p\in P} f_p(X)}{\sum_{p\in P} m_p} - \frac{\sum_{p\in A} \frac{f_p(X)}{m_p}}{|A|} \right| \le 3\epsilon \left(\max_{p\in P} \frac{f_p(X)}{m_p} \right). \tag{15.13}$$

PROOF. By scaling up, we can assume that each m_p is an integer. Let P' be the set constructed by adding m_p copies of each $p \in P$ to P', where each copy of p is assigned the function $\frac{f_p(X)}{m_p}$. By applying Corollary 15.6 to P', there exists a set A such that for all $X \in (\mathbb{R}^d)^k$,

$$\left| \frac{\sum_{p'\in P'} f_{p'}(X)}{|P'|} - \frac{\sum_{p'\in A} f_{p'}(X)}{|A|} \right| \le 3\epsilon \left(\max_{p'\in P'} f_{p'(X)} \right). \tag{15.14}$$

Noting that $\sum_{p'\in P'} f_{p'}(X) = \sum_{p\in P} f_p(X)$, Equation (15.14) is equivalent to Equation (15.13).
Finally, as the VC-dimension is unchanged by adding duplicate elements, Theorem 13.2 implies that $|A| = O\left(\frac{\text{VC-dim}(\mathcal{R})}{\epsilon^2}\right)$. □

In the proof of Theorems 15.10 and 15.11 we will set the parameters m_p, $f_p(\cdot)$ and ϵ' such that an ϵ'-approximation A given by Lemma 15.12 can be used to construct the required ϵ-coreset.

Given our preparation, the proof of our first main result is immediate.

THEOREM 15.10. *Let P be a set of n points in $\mathbb{R}^d$ and $k \in \mathbb{Z}^+$, $\epsilon > 0$ be two given parameters. Define*

$$S = \sum_{p\in P} \sup_{\substack{Y\subseteq\mathbb{R}^d\\ |Y|=k}} \frac{\text{dist}(p, Y)}{\sum_{q\in P} \text{dist}(q, Y)}.$$

Then there exists an ϵ-coreset $A \subseteq P$ of size $O\left(\frac{S^2\, d\, k \log k}{\epsilon^2}\right)$ for the k-median problem on P.

PROOF. Set $f_p(X) = \text{dist}(p, X)$ for each $p \in P$ and let m_p be the weight of $p \in P$. These weights will be set later and normalized so that $\sum_{p\in P} m_p = 1$. Further let $\epsilon' > 0$ be a parameter to be set later.
Let A be an ϵ'-approximation given by Lemma 15.12 applied to P with weights m_p and functions $f_p(\cdot)$. That is, for each set X of k points in $\mathbb{R}^d$, A satisfies

$$\left| \sum_{p\in P} \text{dist}(p, X) - \sum_{p\in A} \text{dist}(p, X) \cdot \frac{1}{|A|\, m_p} \right| \le 3\epsilon' \left(\max_{p\in P} \frac{\text{dist}(p, X)}{m_p} \right). \tag{15.15}$$

For A to be an ϵ-coreset, the right-hand side of the above inequality must be upper bounded by $\epsilon \cdot \sum_{p \in P} \operatorname{dist}(p, X)$. The natural choice is to set $m_p = \frac{\operatorname{dist}(p,X)}{\sum_{q \in P} \operatorname{dist}(q,X)}$ for each $p \in P$ and $\epsilon' = \frac{\epsilon}{3}$. However m_p must be independent of X, as A must work for all choices of X! Considering the worst-case bound for m_p over all choices of X leads to the notion of the *sensitivity* of a point.

DEFINITION 15.16. The sensitivity of each $p \in P$ with respect to $\{f_p : p \in P\}$ is defined to be

$$s(p) = \sup_{\substack{Y \subseteq \mathbb{R}^d \\ |Y|=k}} \frac{f_p(Y)}{\sum_{q \in P} f_q(Y)}.$$

Let $S = \sum_{p \in P} s(p)$ and set

$$f_p(X) = \operatorname{dist}(p, X) \qquad \text{and} \qquad m_p = \frac{s(p)}{S}^{3}.$$

From Equation (15.15),

$$\left| \sum_{p \in P} \operatorname{dist}(p, X) - \sum_{p \in A} \operatorname{dist}(p, X) \cdot \frac{1}{|A|\, m_p} \right| \leq 3\epsilon' \cdot S \cdot \left(\max_{p \in P} \frac{\operatorname{dist}(p, X)}{s(p)} \right)$$

The R.H.S., after substituting for $s(p)$ and multiplying/dividing by $\sum_{q \in P} \operatorname{dist}(q, X)$:

$$3\epsilon' S \sum_{q \in P} \operatorname{dist}(q, X) \cdot \left(\max_{p \in P} \frac{\frac{\operatorname{dist}(p,X)}{\sum_{q \in P} \operatorname{dist}(q,X)}}{\sup_{\substack{Y \subseteq \mathbb{R}^d \\ |Y|=k}} \frac{\operatorname{dist}(p,Y)}{\sum_{q \in P} \operatorname{dist}(q,Y)}} \right) \leq 3\epsilon' S \left(\sum_{q \in P} \operatorname{dist}(q, X) \right).$$

Setting $\epsilon' = \frac{\epsilon}{3S}$ implies that A is an ϵ-coreset where each $p \in A$ is assigned the weight $\frac{1}{|A|\, m_p}$.

Note that A is an ϵ-approximation of the set system induced on P by the union of k balls in $\mathbb{R}^d$; that is, the set system induced by the k-fold union of balls in $\mathbb{R}^d$. Lemma 10.3 and Theorem 11.6 implies that the VC-dimension of this set system is $\Theta(dk \log k)$ and thus $|A| = O\left(\frac{S^2 dk \log k}{\epsilon^2}\right)$ by Lemma 15.12. □

We remark here that to compute A above, we need to compute the weights m_p for each $p \in P$. This, together with the problem of deriving good upper bounds on the total sensitivity S, is a non-trivial algorithmic problem by itself (see discussion).

[3] The division by S is just to get $\sum_p m_p = 1$.

THEOREM 15.11. *Let P be a set of n points in $\mathbb{R}^d$ and $k \in \mathbb{Z}^+, \epsilon > 0$ be two given parameters. Further let $B \subseteq \mathbb{R}^d$ be a set of points and $C \geq 1$ such that*

$$\mathrm{Cost}(P, B) \leq C \cdot \mathrm{Cost}(P, k).$$

Then there exists an ϵ-coreset for the k-median problem on P of size

$$O\left(\frac{C^2\, d\, k \log k}{\epsilon^2} + |B|\right).$$

PROOF. Given B, set the parameters for each $p \in P$ as follows:

$$f_p(X) = \mathrm{dist}(p, X) - \mathrm{dist}(\mathrm{closest}(p, B), X) + \mathrm{dist}\,(p, B)\,,$$
$$m_p = \frac{\mathrm{dist}(p, B)}{\sum_{q \in P} \mathrm{dist}(q, B)},$$

where $\mathrm{closest}\,(p, Q) = \arg\min_{q \in Q} \mathrm{dist}\,(p, q)$ denotes the closest point in Q to p. Note that $f_p\,(X)$ is non-negative[4] due to triangle inequality:

$$\mathrm{dist}(\mathrm{closest}(p, B), X) \leq \mathrm{dist}\,(\mathrm{closest}(p, B), p) + \mathrm{dist}(p, X).$$

Applying Lemma 15.12 with f_p and m_p set above and noting that $\sum_{p \in P} m_p = 1$, we get an ϵ'-approximation A such that for any $X \subseteq \mathbb{R}^d$ of k points,

$$\Bigg| \sum_{p \in P} \Big(\mathrm{dist}(p, X) - \mathrm{dist}(\mathrm{closest}(p, B), X) + \mathrm{dist}\,(p, B) \Big) -$$
$$\sum_{p \in A} \Big(\mathrm{dist}\,(p, X) - \mathrm{dist}\,(\mathrm{closest}\,(p, B)\,, X) + \mathrm{dist}\,(p, B) \Big) \cdot \frac{1}{|A|\, m_p} \Bigg|$$
$$\leq 3\epsilon' \left(\max_{p \in P} \frac{\mathrm{dist}(p, X) - \mathrm{dist}(\mathrm{closest}(p, B), X) + \mathrm{dist}\,(p, B)}{m_p} \right).$$

Using the fact that

$$\sum_{p \in P} \mathrm{dist}(p, B) = \sum_{p \in A} \left(\mathrm{dist}(p, B) \frac{\sum_{q \in P} \mathrm{dist}(q, B)}{|A|\; \mathrm{dist}(p, B)} \right) = \sum_{p \in A} \mathrm{dist}(p, B) \frac{1}{|A|\, m_p},$$

we arrive at

$$\Bigg| \left(\sum_{p \in P} \mathrm{dist}(p, X) \right) - \left(\sum_{p \in P} \mathrm{dist}(\mathrm{closest}(p, B), X) \right)$$
$$- \left(\sum_{p \in A} \mathrm{dist}(p, X) \frac{1}{|A|\, m_p} \right) + \left(\sum_{p \in A} \mathrm{dist}(\mathrm{closest}(p, B), X) \frac{1}{|A|\, m_p} \right) \Bigg|$$
$$\leq 3\epsilon' \left(\max_{p \in P} \frac{\mathrm{dist}(p, X) - \mathrm{dist}(\mathrm{closest}(p, B), X) + \mathrm{dist}\,(p, B)}{m_p} \right). \tag{15.17}$$

R.H.S.: Using a consequence of triangle inequality, that

$$\mathrm{dist}(p, X) - \mathrm{dist}(\mathrm{closest}(p, B), X) \leq \mathrm{dist}(p, \mathrm{closest}(p, B)) = \mathrm{dist}(p, B),$$

[4] Indeed, the additive term $\mathrm{dist}\,(p, B)$ is present in $f_p(X)$ just to make $f_p(X)$ non-negative so that one can apply Lemma 15.12. *Conceptually* we only need $f_p(X) = \mathrm{dist}(p, X) - \mathrm{dist}\,(\mathrm{closest}(p, B), X)$.

as well as substituting the value of m_p, the R.H.S. of Equation (15.17) is at most

$$3\epsilon' \max_{p\in P} \frac{2\,\mathrm{dist}(p,B)}{\frac{\mathrm{dist}(p,B)}{\sum_{q\in P}\mathrm{dist}(q,B)}} = 6\epsilon' \sum_{p\in P} \mathrm{dist}(p,B)$$
$$\leq 6\epsilon' \cdot C \cdot \mathrm{Cost}\,(P,k) \leq 6\epsilon' C \sum_{p\in P} \mathrm{dist}(p,X).$$

L.H.S.: For each $b \in B$, let P_b be the set of points of P whose closest point in B is b. Then the L.H.S. of Equation (15.17) becomes

$$\left| \left(\sum_{p\in P} \mathrm{dist}(p,X)\right) - \left(\sum_{b\in B} |P_b| \cdot \mathrm{dist}(b,X)\right) - \left(\sum_{p\in A} \mathrm{dist}(p,X)\frac{1}{|A|\,m_p}\right) + \left(\sum_{b\in B}\sum_{p\in P_b\cap A} \mathrm{dist}(b,X)\frac{1}{|A|\,m_p}\right)\right|. \tag{15.18}$$

We're done—set $\epsilon' = \frac{\epsilon}{6C}$ and return $A \cup B$ as our ϵ-coreset, with weights dictated by Equation (15.18):

$$p \in A \;:\; w(p) = \frac{1}{|A|\,m_p}.$$
$$b \in B \;:\; w(b) = |P_b| - \sum_{p\in A\cap P_b} \frac{1}{|A|\,m_p}.$$

□

We remark that to compute the set A one again needs to compute the weights m_p for all $p \in P$. However this time it is easier as we are also given the set B.

Bibliography and discussion. The beautiful application of this section is from [**FL11**]. See [**VX12**] for upper bounds on sensitivity for a variety of optimization problems. There are many interesting variations, improvements and applications of the basic ideas presented in this section (e.g., see [**HV20**]). We refer the reader to the surveys [**AHPV07, Phi18**] for more information on coresets.

[AHPV07] P. K. Agarwal, S. Har-Peled, and K. R. Varadarajan, *Geometric approximation via coresets*, Combinatorial and computational geometry, Math. Sci. Res. Inst. Publ., vol. 52, Cambridge Univ. Press, Cambridge, 2005, pp. 1–30, DOI 10.4171/PRIMS/172. MR2178310

[FL11] D. Feldman and M. Langberg, *A unified framework for approximating and clustering data*, STOC'11—Proceedings of the 43rd ACM Symposium on Theory of Computing, ACM, New York, 2011, pp. 569–578, DOI 10.1145/1993636.1993712. MR2932007

[HV20] L. Huang and N. K. Vishnoi, *Coresets for clustering in euclidean spaces: importance sampling is nearly optimal*, STOC '20—Proceedings of the 52nd Annual ACM SIGACT Symposium on Theory of Computing, ACM, New York, 2020, pp. 1416–1429. MR4141850

[Phi18] J. M. Phillips. *Coresets and sketches. Handbook of Discrete and Computational Geometry*. CRC Press, 2018. pp. 1269–1286.

[VX12] K. R. Varadarajan and X. Xiao. *On the sensitivity of shape fitting problems. Proceedings of the IARCS Annual Conference on Foundations of Software Technology and Theoretical Computer Science (FSTTCS)*. Schloss Dagstuhl - Leibniz-Zentrum für Informatik, 2012. pp. 486–497.

CHAPTER 16

A Summary of Known Bounds

This chapter lists the current-best bounds for

(1) **the VC-dimension and the shallow-cell complexity,**
(2) **sizes of ϵ-nets, and**
(3) **sizes of ϵ-approximations**

of many basic geometric set systems.

1. LIST: Complexity of Geometric Set Systems

The art of doing mathematics is finding that special case that contains all the germs of generality.

David Hilbert

This section lists the VC-dimension and shallow-cell complexity of several geometric set systems. First we define/recall some definitions.

Geometric set systems. A given family of geometric objects gives rise to two types of set systems: primal and dual.

DEFINITION 1.11. Given a set P of points in $\mathbb{R}^d$ and a (possibly infinite) family $\mathcal{R}$ of geometric objects in $\mathbb{R}^d$, the primal set system induced on P by $\mathcal{R}$ is

$$\{O \cap P \,:\, O \in \mathcal{R}\}.$$

DEFINITION 1.12. Given a set $\mathcal{R}$ of geometric objects in $\mathbb{R}^d$, the dual set system induced on $\mathcal{R}$ by $\mathbb{R}^d$ is defined as

$$\{\mathcal{R}_p \colon p \in \mathbb{R}^d\}, \quad \text{where} \quad \mathcal{R}_p = \{R \in \mathcal{R} \colon R \ni p\}.$$

Geometric objects. We list some common geometric objects.

Pseudo-disks: A finite set $\mathcal{O}$ of objects in $\mathbb{R}^2$ are called *pseudo-disks* if each object in $\mathcal{O}$ is a simply connected region of $\mathbb{R}^2$ bounded by a simple closed Jordan curve and the boundaries of any two objects in $\mathcal{O}$ cross at most *twice*. The notion of pseudo-disks is a generalization of that of disks. It arises naturally in situations when the geometry of the boundary is irrelevant and one only cares about the combinatorics of the arrangement of objects. The key combinatorial property of pseudo-disks that carries over from disks is that the complexity of the union of any set of n pseudo-disks is at most $6n - 12$ (see Claim 1.10). This implies that the shallow-cell complexity of pseudo-disks and disks is the same within constant factors.

Homothets: Given an object $K \subseteq \mathbb{R}^d$, a *homothet* of K is a set of the form $\alpha \cdot K + a$, where $\alpha \in \mathbb{R}^+$ and $a \in \mathbb{R}^d$. Here α is a scaling factor while a is the translation vector. A *translate* of K is a homothet with $\alpha = 1$. For a convex object K in $\mathbb{R}^2$, a set of n homothets of K form a set of pseudo-disks.

Fat objects: A triangle Δ in $\mathbb{R}^2$ is called α-fat, $\alpha \in [0, \frac{\pi}{3}]$, if each angle of Δ is at least α.

VC-dimension and shallow-cell complexity. Recall these notions.

DEFINITION 4.2. The *VC-dimension* of a set system $(X, \mathcal{F})$, denoted by VC-dim$(\mathcal{F})$, is the size of the largest $Y \subseteq X$ for which $|\mathcal{F}|_Y| = 2^{|Y|}$. We say that such a Y is *shattered* by $\mathcal{F}$.

For a family $\mathcal{R}$ of geometric objects in $\mathbb{R}^d$—e.g., the family of all half-spaces in $\mathbb{R}^d$—the VC-dimension of $\mathcal{R}$ is defined to be the VC-dimension of the primal set system $(\mathbb{R}^d, \mathcal{R})$.

DEFINITION 4.4. A set system $(X, \mathcal{F})$ has shallow-cell complexity $\varphi_{\mathcal{F}}(\cdot, \cdot)$ if for any positive integer k and any finite $Y \subseteq X$, the number of sets in $\mathcal{F}|_Y$ of size at most k is upper bounded by $|Y| \cdot \varphi_{\mathcal{F}}(|Y|, k)$.

For a family $\mathcal{R}$ of geometric objects in $\mathbb{R}^d$—e.g., the family of all half-spaces—the shallow-cell complexity of $\mathcal{R}$ is defined to be the shallow-cell complexity of the primal set system $(\mathbb{R}^d, \mathcal{R})$.

1.1. Dual set systems. The notion of VC-dimension of an abstract set system $(X, \mathcal{F})$ was motivated by considering primal set systems. One can also consider the VC-dimension of the set system dual to $(X, \mathcal{F})$:

DEFINITION 16.1. Given $(X, \mathcal{F})$, its dual set system is defined as $(\mathcal{F}, \mathcal{F}^*)$, with

$$\mathcal{F}^* = \{\mathcal{F}_x \colon x \in X\}, \quad \text{where } \mathcal{F}_x = \{F \in \mathcal{F} \colon F \ni x\}.$$

The VC-dimension of $(\mathcal{F}, \mathcal{F}^*)$ is the size of the largest $\mathcal{F}' \subseteq \mathcal{F}$ such that for all $\mathcal{F}'' \subseteq \mathcal{F}'$, there exists an element of X contained in all sets of $\mathcal{F}''$ and no set of $\mathcal{F}' \setminus \mathcal{F}''$. Pictorially, in the Venn diagram induced by $\mathcal{F}'$ on X, all $2^{|\mathcal{F}'|}$ cells contain at least one point of X.

1.2. k-fold unions and intersections. Finally, recall the notion of k-fold union of set systems.

DEFINITION 10.1. Given a set system $(X, \mathcal{F})$ and an integer $k \geq 1$, the *k-fold union* of $\mathcal{F}$, denoted by $\mathcal{F}^k$, is the set system obtained by adding the union of at most k sets of $\mathcal{F}$ to $\mathcal{F}^k$. That is,

$$\mathcal{F}^k = \Big\{R_{i_1} \cup \cdots \cup R_{i_k} \ : \ R_{i_j} \in \mathcal{F} \text{ for all } 1 \leq j \leq k\Big\}.$$

As the k sets need not be distinct, we have $\mathcal{F} \subseteq \mathcal{F}^k$. Similarly one can define k-fold intersections of a given set system.

We now present two tables summarizing the VC-dimension and shallow-cell complexity for several common geometric set systems.

Primal set systems

	Objects $\mathcal{O}$	$\boldsymbol{m\,\varphi(m,k)}$	**VC-dimension**
$\mathbb{R}$			
	Intervals	$O(mk)$	2
	t-fold union of intervals	$O(m^t k^t)$	$2t$ **[Blu+89]**
$\mathbb{R}^2$			
	Lines	$O(m^2)$ for $k \geq 2$	2
	Halfplanes	$O(mk)$ **[CS8916a]**	3
	Homothets of a convex body	$O(mk^2 + k^3)$ **[AU1616a]**	3 **[NT1016a]**
	Axis-aligned rectangles	$O(m^2k^2)$	4
	Bottomless axis aligned rectangles	$O(mk^2)$	3
	Disks	$O(mk^2)$ **[CS8916a]**	3
	Pseudo-disks	$O(mk^2)$ **[BPR1316a]**	3 **[BPR1316a]**
	t-fold union of half-spaces	$O(m^t k^t)$	$2t+1$ **[Blu+89]**
	t-fold intersection of half-spaces	$O(m^t k^t)$	$2t+1$ **[Blu+89]**
	Triangles ($t = 3$)	$O(m^3k^3)$	7
	Convex sets	$O(m^k)$	∞
$\mathbb{R}^3$			
	Half-spaces	$O(mk^2)$ **[CS8916a]**	4
	Balls	$O(m^2k^2)$	4
	t-fold union of half-spaces	$O(m^t k^{2t})$	$4t$
	t-fold intersection of half-spaces	$O(m^t k^{2t})$	$4t$
	Homothets of a convex body	$O(m^k)$ **[NT1016a]**	∞ **[NT1016a]**
$\mathbb{R}^d$			
	Half-spaces	$O\left(m^{\lfloor d/2 \rfloor} k^{\lceil d/2 \rceil}\right)$ **[CS8916a]**	$d+1$
	Balls	$O\left(m^{\lfloor (d+1)/2 \rfloor} k^{\lceil (d+1)/2 \rceil}\right)$ **[CS8916a]**	$d+1$
	Axis-aligned boxes	$O\left(m^d k^d\right)$	$2d$ **[Blu+89]**
	t-fold union of half-spaces	$O\left(m^{t\lfloor d/2 \rfloor} k^{t\lceil d/2 \rceil}\right)$	$\Theta(d \cdot t \log t)$ **[CMK19]**
	t-fold intersection of half-spaces	$O\left(m^{t\lfloor d/2 \rfloor} k^{t\lceil d/2 \rceil}\right)$	$\Theta(d \cdot t \log t)$ **[CMK19]**

TABLE 1. The table lists the bounds on VC-dimension and shallow-cell complexity for the primal set system induced by common geometric set systems. The constants in the asymptotic notation may depend on d.

Dual set systems

	Objects $\mathcal{O}$	$m\,\varphi(m,k)$	**VC-dimension**
$\mathbb{R}$			
	Intervals	m	2
$\mathbb{R}^2$			
	Lines	$\Theta(m^2)$	2
	Halfplanes	$\Theta(mk)$ **[CS8916a]**	3
	Homothets of a convex body	$O(mk)$	3 **[NT1016a]**
	Axis-aligned rectangles	$\Theta(m^2)$	4
	Disks	$O(mk)$	3
	Pseudo-disks	$O(mk)$ **[BPR1316a]**	$O(1)$ **[BPR1316a]**
	α-fat triangles	$O(mk \log^* \frac{m}{k} + \frac{mk}{\alpha} \log^2 \frac{1}{\alpha})$ **[Aro+14]**	7
	Objects with union complexity $f(\cdot)$	$O\left(f\left(\frac{m}{k}\right)k^2\right)$ **[Sha91]**	$O(1)$
	t-fold union of half-spaces	$O(m^t k^t)$	$2t+1$ **[Blu+89]**
	t-fold intersection of half-spaces	$O(m^t k^t)$	$2t+1$ **[Blu+89]**
$\mathbb{R}^3$			
	Half-spaces	$O(mk^2)$	4
	Balls	$O(m^2 k)$	4
	Axis-aligned cubes	$O(m^2 k)$ **[Boi+98]**	$O(1)$
	t-fold union of half-spaces	$O(m^t k^{2t})$	$4t$
	t-fold intersection of half-spaces	$O(m^t k^{2t})$	$4t$
$\mathbb{R}^d$			
	Half-spaces	$O\left(m^{\lfloor d/2 \rfloor} k^{\lceil d/2 \rceil}\right)$ **[CS8916a]**	$d+1$
	Balls	$O\left(m^{\lceil d/2 \rceil} k^{\lfloor d/2 \rfloor}\right)$	$d+1$
	t-fold union of half-spaces	$O\left(m^{t\lfloor d/2 \rfloor} k^{t\lceil d/2 \rceil}\right)$	$\Theta(d \cdot t \log t)$ **[CMK19]**
	t-fold intersection of half-spaces	$O\left(m^{t\lfloor d/2 \rfloor} k^{t\lceil d/2 \rceil}\right)$	$\Theta(d \cdot t \log t)$ **[CMK19]**

TABLE 2. The table lists the bounds on VC-dimension and shallow-cell complexity for the dual set system induced by common geometric set systems. The constants in the asymptotic notation may depend on d.

Bibliography and discussion. There is a long history of work on improving the various bounds listed in this section. It would be overwhelming to cite all the papers involved in these bounds. We have thus attempted to cite only the latest relevant papers. We refer the reader to the nice 'geometric hypergraph zoo' at Eötvös Loránd University for a database of properties of many geometric objects.

[AU1616a] M. Axenovich and T. Ueckerdt, *Density of range capturing hypergraphs*, J. Comput. Geom. **7** (2016), no. 1, 1–21. MR3455619

[Aro+14] B. Aronov, M. de Berg, E. Ezra, and M. Sharir, *Improved bounds for the union of locally fat objects in the plane*, SIAM J. Comput. **43** (2014), no. 2, 543–572, DOI 10.1137/120891241. MR3188401

[BPR1316a] S. Buzaglo, R. Pinchasi, and G. Rote, *Topological hypergraphs*, Thirty essays on geometric graph theory, Springer, New York, 2013, pp. 71–81, DOI 10.1007/978-1-4614-0110-0_6. MR3205150

[Boi+98] J.-D. Boissonnat, M. Sharir, B. Tagansky, and M. Yvinec, *Voronoi diagrams in higher dimensions under certain polyhedral distance functions*, Discrete Comput. Geom. **19** (1998), no. 4, 485–519, DOI 10.1007/PL00009366. MR1620060

[CMK19] M. Csikós, N. H. Mustafa, and A. Kupavskii, *Tight lower bounds on the VC-dimension of geometric set systems*, J. Mach. Learn. Res. **20** (2019), Paper No. 81, 8. MR3960935

[CS8916a] K. L. Clarkson and P. W. Shor, *Applications of random sampling in computational geometry. II*, Discrete Comput. Geom. **4** (1989), no. 5, 387–421, DOI 10.1007/BF02187740. MR1014736

[NT1016a] M. Naszódi and S. Taschuk, *On the transversal number and VC-dimension of families of positive homothets of a convex body*, Discrete Math. **310** (2010), no. 1, 77–82, DOI 10.1016/j.disc.2009.07.030. MR2558970

[Sha91] M. Sharir, *On k-sets in arrangements of curves and surfaces*, Discrete Comput. Geom. **6** (1991), no. 6, 593–613, DOI 10.1007/BF02574706. MR1121331

2. LIST: Epsilon-Nets

I think it's rarely about what you actually learn in class ... it's mostly about things that you stay motivated to go and continue to do on your own.

Maryam Mirzakhani

As any good teacher knows, the methods of instruction and the range of material covered are matters of small importance as compared with the success in arousing the natural curiosity of the students and stimulating their interest in exploring on their own.

Noam Chomsky

This section lists the state-of-the-art bounds for ϵ-net sizes of geometric set systems. All upper bounds except one follow directly from the bounds on the shallow-cell complexity of the set system together with Theorem 6.2 or Theorem 8.11, both of which we first recall.

THEOREM 6.2. *Let $(X, \mathcal{F})$ be a finite set system with shallow-cell complexity $\varphi_{\mathcal{F}}(\cdot,\cdot)$ and with* VC-dim $(\mathcal{F}) \leq d$. *Then for any $\epsilon \in \left(0, \frac{1}{2}\right)$ there exists an ϵ-net of $\mathcal{F}$ of size*

$$O\left(\frac{d}{\epsilon} + \frac{1}{\epsilon}\log \varphi_{\mathcal{F}}\left(\frac{16d}{\epsilon}, 48d\right)\right).$$

THEOREM 8.11. *Let $(X, \mathcal{F})$ be a set system on n elements and $a \in (1,2)$, $b \in \mathbb{R}^+$ be parameters such that the shallow-cell complexity of $\mathcal{F}$, $\varphi_{\mathcal{F}}(\cdot,\cdot)$, is (a,b)-well-behaved. Let $\epsilon > 0$ be a given parameter and set $c_1 = \exp\left(\frac{\sqrt{2}}{\sqrt{2}-\sqrt{a}}\right)$ and $c_2 = b + 2^{\frac{a}{a-1}}$. Note that $c_1, c_2 \geq 1$. Then there exists a procedure to compute an ϵ-net N of $\mathcal{F}$ such that each $p \in X$ is present in N with probability*

$$O\left(\frac{c_1\, c_2}{\epsilon n} + \frac{c_1}{\epsilon n}\ln \varphi_{\mathcal{F}}\left(\frac{c_1\, n}{2^t}, \frac{\epsilon n}{2^t}\right)\right),$$

where t is the largest integer in $\left[0, \log \frac{\epsilon n}{c_2}\right]$ such that for all $i < t$,

$$\sqrt{\frac{8\ln\left(\frac{\epsilon n}{2^i}\cdot \varphi_{\mathcal{F}}\left(\frac{c_1 n}{2^i}, \frac{\epsilon n}{2^i}\right)\right)}{\frac{\epsilon n}{2^i}}} \leq \frac{1}{2}. \tag{8.11}$$

Furthermore, let $w : X \to \mathbb{R}$ be weights on the elements of X, with $W = \sum_{p\in X} w(p)$. Then there exists an ϵ-net of $\mathcal{F}$ of total weight

$$O\left(W\left(\frac{c_1\, c_2}{\epsilon n} + \frac{c_1}{\epsilon n}\ln \varphi_{\mathcal{F}}\left(\frac{c_1\, n}{2^t}, \frac{\epsilon n}{2^t}\right)\right)\right).$$

The exception is the ϵ-net size bound of $O\left(\frac{1}{\epsilon}\log\log\frac{1}{\epsilon}\right)$ of the primal system induced by axis-aligned rectangles in the plane. This follows immediately by Theorem 6.2 together with Theorem 4.17.

The algorithm in Chapter 7 computes an ϵ-net of these sizes, in expectation.

	Objects $\mathcal{O}$	P/D	Upper bound		Lower bound	
$\mathbb{R}$						
	Intervals	P/D	$\lfloor \frac{1}{\epsilon} \rfloor$		$\lfloor \frac{1}{\epsilon} \rfloor$	
$\mathbb{R}^2$						
	Lines	P/D	$\frac{2}{\epsilon} \log \frac{1}{\epsilon}$	**[HW87]**	$\frac{1}{80\epsilon} \left(\frac{\log \frac{1}{\epsilon}}{\log \log \frac{1}{\epsilon}} \right)^{\frac{1}{2}}$	[BS20]
	Half-spaces	P/D	$\frac{2}{\epsilon} - 1$	**[HW87]**	$\frac{2}{\epsilon} - 2$	**[KPW92]**
	Homothets of a convex body	P/D	$O\left(\frac{1}{\epsilon}\right)$	**[PR08]**	$\Omega\left(\frac{1}{\epsilon}\right)$	
	Axis-aligned rectangles	P	$O\left(\frac{1}{\epsilon} \log \log \frac{1}{\epsilon}\right)$	**[AES10]**	$\Omega\left(\frac{1}{\epsilon} \log \log \frac{1}{\epsilon}\right)$	**[PT13]**
	Axis-aligned rectangles	D	$\frac{5}{\epsilon} \log \frac{1}{\epsilon}$	**[HW87]**	$\frac{1}{9\epsilon} \log \frac{1}{\epsilon}$	**[PT13]**
	Bottomless axis aligned rectangles	P	$O\left(\frac{1}{\epsilon}\right)$		$\Omega\left(\frac{1}{\epsilon}\right)$	
	Disks	P	$\frac{13.4}{\epsilon}$	**[Bus+16]**	$\frac{2}{\epsilon} - 2$	**[HW87]**
	Pseudo-disks	P/D	$O\left(\frac{1}{\epsilon}\right)$	**[PR08]**	$\Omega\left(\frac{1}{\epsilon}\right)$	
	Triangles	P	$\frac{7}{\epsilon} \log \frac{1}{\epsilon}$	**[HW87]**	$\frac{1}{80\epsilon} \left(\frac{\log \frac{1}{\epsilon}}{\log \log \frac{1}{\epsilon}} \right)^{\frac{1}{2}}$	[BS20]
	Fat triangles	D	$O\left(\frac{1}{\epsilon} \log \log^* \frac{1}{\epsilon}\right)$	**[Aro+14]**	$\Omega\left(\frac{1}{\epsilon}\right)$	
	Objects with union complexity $f(\cdot)$	D	$O\left(\frac{1}{\epsilon} \log\left(\epsilon \cdot f\left(\frac{1}{\epsilon}\right)\right)\right)$	**[AES10]**	$\Omega\left(\frac{1}{\epsilon}\right)$	
	Convex sets	P	$\lvert X \rvert - \epsilon \lvert X \rvert$		$\lvert X \rvert - \epsilon \lvert X \rvert$	
$\mathbb{R}^3$						
	Half-spaces	P/D	$O\left(\frac{1}{\epsilon}\right)$	**[Mat92]**	$\frac{2}{\epsilon} - 2$	**[KPW92]**
	Balls	P/D	$O\left(\frac{1}{\epsilon} \log \frac{1}{\epsilon}\right)$		$\Omega\left(\frac{1}{\epsilon}\right)$	
	Homothets of convex bodies	P	$\lvert X \rvert - \epsilon \lvert X \rvert$		$\lvert X \rvert - \epsilon \lvert X \rvert$	**[NT10]**
	Convex sets	P	$\lvert X \rvert - \epsilon \lvert X \rvert$		$\lvert X \rvert - \epsilon \lvert X \rvert$	
$\mathbb{R}^d$						
	Half-spaces	P/D	$\frac{d}{\epsilon}\left(\log \frac{1}{\epsilon} + o(1)\right)$	**[KPW92]**	$\frac{\lfloor d/2 \rfloor - 1}{9} \frac{1}{\epsilon} \log \frac{1}{\epsilon}$	**[KMP16]**
	Balls	P	$\frac{d+1}{\epsilon}\left(\log \frac{1}{\epsilon} + o(1)\right)$	**[KPW92]**	$\frac{\lfloor d/2 \rfloor - 1}{9} \frac{1}{\epsilon} \log \frac{1}{\epsilon}$	**[KMP16]**
	Convex sets	P	$\lvert X \rvert - \epsilon \lvert X \rvert$		$\lvert X \rvert - \epsilon \lvert X \rvert$	

TABLE 3. The table lists ϵ-net sizes for the most common geometric set systems. The column 'P/D' specifies whether the set system is primal or dual. The constants in the asymptotic notation may depend on d.

Bibliography and discussion. We refer the reader to the text **[Cha00]** for efficient algorithms to construct ϵ-nets for several geometric set systems.

[AES10] B. Aronov, E. Ezra, and M. Sharir, *Small-size ϵ-nets for axis-parallel rectangles and boxes*, SIAM J. Comput. **39** (2010), no. 7, 3248–3282, DOI 10.1137/090762968. MR2678074

[Aro+14] B. Aronov, M. de Berg, E. Ezra, and M. Sharir, *Improved bounds for the union of locally fat objects in the plane*, SIAM J. Comput. **43** (2014), no. 2, 543–572, DOI 10.1137/120891241. MR3188401

[BS20] J. Balogh and W. Samotij, *An efficient container lemma*, Discrete Anal., posted on 2020, Paper No. 17, 56, DOI 10.19086/da. MR4186910

[Bus+16] N. Bus, S. Garg, N. H. Mustafa, and S. Ray, *Tighter estimates for ϵ-nets for disks*, Comput. Geom. **53** (2016), 27–35, DOI 10.1016/j.comgeo.2015.12.002. MR3454542

[Cha00] B. Chazelle, *The discrepancy method*, Cambridge University Press, Cambridge, 2000. Randomness and complexity, DOI 10.1017/CBO9780511626371. MR1779341

[HW87] D. Haussler and E. Welzl, *ϵ-nets and simplex range queries*, Discrete Comput. Geom. **2** (1987), no. 2, 127–151, DOI 10.1007/BF02187876. MR884223

[KMP16] A. Kupavskii, N. H. Mustafa, and J. Pach, *New lower bounds for ϵ-nets*, 32nd International Symposium on Computational Geometry, LIPIcs. Leibniz Int. Proc. Inform., vol. 51, Schloss Dagstuhl. Leibniz-Zent. Inform., Wadern, 2016, pp. Art. 54, 16. MR3540896

[KPW92] J. Komlós, J. Pach, and G. Woeginger, *Almost tight bounds for ϵ-nets*, Discrete Comput. Geom. **7** (1992), no. 2, 163–173, DOI 10.1007/BF02187833. MR1139078

[Mat92] J. Matoušek, *Reporting points in halfspaces*, Comput. Geom. **2** (1992), no. 3, 169–186, DOI 10.1016/0925-7721(92)90006-E. MR1190599

[NT10] M. Naszódi and S. Taschuk, *On the transversal number and VC-dimension of families of positive homothets of a convex body*, Discrete Math. **310** (2010), no. 1, 77–82, DOI 10.1016/j.disc.2009.07.030. MR2558970

[PR08] E. Pyrga and S. Ray, *New existence proofs for ϵ-nets*, Computational geometry (SCG'08), ACM, New York, 2008, pp. 199–207, DOI 10.1145/1377676.1377708. MR2504287

[PT13] J. Pach and G. Tardos, *Tight lower bounds for the size of epsilon-nets*, J. Amer. Math. Soc. **26** (2013), no. 3, 645–658, DOI 10.1090/S0894-0347-2012-00759-0. MR3037784

3. LIST: Epsilon-Approximations

It reminds me of an anecdote about André Weil who at some point had some problems with elliptic operators so he invited a great expert in the field and he gave him the problem. The expert sat at the kitchen table and solved the problem after several hours. To thank him, André Weil said 'when I have a problem with electricity I call an electrician, when I have a problem with ellipticity I use an elliptician'.

Alain Connes

We now turn towards sizes of ϵ-approximations for combinatorial and geometric set systems.

Upper bounds. We first recall the notion of discrepancy, and restate Theorem 12.9.

DEFINITION 12.7. Given a set system $(X, \mathcal{F})$ and a two-coloring $\chi\colon X \to \{-1, 1\}$, define the discrepancy of a set $R \subseteq X$ with respect to χ as

$$\mathrm{disc}_\chi(R) = \left| \sum_{p \in R} \chi(p) \right|,$$

and the discrepancy of $\mathcal{F}$ with respect to χ as

$$\mathrm{disc}_\chi(\mathcal{F}) = \max_{R \in \mathcal{F}} \mathrm{disc}_\chi(R).$$

The discrepancy of $\mathcal{F}$ is then defined to be

$$\mathrm{disc}(\mathcal{F}) = \min_{\chi: X \to \{-1,1\}} \mathrm{disc}_\chi(\mathcal{F}).$$

THEOREM 12.9. *Let $(X, \mathcal{F})$ be a finite set system with $X \in \mathcal{F}$ and let $f(\cdot)$ be a function such that $\mathrm{disc}(\mathcal{F}|_Y) \leq f(|Y|)$ for all $Y \subseteq X$. Then for every integer $t \geq 0$, there exists a set $A \subseteq X$ of size $\lceil \frac{n}{2^t} \rceil$ such that A is an ϵ-approximation of $\mathcal{F}$, where*

$$\epsilon = \frac{2}{n}\left(f(n) + 2f\left(\left\lceil \frac{n}{2} \right\rceil\right) + \cdots + 2^{t-1} f\left(\left\lceil \frac{n}{2^{t-1}} \right\rceil\right)\right).$$

In particular, if there exists a constant $c > 1$ such that $f(2m) \leq \frac{2}{c} f(m)$ for all $m \geq \lceil \frac{n}{2^t} \rceil$, then there exists an ϵ-approximation of $\mathcal{F}$ of size $\lceil \frac{n}{2^t} \rceil$, and where

$$\epsilon = \Theta\left(\frac{2^t}{n} f\left(\left\lceil \frac{n}{2^t} \right\rceil\right)\right).$$

The best upper bounds on sizes of ϵ-approximations are derived by applying Theorem 12.9 using upper bounds on combinatorial discrepancy of the corresponding set system.

- For a set system $(X, \mathcal{F})$ with $|X| = n$ and VC-dim$(\mathcal{F}) \leq d$, we have

  $$\mathrm{disc}(\mathcal{F}) = O\left(n^{\frac{1}{2} - \frac{1}{2d}}\right).$$

  Thus we apply Theorem 12.9, with $f(n) = O\left(n^{\frac{1}{2} - \frac{1}{2d}}\right)$ and the value of t with

  $$\frac{2^t}{n}\left(\frac{n}{2^t}\right)^{\frac{1}{2} - \frac{1}{2d}} = \Theta(\epsilon) \implies 2^t = \Theta\left(\epsilon^{\frac{2d}{d+1}} n\right),$$

to get the existence of an $O(\epsilon)$-approximation of size

$$O\left(\frac{n}{2^t}\right) = O\left(\frac{1}{\epsilon^{\frac{2d}{d+1}}}\right).$$

- For a set system $(X, \mathcal{F})$ with $|X| = n$, such that the VC-dimension of the *dual* set system $(\mathcal{F}, \mathcal{F}^*)$ is at most d, we have

$$\operatorname{disc}(\mathcal{F}^*) = O\left(n^{\frac{1}{2}-\frac{1}{2d}}\sqrt{\log n}\right).$$

  Thus we apply Theorem 12.9, with $f(n) = O\left(n^{\frac{1}{2}-\frac{1}{2d}}\sqrt{\log n}\right)$ and the value of t such that

$$\frac{2^t}{n}\left(\frac{n}{2^t}\right)^{\frac{1}{2}-\frac{1}{2d}}\sqrt{\log n} = \Theta(\epsilon) \implies 2^t = \Theta\left(\frac{\epsilon^{\frac{2d}{d+1}}}{\left(\log\frac{1}{\epsilon}\right)^{\frac{d}{d+1}}}n\right),$$

  to get the existence of an $O(\epsilon)$-approximation of size

$$O\left(\frac{n}{2^t}\right) = O\left(\frac{1}{\epsilon^{\frac{2d}{d+1}}}\left(\log\frac{1}{\epsilon}\right)^{\frac{d}{d+1}}\right).$$

- The primal set system induced by half-spaces in $\mathbb{R}^d$ has VC-dimension at most d, and thus the previous case implies an ϵ-approximation of size $O\left(\frac{n}{2^t}\right) = O\left(\frac{1}{\epsilon^{\frac{2d}{d+1}}}\right)$.
- The primal set system induced by balls in $\mathbb{R}^d$ has VC-dimension $d+1$, and thus applying the discrepancy bound based on its VC-dimension would give an ϵ-approximation of sub-optimal size $O\left(1/\epsilon^{2(d+1)/(d+2)}\right)$. Instead, as the VC-dimension of the dual set system is d, applying the discrepancy bound for the dual set system gives an ϵ-approximation of size

$$O\left(\frac{n}{2^t}\right) = O\left(\frac{1}{\epsilon^{\frac{2d}{d+1}}}\left(\log\frac{1}{\epsilon}\right)^{\frac{d}{d+1}}\right).$$

Lower bounds. The upper bounds on ϵ-approximation sizes given in the table are tight except for the cases where there is a logarithmic term present, in which case it is known to be tight up to some logarithmic term.

The lower bounds for geometric set systems follow from lower bounds on the so-called *Lebesgue discrepancy* for the corresponding systems. We sketch a proof for the case of half-spaces in $\mathbb{R}^d$, which relies on the following beautiful result of Alexander [**Ale90**], stated without proof, on the Lebesgue discrepancy of half-spaces.

THEOREM 16.2. *Let $\mathcal{C} = [0,1]^d$ denote the unit hypercube in $\mathbb{R}^d$. Then for any set $A \subset \mathcal{C}$ of size t, there exists a half-space h^+ in $\mathbb{R}^d$ such that*

$$\left||h^+ \cap A| - t \operatorname{vol}(h^+ \cap \mathcal{C})\right| = \Omega\left(t^{\frac{1}{2}-\frac{1}{2d}}\right).$$

	Objects	**P/D**	**ϵ-Approximations**		**Discrepancy**	
	$(X, \mathcal{F})$ VC-dim$(\mathcal{F}) \leq d$		$\frac{1}{\epsilon^{\frac{2d}{d+1}}}$	**[MWW9316c]**	$n^{\frac{1}{2}-\frac{1}{2d}}$	**[Mat95a]**
	$(X, \mathcal{F})$ VC-dim$(\mathcal{F}^*) \leq d$		$\frac{1}{\epsilon^{\frac{2d}{d+1}}}(\log \frac{1}{\epsilon})^{\frac{d}{d+1}}$	**[MWW9316c]**	$n^{\frac{1}{2}-\frac{1}{2d}}\sqrt{\log n}$	**[MWW9316c]**
$\mathbb{R}^d$						
	Half-spaces	P/D	$\frac{1}{\epsilon^{\frac{2d}{d+1}}}$	**[MWW9316c]**	$n^{\frac{1}{2}-\frac{1}{2d}}$	**[Mat95a]**
	Balls	P	$\frac{1}{\epsilon^{\frac{2d}{d+1}}}(\log \frac{1}{\epsilon})^{\frac{d}{d+1}}$	**[MWW9316c]**	$n^{\frac{1}{2}-\frac{1}{2d}}\sqrt{\log n}$	**[MWW9316c]**
	Balls	D	$\frac{1}{\epsilon^{\frac{2d}{d+1}}}$	**[MWW9316c]**	$n^{\frac{1}{2}-\frac{1}{2d}}$	**[Mat95a]**
	Axis-aligned boxes	P	$\frac{1}{\epsilon} \cdot \log^{d-\frac{1}{2}} \frac{1}{\epsilon}$	**[MWW9316c]**	$\log^{d-\frac{1}{2}} n$	**[Nik17]**

TABLE 4. The table lists the approximation sizes for the most common geometric set systems. The column 'P/D' specifies whether the set system is primal or dual. All bounds are in asymptotic notation, though to save space we omit the '$O(\cdot)$' notation (the omitted constants can depend on d).

Equivalently,

$$\left| \frac{|h^+ \cap A|}{t} - \mathrm{vol}\left(h^+ \cap \mathcal{C}\right) \right| = \Omega\left(\frac{1}{t^{\frac{1}{2}+\frac{1}{2d}}} \right). \tag{16.3}$$

Theorem 16.2 immediately implies a lower bound on ϵ-approximations for the primal set system $\mathcal{R}$ induced on $\mathcal{C}$ by half-spaces in $\mathbb{R}^d$: for the R.H.S. of Equation (16.3) to be at most ϵ, it must be that

$$\frac{1}{t^{\frac{1}{2}+\frac{1}{2d}}} = O(\epsilon) \quad \Longrightarrow \quad t = \Omega\left(\left(\frac{1}{\epsilon}\right)^{\frac{2d}{d+1}} \right).$$

By choosing points from a fine-enough grid in $\mathcal{C}$, the same lower bound applies for the finite case, as we now outline.

Fix a real

$$\delta = o\left(\frac{1}{t^{\frac{1}{2}+\frac{1}{2d}}} \right),$$

and let $\mathcal{G}$ be a uniform grid in $\mathcal{C} = [0,1]^d$ where each cell has side-length δ and let X be a set of $n = \frac{1}{\delta^d}$ points, one from each cell of $\mathcal{G}$ (arbitrarily). Let $A \subseteq X$ be an ϵ-approximation of the primal set system induced on X by half-spaces in $\mathbb{R}^d$, and set $t = |A|$.

Let h^+ be the half-space obtained by applying Theorem 16.2 to A, and so

$$\left| \frac{|h^+ \cap A|}{t} - \mathrm{vol}\left(h^+ \cap \mathcal{C}\right) \right| = \Omega\left(\frac{1}{t^{\frac{1}{2}+\frac{1}{2d}}} \right). \tag{16.4}$$

As each hyperplane in $\mathbb{R}^d$ intersects $O\left(\frac{1}{\delta^{d-1}}\right)$ cells of $\mathcal{G}$, we have

$$\left|h^+ \cap X\right| = n \cdot \mathrm{vol}\left(h^+ \cap \mathcal{C}\right) \pm O\left(\frac{1}{\delta^{d-1}} \right),$$

$$\Longrightarrow \mathrm{vol}\left(h^+ \cap \mathcal{C}\right) = \frac{\left|h^+ \cap X\right|}{n} \pm O(\delta).$$

The above, together with Equation (16.4) and our choice of δ, implies that

$$\left|\frac{|h^+ \cap A|}{t} - \frac{|h^+ \cap X|}{n}\right| = \Omega\left(\frac{1}{t^{\frac{1}{2}+\frac{1}{2d}}}\right) - O(\delta) = \Omega\left(\frac{1}{t^{\frac{1}{2}+\frac{1}{2d}}}\right),$$

and the required lower bound on the size of A follows as earlier.

The other lower bounds follow in a similar manner.

Bibliography and discussion. A nice table with many upper and lower bounds on both combinatorial and Lebesgue discrepancy can be found in [**Mat99**, Appendix A].

[Ale90] R. Alexander, *Geometric methods in the study of irregularities of distribution*, Combinatorica **10** (1990), no. 2, 115–136, DOI 10.1007/BF02123006. MR1082645

[MWW9316c] J. Matoušek, E. Welzl, and L. Wernisch, *Discrepancy and approximations for bounded VC-dimension*, Combinatorica **13** (1993), no. 4, 455–466, DOI 10.1007/BF01303517. MR1262921

[Mat95a] J. Matoušek, *Tight upper bounds for the discrepancy of half-spaces*, Discrete Comput. Geom. **13** (1995), no. 3-4, 593–601, DOI 10.1007/BF02574066. MR1318799

[Mat99] J. Matoušek. *Geometric Discrepancy: An Illustrated Guide*. Springer, 1999.

[Nik17] A. Nikolov, *Tighter bounds for the discrepancy of boxes and polytopes*, Mathematika **63** (2017), no. 3, 1091–1113, DOI 10.1112/S0025579317000250. MR3731316

Bibliography

[AES10] B. Aronov, E. Ezra, and M. Sharir, *Small-size ϵ-nets for axis-parallel rectangles and boxes*, SIAM J. Comput. **39** (2010), no. 7, 3248–3282, DOI 10.1137/090762968. MR2678074

[AHPV07] P. K. Agarwal, S. Har-Peled, and K. R. Varadarajan, *Geometric approximation via coresets*, Combinatorial and computational geometry, Math. Sci. Res. Inst. Publ., vol. 52, Cambridge Univ. Press, Cambridge, 2005, pp. 1–30, DOI 10.4171/PRIMS/172. MR2178310

[APS08] P. K. Agarwal, J. Pach, and M. Sharir, *State of the union (of geometric objects)*, Surveys on discrete and computational geometry, Contemp. Math., vol. 453, Amer. Math. Soc., Providence, RI, 2008, pp. 9–48, DOI 10.1090/conm/453/08794. MR2405676

[AS00] P. K. Agarwal and M. Sharir, *Arrangements and their applications*, Handbook of computational geometry, North-Holland, Amsterdam, 2000, pp. 49–119, DOI 10.1016/B978-044482537-7/50003-6. MR1746675

[AS16] N. Alon and J. H. Spencer, *The probabilistic method*, 4th ed., Wiley Series in Discrete Mathematics and Optimization, John Wiley & Sons, Inc., Hoboken, NJ, 2016. MR3524748

[AU16] M. Axenovich and T. Ueckerdt, *Density of range capturing hypergraphs*, J. Comput. Geom. **7** (2016), no. 1, 1–21. MR3455619

[Aga18] P. K. Agarwal, *Range searching*, Handbook of discrete and computational geometry, CRC Press Ser. Discrete Math. Appl., CRC, Boca Raton, FL, 1997, pp. 575–598. MR1730187

[Ale90] R. Alexander, *Geometric methods in the study of irregularities of distribution*, Combinatorica **10** (1990), no. 2, 115–136, DOI 10.1007/BF02123006. MR1082645

[Alo+08] N. Alon, H. Kaplan, G. Nivasch, M. Sharir, and S. Smorodinsky, *Weak ϵ-nets and interval chains*, J. ACM **55** (2008), no. 6, Art. 28, 32, DOI 10.1145/1455248.1455252. MR2477488

[Alo+92] N. Alon, I. Bárány, Z. Füredi, and D. J. Kleitman, *Point selections and weak ϵ-nets for convex hulls*, Combin. Probab. Comput. **1** (1992), no. 3, 189–200, DOI 10.1017/S0963548300000225. MR1208800

[Aro+14] B. Aronov, M. de Berg, E. Ezra, and M. Sharir, *Improved bounds for the union of locally fat objects in the plane*, SIAM J. Comput. **43** (2014), no. 2, 543–572, DOI 10.1137/120891241. MR3188401

[Ary+20] R. Arya, S. Arya, G. D. da Fonseca, and D. M. Mount, *Optimal bound on the combinatorial complexity of approximating polytopes*, Proceedings of the 2020 ACM-SIAM Symposium on Discrete Algorithms, SIAM, Philadelphia, PA, 2020, pp. 786–805. MR4141230

[BC87] J. Beck and W. W. L. Chen, *Irregularities of distribution*, Cambridge Tracts in Mathematics, vol. 89, Cambridge University Press, Cambridge, 1987, DOI 10.1017/CBO9780511565984. MR903025

[BCM99] H. Brönnimann, B. Chazelle, and J. Matoušek, *Product range spaces, sensitive sampling, and derandomization*, SIAM J. Comput. **28** (1999), no. 5, 1552–1575, DOI 10.1137/S0097539796260321. MR1694192

[BLM13] S. Boucheron, G. Lugosi, and P. Massart, *Concentration inequalities*, Oxford University Press, Oxford, 2013. A nonasymptotic theory of independence; With a foreword by Michel Ledoux, DOI 10.1093/acprof:oso/9780199535255.001.0001. MR3185193

[BMN11] B. Bukh, J. Matoušek, and G. Nivasch, *Lower bounds for weak epsilon-nets and stair-convexity*, Israel J. Math. **182** (2011), 199–208, DOI 10.1007/s11856-011-0029-1. MR2783971

[BPR13] S. Buzaglo, R. Pinchasi, and G. Rote, *Topological hypergraphs*, Thirty essays on geometric graph theory, Springer, New York, 2013, pp. 71–81, DOI 10.1007/978-1-4614-0110-0_6. MR3205150

[BS20] J. Balogh and W. Samotij, *An efficient container lemma*, Discrete Anal., posted on 2020, Paper No. 17, 56, DOI 10.19086/da. MR4186910

[BS95] M. de Berg and O. Schwarzkopf, *Cuttings and applications*, Internat. J. Comput. Geom. Appl. **5** (1995), no. 4, 343–355, DOI 10.1142/S0218195995000210. MR1359425

[Blu+89] A. Blumer, A. Ehrenfeucht, D. Haussler, and M. K. Warmuth, *Learnability and the Vapnik-Chervonenkis dimension*, J. Assoc. Comput. Mach. **36** (1989), no. 4, 929–965, DOI 10.1145/76359.76371. MR1072253

[Boi+98] J.-D. Boissonnat, M. Sharir, B. Tagansky, and M. Yvinec, *Voronoi diagrams in higher dimensions under certain polyhedral distance functions*, Discrete Comput. Geom. **19** (1998), no. 4, 485–519, DOI 10.1007/PL00009366. MR1620060

[Bol86] B. Bollobás, *Combinatorics*, Cambridge University Press, Cambridge, 1986. Set systems, hypergraphs, families of vectors and combinatorial probability. MR866142

[Bus+16] N. Bus, S. Garg, N. H. Mustafa, and S. Ray, *Tighter estimates for ϵ-nets for disks*, Comput. Geom. **53** (2016), 27–35, DOI 10.1016/j.comgeo.2015.12.002. MR3454542

[CF90] B. Chazelle and J. Friedman, *A deterministic view of random sampling and its use in geometry*, Combinatorica **10** (1990), no. 3, 229–249, DOI 10.1007/BF02122778. MR1092541

[CKP20] V. Chepoi, K. Knauer, and M. Philibert, *Two-dimensional partial cubes*, Electron. J. Combin. **27** (2020), no. 3, Paper No. 3.29, 40, DOI 10.37236/8934. MR4245142

[CM21] M. Csikós and N. H. Mustafa. *Optimal approximations made easy. CoRR* abs/2008.08970, 2021. `arXiv:2008.08970`. URL: `https://arxiv.org/abs/2008.08970`.

[CM96] B. Chazelle and J. Matoušek, *On linear-time deterministic algorithms for optimization problems in fixed dimension*, J. Algorithms **21** (1996), no. 3, 579–597, DOI 10.1006/jagm.1996.0060. MR1417665

[CMK19] M. Csikós, N. H. Mustafa, and A. Kupavskii, *Tight lower bounds on the VC-dimension of geometric set systems*, J. Mach. Learn. Res. **20** (2019), Paper No. 81, 8. MR3960935

[CS89] K. L. Clarkson and P. W. Shor, *Applications of random sampling in computational geometry. II*, Discrete Comput. Geom. **4** (1989), no. 5, 387–421, DOI 10.1007/BF02187740. MR1014736

[CST14] N. Srivastava, *Discrepancy, graphs, and the Kadison-Singer problem*, Asia Pac. Math. Newsl. **3** (2013), no. 4, 15–20. MR3156130

[CV07] K. L. Clarkson and K. Varadarajan, *Improved approximation algorithms for geometric set cover*, Discrete Comput. Geom. **37** (2007), no. 1, 43–58, DOI 10.1007/s00454-006-1273-8. MR2279863

[CW89] B. Chazelle and E. Welzl, *Quasi-optimal range searching in spaces of finite VC-dimension*, Discrete Comput. Geom. **4** (1989), no. 5, 467–489, DOI 10.1007/BF02187743. MR1014739

[Cha+12] T. M. Chan, E. Grant, J. Könemann, and M. Sharpe, *Weighted capacitated, priority, and geometric set cover via improved quasi-uniform sampling*, Proceedings of the Twenty-Third Annual ACM-SIAM Symposium on Discrete Algorithms, ACM, New York, 2012, pp. 1576–1585. MR3205315

[Cha+95] B. Chazelle, H. Edelsbrunner, M. Grigni, L. Guibas, M. Sharir, and E. Welzl, *Improved bounds on weak ϵ-nets for convex sets*, Discrete Comput. Geom. **13** (1995), no. 1, 1–15, DOI 10.1007/BF02574025. MR1300506

[Cha00] B. Chazelle, *The discrepancy method*, Cambridge University Press, Cambridge, 2000. Randomness and complexity, DOI 10.1017/CBO9780511626371. MR1779341

[Cha12] T. M. Chan, *Optimal partition trees*, Discrete Comput. Geom. **47** (2012), no. 4, 661–690, DOI 10.1007/s00454-012-9410-z. MR2901245

[Cha93] B. Chazelle, *Cutting hyperplanes for divide-and-conquer*, Discrete Comput. Geom. **9** (1993), no. 2, 145–158, DOI 10.1007/BF02189314. MR1194032

[Cla+90] K. L. Clarkson, H. Edelsbrunner, L. J. Guibas, M. Sharir, and E. Welzl, *Combinatorial complexity bounds for arrangements of curves and spheres*, Discrete Comput. Geom. **5** (1990), no. 2, 99–160, DOI 10.1007/BF02187783. MR1032370

[Cla87] K. L. Clarkson, *New applications of random sampling in computational geometry*, Discrete Comput. Geom. **2** (1987), no. 2, 195–222, DOI 10.1007/BF02187879. MR884226

[Cla88] K. L. Clarkson, *A randomized algorithm for closest-point queries*, SIAM J. Comput. **17** (1988), no. 4, 830–847, DOI 10.1137/0217052. MR953296

[Csi16] M. Csikós. Personal communication. 2016.

[DEG16] K. Dutta, E. Ezra, and A. Ghosh, *Two proofs for shallow packings*, Discrete Comput. Geom. **56** (2016), no. 4, 910–939, DOI 10.1007/s00454-016-9824-0. MR3561795

[DGL96] L. Devroye, Györfi and G. Lugosi. *A Probabilistic Theory of Pattern Recognition*. Springer, 1996.

[DP09] D. P. Dubhashi and A. Panconesi, *Concentration of measure for the analysis of randomized algorithms*, Cambridge University Press, Cambridge, 2009, DOI 10.1017/CBO9780511581274. MR2547432

[Dye+18] M. Dyer, B. Gärtner, N. Megiddo and E. Welzl. *Linear programming. Handbook of Discrete and Computational Geometry*. CRC Press, 2018. pp. 1291–1310.

[EA07] D. Eisenstat and D. Angluin, *The VC dimension of k-fold union*, Inform. Process. Lett. **101** (2007), no. 5, 181–184, DOI 10.1016/j.ipl.2006.10.004. MR2291190

[ERS05] G. Even, D. Rawitz, and S. Shahar, *Hitting sets when the VC-dimension is small*, Inform. Process. Lett. **95** (2005), no. 2, 358–362, DOI 10.1016/j.ipl.2005.03.010. MR2149262

[Eis09] D. Eisenstat, *k-fold unions of low-dimensional concept classes*, Inform. Process. Lett. **109** (2009), no. 23-24, 1232–1234, DOI 10.1016/j.ipl.2009.09.005. MR2571754

[Ezr+20] E. Ezra, S. Har-Peled, H. Kaplan, and M. Sharir, *Decomposing arrangements of hyperplanes: VC-dimension, combinatorial dimension, and point location*, Discrete Comput. Geom. **64** (2020), no. 1, 109–173, DOI 10.1007/s00454-019-00141-7. MR4110531

[Ezr16] E. Ezra, *A size-sensitive discrepancy bound for set systems of bounded primal shatter dimension*, SIAM J. Comput. **45** (2016), no. 1, 84–101, DOI 10.1137/140977746. MR3452262

[FK15] A. Frieze and M. Karoński, *Introduction to random graphs*, Cambridge University Press, Cambridge, 2016, DOI 10.1017/CBO9781316339831. MR3675279

[FL11] D. Feldman and M. Langberg, *A unified framework for approximating and clustering data*, STOC'11—Proceedings of the 43rd ACM Symposium on Theory of Computing, ACM, New York, 2011, pp. 569–578, DOI 10.1145/1993636.1993712. MR2932007

[FP08] J. Fox and J. Pach, *Separator theorems and Turán-type results for planar intersection graphs*, Adv. Math. **219** (2008), no. 3, 1070–1080, DOI 10.1016/j.aim.2008.06.002. MR2442062

[FP83] P. Frankl and J. Pach, *On the number of sets in a null t-design*, European J. Combin. **4** (1983), no. 1, 21–23, DOI 10.1016/S0195-6698(83)80004-3. MR694464

[FS13] Z. Füredi and M. Simonovits, *The history of degenerate (bipartite) extremal graph problems*, Erdös centennial, Bolyai Soc. Math. Stud., vol. 25, János Bolyai Math. Soc., Budapest, 2013, pp. 169–264, DOI 10.1007/978-3-642-39286-3_7. MR3203598

[Fra87] P. Frankl, *The shifting technique in extremal set theory*, Surveys in combinatorics 1987 (New Cross, 1987), London Math. Soc. Lecture Note Ser., vol. 123, Cambridge Univ. Press, Cambridge, 1987, pp. 81–110. MR905277

[GKP94] R. L. Graham, D. E. Knuth, and O. Patashnik, *Concrete mathematics*, 2nd ed., Addison-Wesley Publishing Company, Reading, MA, 1994. A foundation for computer science. MR1397498

[Gär+08] B. Gärtner, J. Matoušek, L. Rüst, and P. Škovroň, *Violator spaces: structure and algorithms*, Discrete Appl. Math. **156** (2008), no. 11, 2124–2141, DOI 10.1016/j.dam.2007.08.048. MR2437006

[HLW94] D. Haussler, N. Littlestone, and M. K. Warmuth, *Predicting $\{0,1\}$-functions on randomly drawn points*, Inform. and Comput. **115** (1994), no. 2, 248–292, DOI 10.1006/inco.1994.1097. MR1304811

[HP11] S. Har-Peled, *Geometric approximation algorithms*, Mathematical Surveys and Monographs, vol. 173, American Mathematical Society, Providence, RI, 2011, DOI 10.1090/surv/173. MR2760023

[HPS11] S. Har-Peled and M. Sharir, *Relative (p, ϵ)-approximations in geometry*, Discrete Comput. Geom. **45** (2011), no. 3, 462–496, DOI 10.1007/s00454-010-9248-1. MR2770547

[HV20] L. Huang and N. K. Vishnoi, *Coresets for clustering in euclidean spaces: importance sampling is nearly optimal*, STOC '20—Proceedings of the 52nd Annual ACM SIGACT Symposium on Theory of Computing, ACM, New York, 2020, pp. 1416–1429. MR4141850

[HW87] D. Haussler and E. Welzl, *ϵ-nets and simplex range queries*, Discrete Comput. Geom. **2** (1987), no. 2, 127–151, DOI 10.1007/BF02187876. MR884223

[Hau95] D. Haussler, *Sphere packing numbers for subsets of the Boolean n-cube with bounded Vapnik-Chervonenkis dimension*, J. Combin. Theory Ser. A **69** (1995), no. 2, 217–232, DOI 10.1016/0097-3165(95)90052-7. MR1313896

[IK10] R. Impagliazzo and V. Kabanets, *Constructive proofs of concentration bounds*, Approximation, randomization, and combinatorial optimization, Lecture Notes in Comput. Sci., vol. 6302, Springer, Berlin, 2010, pp. 617–631, DOI 10.1007/978-3-642-15369-3_46. MR2755867

[KMP16] A. Kupavskii, N. H. Mustafa, and J. Pach, *New lower bounds for ϵ-nets*, 32nd International Symposium on Computational Geometry, LIPIcs. Leibniz Int. Proc. Inform., vol. 51, Schloss Dagstuhl. Leibniz-Zent. Inform., Wadern, 2016, pp. Art. 54, 16. MR3540896

[KMP17] A. Kupavskii, N. H. Mustafa, and J. Pach, *Near-optimal lower bounds for ϵ-nets for half-spaces and low complexity set systems*, A journey through discrete mathematics, Springer, Cham, 2017, pp. 527–541. MR3726612

[KPW92] J. Komlós, J. Pach, and G. Woeginger, *Almost tight bounds for ϵ-nets*, Discrete Comput. Geom. **7** (1992), no. 2, 163–173, DOI 10.1007/BF02187833. MR1139078

[KY15] P. Klein and N. E. Young, *On the number of iterations for Dantzig-Wolfe optimization and packing-covering approximation algorithms*, SIAM J. Comput. **44** (2015), no. 4, 1154–1172, DOI 10.1137/12087222X. MR3390154

[LLS01] Y. Li, P. M. Long, and A. Srinivasan, *Improved bounds on the sample complexity of learning*, J. Comput. System Sci. **62** (2001), no. 3, 516–527, DOI 10.1006/jcss.2000.1741. MR1824457

[Loe+19] J. A. De Loera, X. Goaoc, F. Meunier, and N. H. Mustafa, *The discrete yet ubiquitous theorems of Carathéodory, Helly, Sperner, Tucker, and Tverberg*, Bull. Amer. Math. Soc. (N.S.) **56** (2019), no. 3, 415–511, DOI 10.1090/bull/1653. MR3974609

[Lon01] P. M. Long, *Using the pseudo-dimension to analyze approximation algorithms for integer programming*, Algorithms and data structures (Providence, RI, 2001), Lecture Notes in Comput. Sci., vol. 2125, Springer, Berlin, 2001, pp. 26–37, DOI 10.1007/3-540-44634-6_4. MR1936398

[MDG18] N. H. Mustafa, K. Dutta, and A. Ghosh, *A simple proof of optimal epsilon nets*, Combinatorica **38** (2018), no. 5, 1269–1277, DOI 10.1007/s00493-017-3564-5. MR3884789

[MG07] J. Matoušek and B. Gärtner. *Understanding and Using Linear Programming.* Springer, 2007.

[MP16] N. H. Mustafa and J. Pach, *On the Zarankiewicz problem for intersection hypergraphs*, J. Combin. Theory Ser. A **141** (2016), 1–7, DOI 10.1016/j.jcta.2016.02.001. MR3479234

[MP18] M. Matheny and J. M. Phillips, *Practical low-dimensional halfspace range space sampling*, 26th European Symposium on Algorithms, LIPIcs. Leibniz Int. Proc. Inform., vol. 112, Schloss Dagstuhl. Leibniz-Zent. Inform., Wadern, 2018, pp. Art. No. 62, 14. MR3847688

[MR09] N. H. Mustafa and S. Ray, *An optimal extension of the centerpoint theorem*, Comput. Geom. **42** (2009), no. 6-7, 505–510, DOI 10.1016/j.comgeo.2007.10.004. MR2519371

[MR17] N. H. Mustafa and S. Ray, *ε-Mnets: hitting geometric set systems with subsets*, Discrete Comput. Geom. **57** (2017), no. 3, 625–640, DOI 10.1007/s00454-016-9845-8. MR3614776

[MW04] J. Matoušek and U. Wagner, *New constructions of weak ϵ-nets*, Discrete Comput. Geom. **32** (2004), no. 2, 195–206, DOI 10.1007/s00454-004-1116-4. MR2074337

[MWW93] J. Matoušek, E. Welzl, and L. Wernisch, *Discrepancy and approximations for bounded VC-dimension*, Combinatorica **13** (1993), no. 4, 455–466, DOI 10.1007/BF01303517. MR1262921

[Mat02] J. Matoušek. *Lectures in Discrete Geometry.* Springer, 2002.

[Mat91] J. Matoušek, *Cutting hyperplane arrangements*, Discrete Comput. Geom. **6** (1991), no. 5, 385–406, DOI 10.1007/BF02574697. MR1115098

[Mat92] J. Matoušek, *Efficient partition trees*, Discrete Comput. Geom. **8** (1992), no. 3, 315–334, DOI 10.1007/BF02293051. ACM Symposium on Computational Geometry (North Conway, NH, 1991). MR1174360

[Mat92] J. Matoušek, *Reporting points in halfspaces*, Comput. Geom. **2** (1992), no. 3, 169–186, DOI 10.1016/0925-7721(92)90006-E. MR1190599

[Mat95a] J. Matoušek, *Tight upper bounds for the discrepancy of half-spaces*, Discrete Comput. Geom. **13** (1995), no. 3-4, 593–601, DOI 10.1007/BF02574066. MR1318799

[Mat95b] J. Matoušek, *Approximations and optimal geometric divide-and-conquer*, J. Comput. System Sci. **50** (1995), no. 2, 203–208, DOI 10.1006/jcss.1995.1018. 23rd Symposium on the Theory of Computing (New Orleans, LA, 1991). MR1330253

[Mat99] J. Matoušek. *Geometric Discrepancy: An Illustrated Guide.* Springer, 1999.

[McM70] P. McMullen, *The maximum numbers of faces of a convex polytope*, Mathematika **17** (1970), 179–184, DOI 10.1112/S0025579300002850. MR283691

[Mul18] W. Mulzer, *Five proofs of Chernoff's bound with applications*, Bull. Eur. Assoc. Theor. Comput. Sci. EATCS **124** (2018), 59–76. MR3793013

[Mus16] N. H. Mustafa, *A simple proof of the shallow packing lemma*, Discrete Comput. Geom. **55** (2016), no. 3, 739–743, DOI 10.1007/s00454-016-9767-5. MR3473678

[Mus19] N. H. Mustafa, *Computing optimal epsilon-nets is as easy as finding an unhit set*, 46th International Colloquium on Automata, Languages, and Programming, LIPIcs. Leibniz Int. Proc. Inform., vol. 132, Schloss Dagstuhl. Leibniz-Zent. Inform., Wadern, 2019, pp. Art. No. 87, 12. MR3984904

[NT10] M. Naszódi and S. Taschuk, *On the transversal number and VC-dimension of families of positive homothets of a convex body*, Discrete Math. **310** (2010), no. 1, 77–82, DOI 10.1016/j.disc.2009.07.030. MR2558970

[Nas19] M. Naszódi, *Approximating a convex body by a polytope using the epsilon-net theorem*, Discrete Comput. Geom. **61** (2019), no. 3, 686–693, DOI 10.1007/s00454-018-9977-0. MR3918553

[Nik17] A. Nikolov, *Tighter bounds for the discrepancy of boxes and polytopes*, Mathematika **63** (2017), no. 3, 1091–1113, DOI 10.1112/S0025579317000250. MR3731316

[PA95] J. Pach and P. K. Agarwal, *Combinatorial geometry*, Wiley-Interscience Series in Discrete Mathematics and Optimization, John Wiley & Sons, Inc., New York, 1995. A Wiley-Interscience Publication, DOI 10.1002/9781118033203. MR1354145

[PR08] E. Pyrga and S. Ray, *New existence proofs for ϵ-nets*, Computational geometry (SCG'08), ACM, New York, 2008, pp. 199–207, DOI 10.1145/1377676.1377708. MR2504287

[PT13] J. Pach and G. Tardos, *Tight lower bounds for the size of epsilon-nets*, J. Amer. Math. Soc. **26** (2013), no. 3, 645–658, DOI 10.1090/S0894-0347-2012-00759-0. MR3037784

[Pel97] M. Pellegrini, *On counting pairs of intersecting segments and off-line triangle range searching*, Algorithmica **17** (1997), no. 4, 380–398, DOI 10.1007/BF02523679. MR1429210

[Phi18] J. M. Phillips. *Coresets and sketches. Handbook of Discrete and Computational Geometry.* CRC Press, 2018. pp. 1269–1286.

[Rub18] N. Rubin, *An improved bound for weak epsilon-nets in the plane*, 59th Annual IEEE Symposium on Foundations of Computer Science—FOCS 2018, IEEE Computer Soc., Los Alamitos, CA, 2018, pp. 224–235, DOI 10.1109/FOCS.2018.00030. MR3899592

[STZ06] S. Suri, C. D. Tóth, and Y. Zhou, *Range counting over multidimensional data streams*, Discrete Comput. Geom. **36** (2006), no. 4, 633–655, DOI 10.1007/s00454-006-1269-4. MR2267550

[SW92] M. Sharir and E. Welzl, *A combinatorial bound for linear programming and related problems*, STACS 92 (Cachan, 1992), Lecture Notes in Comput. Sci., vol. 577, Springer, Berlin, 1992, pp. 569–579. MR1255620

[Sha91] M. Sharir, *On k-sets in arrangements of curves and surfaces*, Discrete Comput. Geom. **6** (1991), no. 6, 593–613, DOI 10.1007/BF02574706. MR1121331

[Spi19] M. Z. Spivey, *The art of proving binomial identities*, Discrete Mathematics and its Applications (Boca Raton), CRC Press, Boca Raton, FL, 2019, DOI 10.1201/9781351215824. MR3931743

[Tal94] M. Talagrand, *Sharper bounds for Gaussian and empirical processes*, Ann. Probab. **22** (1994), no. 1, 28–76. MR1258865

[VC71] V. N. Vapnik and A. Ja. Červonenkis, *The uniform convergence of frequencies of the appearance of events to their probabilities* (Russian, with English summary), Teor. Verojatnost. i Primenen. **16** (1971), 264–279. MR0288823

[VX12] K. R. Varadarajan and X. Xiao. *On the sensitivity of shape fitting problems. Proceedings of the IARCS Annual Conference on Foundations of Software Technology and Theoretical Computer Science (FSTTCS).* Schloss Dagstuhl - Leibniz-Zentrum für Informatik, 2012. pp. 486–497.

[Var10] K. Varadarajan, *Weighted geometric set cover via quasi-uniform sampling*, STOC'10—Proceedings of the 2010 ACM International Symposium on Theory of Computing, ACM, New York, 2010, pp. 641–647. MR2743313

[Wag08] U. Wagner, *k-sets and k-facets*, Surveys on discrete and computational geometry, Contemp. Math., vol. 453, Amer. Math. Soc., Providence, RI, 2008, pp. 443–513, DOI 10.1090/conm/453/08810. MR2405692

[Zie95] G. M. Ziegler, *Lectures on polytopes*, Graduate Texts in Mathematics, vol. 152, Springer-Verlag, New York, 1995, DOI 10.1007/978-1-4613-8431-1. MR1311028

Index

$(\leq k)$-sets, 4
k-fold union, intersection, 144, 158, 229
  balls in $\mathbb{R}^d$, 223
  lower bound for general systems, 158
  relation to epsilon-nets, 157
  VC-dimension upper bound, 145
k-median
  cost function, 220
  definition, 220
k-median problem, 215
k-tuples of points, 216

Alexander's theorem, 237
algorithms
  epsilon-approximations for general case, 173
  epsilon-approximations for half-spaces, 195
  epsilon-approximations for VC set systems, 177
  epsilon-nets for shallow-cell complexity, 101, 108
  iterative for epsilon-nets, 94
  linear-sized epsilon-nets, 102
  merge-and-reduce, 183
  weak epsilon-nets for convex sets, 134
  weighted general shallow-cell complexity, 122
  weighted linear shallow-cell complexity, 116
alterations technique, 37, 42
  cuttings, 45, 49
  disks, 40
  epsilon-net algorithm, 101
  epsilon-nets for shallow-cell complexity, 97
AM-GM inequality, 17
approximation of distances, 213
arithmetic progressions, 143
averaging technique
  k-Delaunay graphs, 1
  anchored rectangles, 72
  epsilon-nets for VC systems, 91
  level sets for disks (dual), 7
  level sets for disks (primal), 5
  level sets for half-spaces, 9
  Markov's inequality, 15
  packing theorem, 80
  tail bounds, 16
  union complexity, 68

balls in $\mathbb{R}^d$, 213
binomial theorem, 17
  approximations, 19
boxes
  k-fold union, 158
  dual set system, 159

canonical objects, 35, 96
  disks, 4, 34, 39, 47
  halfplanes, 34, 39
  rectangles, 31
  trapezoids, 46
centerpoints, 133
  centerpolytope, 133
  for all subsets, 134
  use for weak epsilon-nets, 134
centroids, 140
chaining, 199
  intuition, 185
charging scheme, 116
  with permutations, 120
Chernoff's bound, 13, 92
  asymmetry, 19
  proof, 16
clustering, 215
combinatorial approximation of points, 213
convex-hull, 135
convexity
  center polytope, 133
  points in convex position, 137
  polytopes, 131
  shrinking, 138
  volume, 139
coresets, 220
  definition, 220
  literature, 225
  sensitivity, 223

cuttings
  inside a simplex, 50
  optimality, 44
  segments, 44, 47, 54
  shallow cuttings, 51, 78
  simplices, 50
  uses for divide-and-conquer, 55

decomposition
  trapezoidal, 46
  triangles, 45
discrepancy
  balls (dual), 238
  balls (primal), 238
  books, 183
  bound for VC systems, 179
  boxes, 238
  chaining, 192
  combinatorial, definition, 178, 236
  combinatorial, for half-spaces, 182
  connection to lower bounds for approximations, 239
  dual set systems, 237
  geometric, 155, 237
  half-spaces, 238
  Lebesgue, 237
  optimal bounds, 236
discretization, 94
distance decomposition, 214
distributions
  negatively associated, 12
duality, 77

epsilon-approximation
  a first calculation, 172
  algorithm for general case, 173
  definition, 172, 185
  for general functions, 216
  functional, 216
  general case, 172
  half-spaces, 193
  intuition, 171
  VC-dimension, 186
  weighted, 221
epsilon-approximations
  additivity lemma, 174, 186
  chaining for VC set systems, 186
  composition properties, 189
  discrepancy proof for VC bound, 180
  experimental evaluation, 198
  inductive proof, 174
  lower bounds, 237
  optimal bound for half-spaces, 182
  relation to discrepancy, 178, 236
  using jittered sampling, 193
  VC bound via discrepancy, 178
  VC original theorem, 174
epsilon-net
  anchored rectangles, 24
  bounded degree, 38
  boxes in $\mathbb{R}^d$, 26, 29
  disks, 31, 37, 39, 57, 58
  general set system, 26, 38
  general set systems, 38
  half-spaces, 139
  halfplanes, 24
  history, 89
  intervals, 23
  linear shallow-cell complexity, 115
  lower bound abstract case, 167
  lower bound dual boxes, 168
  lower bound half-spaces, 168
  optimality of VC-dimension bound, 89
  rectangles in $\mathbb{R}^2$, 31
  shallow-cell complexity, 96
  shallow-cell complexity theorem, 90, 96, 233
  triangles, 46
  VC-dimension, 91
  weak convex, 131
  weak convex in $\mathbb{R}^2$, 132
  weak convex theorem, 132
  weighted, 22, 113, 115
  weighted disks (primal), 115
epsilon-nets
  balls, 234
  bottomless rectangles (primal), 234
  convex sets, 234
  disks (primal), 234
  fat triangles (primal), 234
  half-spaces, 234
  homothets, 234
  inductive proof for VC-dimension, 176
  intervals, 234
  iterative constructions, 94
  lines, 234
  pseudo-disks (primal), 234
  rectangles (dual), 234
  rectangles (primal), 234
  triangles (primal), 234
  union complexity (dual), 234

fat triangles, 228
finite differencing, 218
forbidden induced subgraph problem, 52

ghost sampling, 94
graphs
  k-Delaunay, 1
  boundary of union of disks, 7
  Delaunay, 4, 5
  planar, 1, 7, 46
  unit distance, 61, 82
grid
  non-uniform, 29
  uniform, 238
Grünbaum's inequality, 140

half-space depth, 139
half-spaces
  arrangement, 9
Haussler's packing theorem, 75, 80
Helly's theorem
  convex sets, 133
  rectangles, 152
homothets, 228

jittered sampling, 193

Kővári-Sós-Turán theorem, 52

level of points
  half-spaces, 9
level sets, 4
  disks (dual), 7
  disks (primal), 4, 40
  functions, 216
  half-spaces (dual), 9, 50
  half-spaces (primal), 9
  rectangles (primal), 70
  trapezoids, 47, 49
linear programming, 21, 35
  books, 129
  rounding, 128
  weighted hitting set, 128
list
  epsilon-approximations, 236
  epsilon-nets, 233
  properties of geometric set systems, 228
lower bounds
  k-fold union, intersection for general systems, 145
  abstract epsilon-nets, 167
  basic idea for epsilon-nets, 143, 167
  direct probabilistic construction, 168
  dual boxes epsilon-nets, 168
  half-spaces epsilon-nets, 168
  weak epsilon-nets, 137

Markov's inequality, 15, 33, 35
  proof, 15
minimum hitting set problem
  definition, 21
  NP-hard, 21
  weighted, 128

nearest-neighbor queries, 89
non-uniform sampling, 114

oracles, 107, 108
orthogonal functions, 155

packing statements
  geometric set systems, 207
  half-spaces in $\mathbb{R}^d$, 77
  Haussler's packing theorem, 80
  maximal, 98
  maximal packings, 103
  shallow-cell complexity, 85
  use for chaining epsilon-approximations, 188
  use for epsilon-nets, 96
  use for relative epsilon-approximations, 210
  VC-dimension, 85, 188
Pascal's identity, 53
polytope approximation, 138
  history, 140
  proof, 138
prefix, 151
probability distribution
  Bernoulli, 106
  binomial, 11
  negative binomial, 106
  without replacement, 12
pseudo-dimension of functions, 219
pseudo-disks, 228

quotes
  Alain Connes, 236
  Alexandre Grothendieck, 96
  Andrew Wiles, 202
  David Hilbert, 228
  David Mumford, 138
  Elizabeth Anscombe, 158
  Franz Kafka, 167
  George Pólya, 121
  Gian-Carlo Rota, 31, 103, 216
  Gil Kalai, 4
  Hans Frauenfelder, 77
  Henri Poincaré, 108
  Herman Chernoff, 11
  Hermann Weyl, 26
  Israel Gelfand, 39
  Jiří Matoušek, 132
  Joel Spencer, 80
  John Barrow, 162
  Maryam Mirzakhani, 233
  Matt Stoller, 178
  Michael Atiyah, 128, 186
  Noam Chomsky, 233
  Noam Elkies, 91
  Norbert Wiener, 44
  Oded Goldreich, 67
  Philip Anderson, 174
  Pierre Deligne, 193
  Robert Tarjan, 220
  Timothy Gowers, 145
  William Thurston, 23, 52, 60, 85, 115, 208

Ramsey theory, 143
rectangles
  k-fold union (dual), 150
  binary representation, 151
  lifting, 160
  lower bound for epsilon-nets, 150

orthogonal system, 152
parity argument, 153
relative epsilon-approximation
chaining for VC set systems, 208
composition properties, 202, 209
definition, 200, 202, 208
first bound, 203
relation to (ν, α)-samples, 207
relation to epsilon-approximations, 201
relation to epsilon-nets, 200

sensitive epsilon-approximation
bound for VC set systems, 205
definition, 205
intuition, 204
lower bound, 205
separators, 55
sequence, 91
set system
anchored rectangles in $\mathbb{R}^2$ (primal), 24
disks (primal), 31, 90
disks in $\mathbb{R}^2$ (primal), 21, 23
downward-facing half-spaces, 9
half-spaces in $\mathbb{R}^2$ (primal), 24
half-spaces in $\mathbb{R}^d$ (dual), 9
half-spaces in $\mathbb{R}^d$ (primal), 8
induced by simplices, 163
intervals (primal), 24
products, 159
rectangles in $\mathbb{R}^2$ (dual), 150
rectangles in $\mathbb{R}^2$ (primal), 31
rectangles in $\mathbb{R}^d$ (primal), 26, 29
sine curves (primal), 143
triangles, 44
set systems
dual, 8, 228
primal, 8, 228
projection, 57
shallow-cell complexity, 58, 67, 85, 89, 96, 115, 228
k-fold union of intervals, 230
anchored rectangles, 70, 72
balls, 230
bottomless rectangles, 230
convex sets, 230
disks (dual), 67
disks (primal), 67, 230
half-spaces k-fold intersection, 230
halfplanes, 230
halfspaces k-fold union, 230
homothets of a convex body, 230
intervals, 230
lines, 230
pseudo-disks, 228
pseudo-disks (primal), 230
rectangles, 70, 230
set systems (dual), 67
triangles, 230
well-behaved, 59
shifting, 61, 63
simplicial partition theorem, 134, 194
construction algorithms, 198
history, 137
use for epsilon-approximations, 194
spatial partitioning, 89
symmetric difference, 61, 75

tail bounds, 11, 199
threshold structures, 171
transformations
dualizing points, 163
lifting boxes, 162
paraboloid lift of points, 9
rectangle products, 160
stretching points, 162

union
balls, 215
union complexity, 68
disks, 7, 69
dual set systems, 68
epsilon-nets, 103
rectangles, 69
upper bound theorem
convex polytopes, 9

VC-dimension, 58, 63, 157, 228
k-fold union of intervals, 230
abstract k-fold union, 145
balls, 65, 230
bottomless rectangles (primal), 230
convex sets, 131, 230
disks, 230
dual, 229
dual boxes k-fold union, 158
dual rectangles k-fold union, 150
half-spaces, 65
half-spaces k-fold intersection, 230
half-spaces k-fold union, 162, 230
halfplanes, 230
history, 66
homothets of a convex body, 230
hyperplanes, 64
induced by simplices, 163
intervals, 230
lines, 230
orthants k-fold union, 162
polynomials, 65
pseudo-disks, 230
rectangles (primal), 230
triangles, 230
Venn diagram, 229

weighted
epsilon-approximations, 221
epsilon-nets, 113, 115

Zarankiewicz problem, 52
$K_{2,2}$ proof, 53
inductive proof, 53
segments in $\mathbb{R}^2$, 52
simplices in $\mathbb{R}^d$, 54

265 **Nabil H. Mustafa,** Sampling in Combinatorial and Geometric Set Systems, 2022

264 **J. Scott Carter and Seiichi Kamada,** Diagrammatic Algebra, 2021

263 **Vugar E. Ismailov,** Ridge Functions and Applications in Neural Networks, 2021

262 **Ragnar-Olaf Buchweitz,** Maximal Cohen–Macaulay Modules and Tate Cohomology, 2021

261 **Shiri Artstein-Avidan, Apostolos Giannopoulos, and Vitali D. Milman,** Asymptotic Geometric Analysis, Part II, 2021

260 **Lindsay N. Childs, Cornelius Greither, Kevin P. Keating, Alan Koch, Timothy Kohl, Paul J. Truman, and Robert G. Underwood,** Hopf Algebras and Galois Module Theory, 2021

259 **William Heinzer, Christel Rotthaus, and Sylvia Wiegand,** Integral Domains Inside Noetherian Power Series Rings, 2021

258 **Pramod N. Achar,** Perverse Sheaves and Applications to Representation Theory, 2021

257 **Juha Kinnunen, Juha Lehrbäck, and Antti Vähäkangas,** Maximal Function Methods for Sobolev Spaces, 2021

256 **Michio Jimbo, Tetsuji Miwa, and Fedor Smirnov,** Local Operators in Integrable Models I, 2021

255 **Alexandre Boritchev and Sergei Kuksin,** One-Dimensional Turbulence and the Stochastic Burgers Equation, 2021

254 **Karim Belabas and Henri Cohen,** Numerical Algorithms for Number Theory, 2021

253 **Robert R. Bruner and John Rognes,** The Adams Spectral Sequence for Topological Modular Forms, 2021

252 **Julie Déserti,** The Cremona Group and Its Subgroups, 2021

251 **David Hoff,** Linear and Quasilinear Parabolic Systems, 2020

250 **Bachir Bekka and Pierre de la Harpe,** Unitary Representations of Groups, Duals, and Characters, 2020

249 **Nikolai M. Adrianov, Fedor Pakovich, and Alexander K. Zvonkin,** Davenport–Zannier Polynomials and Dessins d'Enfants, 2020

248 **Paul B. Larson and Jindrich Zapletal,** Geometric Set Theory, 2020

247 **István Heckenberger and Hans-Jürgen Schneider,** Hopf Algebras and Root Systems, 2020

246 **Matheus C. Bortolan, Alexandre N. Carvalho, and José A. Langa,** Attractors Under Autonomous and Non-autonomous Perturbations, 2020

245 **Aiping Wang and Anton Zettl,** Ordinary Differential Operators, 2019

244 **Nabile Boussaïd and Andrew Comech,** Nonlinear Dirac Equation, 2019

243 **José M. Isidro,** Jordan Triple Systems in Complex and Functional Analysis, 2019

242 **Bhargav Bhatt, Ana Caraiani, Kiran S. Kedlaya, Peter Scholze, and Jared Weinstein,** Perfectoid Spaces, 2019

241 **Dana P. Williams,** A Tool Kit for Groupoid C^*-Algebras, 2019

240 **Antonio Fernández López,** Jordan Structures in Lie Algebras, 2019

239 **Nicola Arcozzi, Richard Rochberg, Eric T. Sawyer, and Brett D. Wick,** The Dirichlet Space and Related Function Spaces, 2019

238 **Michael Tsfasman, Serge Vlăduţ, and Dmitry Nogin,** Algebraic Geometry Codes: Advanced Chapters, 2019

237 **Dusa McDuff, Mohammad Tehrani, Kenji Fukaya, and Dominic Joyce,** Virtual Fundamental Cycles in Symplectic Topology, 2019

236 **Bernard Host and Bryna Kra,** Nilpotent Structures in Ergodic Theory, 2018

For a complete list of titles in this series, visit the
AMS Bookstore at **www.ams.org/bookstore/survseries/**.